Objective
Environmental Studies
with Glossary

Objective Environmental Studies with Glossary

(for Undergraduate Students of All Indian Universities JRF, SRF, ARS, NET, SLET, Civil Services & Other Competitive Examinations)

Dr. C.S. Bohra
Assistant Professor & Coordinator
Department of Environmental & Mechanical Engineering
Amrapali Institute of Technology & Sciences (AITS)
Haldwani – 263 139, Uttarakhand

Dr. B.S. Jina
Senior Lecturer
Department of Biotechnology
Infusion College of Science and Technology
Haldwani – 263 139, Uttarakhand

2010
DAYA PUBLISHING HOUSE
Delhi - 110 035

ISBN 81-7035-661-X
ISBN 978-81-7035-661-5

Published by : **Daya Publishing House**
1123/74, Deva Ram Park
Tri Nagar, Delhi - 110 035
Phone: 27383999
Fax: (011) 23260116
e-mail : dayabooks@vsnl.com
website : www.dayabooks.com

Showroom : 4760-61/23, Ansari Road, Darya Ganj,
New Delhi - 110 002
Phone: 23245578, 23244987

Laser Typesetting : **Classic Computer Services**
Delhi - 110 035

Printed at : **Chawla Offset Printers**
Delhi - 110 052

PRINTED IN INDIA

Dedicated to
my uncle
Late Shri Chandan Singh Bohra

Preface

Today, conservation of ecological habitat has acquired significance due to the burning problems of the pollution and fast depletion of natural habitat. Hence, based on the directions of the Honorable Supreme Court of India, UGC had made it compulsory for all the Indian Universities to introduce a course in Environmental Studies at undergraduate levels in higher education.

'Environmental Studies' is a fast growing area which relates to the study of environment in its totality with special emphasis on man and his adaptability.

The present work entitled "*Objective Environmental Studies with Glossary*" has been prepared to fulfill the needs of undergraduate classes. The Multiple Choice Questions (MCQs) test a candidate's ability to apply his knowledge gained during class work in the subject. Environmental Studies is not a new subject but an extension of ecological approach that stresses upon the study of the environment in its totality. The MCQs in the book have been with great care and after an exhaustive and extensive study of pattern of multiple choice questions in the subject. The questions are precise and clear enabling the reader to make correct choice based on his understanding of the subject.

Several books are published on this subject but a glossary of terms frequently encountered in environmental science in a simple to understand and comprehensive meaning to capture the understanding of the subject by the young brains, academicians, policy makers and anyone else will unveil the substance of the terminology instant. With this background a humble effort is made to collect and coalesce the terms frequently used in the form of ready to hand material. This compilation will not only open new vistas in better and fast understanding but also open new criticisms for further refinement.

We are confidant that the present book will prove useful to the students of undergraduate classes of Indian Universities. Constructive criticism and valuable suggestions from the readers for the improvement of the book will be greatly appreciated.

Dr. C.S. Bohra

Dr. B.S. Jina

Contents

Syllabus

For undergraduate courses of all branches of Higher Education

UNIT-I: The multidisciplinary nature of Environmental Studies

Definition, Scope and Importance, Need for Public Awareness.

UNIT-II: Natural Resources

Renewable and Non-renewable Resources

1. **Forest Resources:** Use and over exploitation, deforestation, case studies, timber extraction, mining, dams and their effects on forests and tribal people.
2. **Water Resources:** Use and over exploitation of surface and ground water, floods, drought, conflicts over water, dams; benefits and related problems.
3. **Mineral Resources:** Use and over exploitation, environmental effects of extracting and using mineral resources, case studies.
4. **Food Resources:** World food problems, changes caused by agriculture and over grazing, effects of modern

agriculture, fertilizers, pesticide problems, water logging, salinity, related case studies.

5. **Energy Resources:** Growing energy needs, renewable and non- renewable energy resources, use of alternate energy resources, case studies.
6. **Land Resources:** Land as a resource, land degradation, man induced land slides, soil erosion and desertification.

   Role of an individual in conservation of natural resources.

   Equitable use of resources for sustainable life style.

UNIT-III: Ecosystem

Concept of an ecosystem, structure and function of an ecosystem, producers, consumers, decomposers, energy flow in the ecosystem, ecological succession, food chain, food webs and ecological pyramids. Introduction, types, characteristic features, structure and function of the following ecosystems; Forest, grassland, desert, aquatic ecosystem (pond, lakes, streams, rivers, oceans and estuaries).

UNIT-IV: Biodiversity and its Conservation

Introduction, definition, genetic, species and ecosystem diversity; Biogeographical classification of India, value of biodiversity; consumptive use, productive use, social, ethical and esthetic values, biodiversity at global, national and local levels, India as a mega diversity nation, hot spots of biodiversity, threats to biodiversity; habitat loss, poaching of wildlife, man-wildlife conflicts, endangered and endemic species of India, conservation of biodiversity; *In- situ* and *Ex-situ* conservation of biodiversity.

UNIT-V: Environmental Pollution

Definition, causes, effects and control measures of; Air pollution, Noise pollution, Water pollution, Soil pollution, Marine pollution, Thermal pollution, Nuclear pollution.

- ☆ **Solid waste management:** Causes, effects and control measures of urban and industrial wastes; role of an individual in prevention of pollution, pollution case studies, disaster management; floods, earthquakes, cyclones and land slides.

UNIT-VI: Social Issues and the Environment

From unsustainable to sustainable development, urban problems related to energy, water conservation, rain water harvesting, watershed management, resettlement and rehabilitation of the people, its problems and concerns, case studies, environmental ethics; issues and possible solutions, climate change, global warming, acid rain, ozone layer depletion, nuclear accidents and holocaust, case studies. Wasteland reclamation, consumerism and waste products. Environmental Protection Acts; Air (prevention and control of pollution) Act, Water (prevention and control of pollution) Act, Wild Life Protection Act, Forest Conservation Act, issues involved in enforcement of environmental legislation, public awareness.

UNIT-VII: Human Population and Environment

Population growth, variations among nations, population explosion, family welfare programmes, environmental and human health, human rights, value education, HIV / AIDS; women and child welfare, role of information technology in environment and human health, care studies.

Part I

Objective Questions

Chapter 1

Environmental Studies: Scope and Importance

1. **The world environmental day is**

   (*a*) 10th June (*b*) 5th June

   (*b*) 6th June (*d*) 4th June

2. **The world health day is on**

   (*a*) 7th April (*b*) 8th April

   (*c*) 6th April (*d*) 10th April

3. **WWF stands for**

   (*a*) Worldwide Fund for Nature (*b*) World Wildlife Fund

   (*c*) World Wide Forest (*d*) Both (*a*) and (*b*)

4. **PCB stands for**

   (*a*) Plant Control Board

   (*b*) Pollution Control Board

   (*d*) People's Control Board

   (*d*) None of the above

5. Organization means

(*a*) Co-ordination (*b*) Common goal
(*c*) Hierarchy (*d*) All the above

6. The following are a part of population except

(*a*) Genes (*b*) Cells
(*c*) Organisms (*d*) Organs
(*e*) Biome

7. The following are a part of biosphere except

(*a*) Atmosphere (*b*) Biome
(*c*) Landscape (*d*) Ecosystem
(*e*) Population

8. A biological community is sustained due to

(*a*) Cycling of nutrients (*b*) Flow of energy
(*c*) Wind (*d*) Both (*a*) and (*b*)

9. An organism with other individual of the same species is known as

(*a*) Population (*b*) Extrapolation
(*c*) Biome (*d*) Genome

10. A distinct unit of life in nature is

(*a*) Microbes (*b*) Organisms
(*c*) Ecology (*d*) Biome

11. The Environment is mutually related with

(*a*) The organism (*b*) Nature
(*c*) Human beings (*d*) All the above

12. The organisms and the environment are interdependent entities of

(*a*) Atmosphere (*b*) Nature
(*c*) Earth (*d*) None of the above

13. The relationship between the organisms and environment is known as

(*a*) Ecology (*b*) Ecosystem

(*c*) Anthropology (*d*) Phytogeography

14. The word 'Ecology' is derived from

(*a*) Latin (*b*) Greek

(*c*) French (*d*) Spanish

15. The word 'Oikos' in Greek means

(*a*) House (*b*) Land

(*c*) Monkey (*d*) Area

16. The term' Ecology' was first coined by

(*a*) Charles Darwin (*b*) Ernst Haeckel

(*c*) Thomas Edison (*d*) Baba Amte

17. The branch of biology that deals with environment

(*a*) Environmentalogy (*b*) Ecology

(*c*) Elementology (*d*) None

18. Ernst Haeckel belonged to

(*a*) Portugal (*b*) Russia

(*c*) Germany (*d*) England

19. Ernst Haeckel was a

(*a*) Doctor (*b*) Engineer

(*c*) Biologist (*d*) Dentist

20. The physical environment has the following except

(*a*) Ionosphere (*b*) Lithosphere

(*c*) Hydrosphere (*d*) Atmosphere

21. Basic unit of Ecological hierarchy is

(*a*) Biome (*b*) Genome

(*c*) Organism (*d*) Population

22. The characters of species are except

(*a*) Breeding (*b*) Similar organisms
(*c*) Produce off springs (*d*) Colony

23. Group of individuals inhabiting a given area at a particular time are called

(*a*) Genome (*b*) Biome
(*c*) Population (*d*) Ecome

24. Ecosphere is also known as

(*a*) Biosphere (*b*) Stratosphere
(*c*) Nanosphere (*d*) Both (*a*) and (*b*)

25. The different populations of the same kind of organisms are called

(*a*) Local population (*b*) Foreign population
(*c*) Discrete population (*d*) None of the above

26. The earth's terrestrial biomes and aquatic systems constitute

(*a*) Biosphere (*b*) Biome
(*c*) Biological community (*d*) Genome

27. A population genetically adapted to its particular environment is known as

(*a*) Phenotype (*b*) Genotype
(*c*) Ecotype (*d*) Biotype

28. Organisms of the same kind may form____population

(*a*) Single (*b*) Several
(*c*) Both (*a*) and (*b*) (*d*) Unique

29. Interaction between different populations in an area are following

(*a*) Competition (*b*) Mutualism
(*c*) Predation (*d*) Synergy

30. Thin life supporting belt on and around earth

(*a*) Lithosphere (*b*) Biosphere

(*c*) Atmosphere (*d*) Ecology

31. The microscopic, photosynthetic organisms that float on water surface are called

(*a*) Grinnel (*b*) Phytoplankton

(*c*) Niche (*d*) Fungi

32. All the dogs living in a colony form

(*a*) Group (*b*) Population

(*c*) Species (*d*) Biome

33. Assemblage of populations of plants, animals, bacteria and fungi living in an area is known as

(*a*) Biotic community

(*b*) Biological community

(*c*) Genome

(*d*) Both (*a*) and (*b*)

34. A unit of land with a natural boundary having a mosaic of patches is known as

(*a*) Landscape (*b*) Biome

(*c*) Biosphere (*d*) Ecosystem

35. A landscape has____ecosystem

(*a*) No (*b*) Several

(*c*) Only one (*d*) None of the above

36. Desert is an example of

(*a*) Terrestrial biome (*b*) Aquatic biome

(*c*) Landscape (*d*) Biosphere

37. A large regional unit characterized by a major vegetation type and associated fauna in a specific climate zone is

(*a*) Landscape (*b*) Biome

(*c*) Biosphere (*d*) Genome

38. **Solid earth crust is known as**

(*a*) Stratosphere (*b*) Lithosphere
(*c*) Hydrosphere (*d*) Atmosphere

39. **Earth's terrestrial and aquatic systems constitute a**

(*a*) Biosphere (*b*) Lithosphere
(*c*) Stratosphere (*d*) Ecosystem

40. **Biotic and abiotic components are the parts of**

(*a*) Biome (*b*) Biosphere
(*c*) Ecosystem (*d*) All of the above

41. **The extent of Biosphere from ocean floor to mountain top is**

(*a*) 40 kms (*b*) 60 kms
(*c*) 22.5 kms (*d*) 80 kms

42. **Tropical rain forests are examples of**

(*a*) Biosphere (*b*) Biome
(*c*) Landscape (*d*) Both (*a*) and (*b*)

43. **Similar organisms which interact freely and produce fertile off springs constitute a**

(*a*) Population (*b*) Community
(*c*) Species (*d*) Ecotype

44. **The role of an organism in the ecological system is known as**

(*a*) Habitat (*b*) Herbivory
(*c*) Niche (*d*) Interaction

45. **The environment which has been modified by human activities is called**

(*a*) Natural environment
(*b*) Anthropogenic environment
(*c*) Urban environment
(*d*) Modern environment

46. The surroundings in which the organisms live is known as

(*a*) Atmosphere (*b*) Environment
(*c*) Ecosystem (*d*) All the above

47. Physical space occupied by an organism is known as

(*a*) Spacial (*b*) Special
(*c*) Racial (*d*) Colossal

48. Short term properties of the atmosphere is

(*a*) Climate (*b*) Weather
(*c*) Environment (*d*) None of the above

49. The climate conditions in an area of limited size

(*a*) Macro climate (*b*) Micro climate
(*c*) Weather (*d*) All of the above

50. Sum total of conditions which surround man at a given point in space and time

(*a*) Atmosphere (*b*) Climate
(*c*) Environment (*d*) Nature

51. The regulatory factor of environment is

(*a*) Temperature (*b*) Light
(*c*) Soil (*d*) Both (*a*) and (*b*)

52. Resources of the environment are

(*a*) Soil (*c*) Water
(*b*) Temperature (*d*) Both (*a*) and (*c*)

53. The average weather of an area is

(*a*) Temperature (*b*) Climate
(*c*) Environment (*d*) Both (*b*) and (*c*)

54. _____ is determined by the temperature and rainfall of an area

(*a*) Climate (*b*) Weather
(*c*) Environment (*d*) Both (*a*) and (*b*)

55. Hourly and weekly change in temperature and pressure etc. refer to

(*a*) Weather (*b*) Climate

(*c*) Nature (*d*) Both (*a*) and (*b*)

56. Global climate patterns are governed by variations in

(*a*) Temperature (*b*) Rainfall

(*c*) Humidity (*d*) All of the above

57. The regions in 20°–40° latitude are known as

(*a*) Tropical region (*b*) Subtropical region

(*c*) Temperate region (*d*) Arctic region

58. Temperature and pressure changes and variations over a season or for longer periods is known as

(*a*) Climate (*b*) Atmosphere

(*c*) Weather (*d*) Both (*a*) and (*b*)

59. Global variation in temperature and rainfall is due to

(*a*) Difference in solar radiation inputs

(*b*) Heat energy redistribution differences

(*c*) Ocean currents

(*d*) All of the above

60. Arctic region is

(*a*) 0°–20° latitude (*b*) 20°–40° latitude

(*c*) 40°–60° latitude (*d*) 60°–80° latitude

61. Temperature varies with changes in

(*a*) Latitude (*b*) Altitude

(*c*) Pressure (*d*) Both (*a*) and (*b*)

62. The immediate surroundings of plants and animals is

(*a*) Macroclimate (*b*) Microclimate

(*c*) Environment (*d*) Atmosphere

63. The main temperature decreases as we move from

(*a*) Tropical to arctic region

(*b*) Arctic region to tropical region

(*c*) Both (*a*) and (*b*)

(*d*) No change

64. Vegetation and Soil type of a region are determined by

(*a*) Temperature (*b*) Precipitation

(*c*) Pressure (*d*) Both (*a*) and (*b*)

65. Temperate region is

(*a*) 0–20 (*b*) 20°–40° latitude

(*c*) 40°–60° latitude (*d*) 60°–80° latitude

66. As the height of the mountain increases, the temperature

(*a*) Increases (*b*) Decreases

(*c*) No Change (*d*) Cannot predict

67. The high altitudes in temperate zone will have____climatic conditions

(*a*) Alpine (*b*) Tropical

(*c*) Subtropical (*d*) None of the above

68. The ecological factors related to soil and substratum are called____factors

(*a*) Endemic (*b*) Edaphic

(*c*) Mutual (*d*) Both (*a*) and (*b*)

69. The components of environment are

(*a*) Biotic (*b*) Abiotic

(*c*) Symbiotic (*d*) Both (*a*) and (*b*)

70. The water mantle of the earth is

(*a*) Hydrosphere (*b*) Hydrocele

(*c*) Hydrocarbon (*d*) Hydrogen

71. **Place occupied by an organism in relation to environment is**

(*a*) Habit (*b*) Habitat

(*c*) Edaphic (*d*) All of the above

72. **Capability of an organism to survive and reproduce in a particular environment is known as**

(*a*) Adaptation (*b*) Adoption

(*c*) Adjustment (*d*) None

73. **The ability of some animals to blend with the surroundings is**

(*a*) Disguise (*b*) Camouflage

(*c*) Conceal (*d*) None

74. **During the daytime, temperature inside the forest is lower because**

(*a*) Less light comes inside (*b*) Dense foliage of trees

(*c*) More humidity (*d*) All of the above

75. **Minor variations in the habitat due to changes in the environmental factors are known as**

(*a*) Microhabitat (*b*) Partial habitat

(*c*) Both (*a*) and (*b*) (*d*) Impartial habitat

76. **The term 'Nicbe' was first coined by**

(*a*) Gauss (*b*) Odum

(*c*) Grinnel (*d*) Lack

77. **The scientist who differentiated between habitat and ecological niche is**

(*a*) Grinnel (*b*) Odum

(*c*) Gauss (*d*) Lack

78. **The habitat together with the functions of a species is called**

(*a*) Niche (*b*) Biche

(*c*) Creche (*d*) None of the above

79. Groups of species exploiting a common resource are called

(*a*) Builds (*b*) Guilds

(*c*) Community (*d*) Population

80. Who showed that closely related birds occur together due to differences in their nesting and feeding habits

(*a*) Grinnel (*b*) Odum

(*c*) Lack (*d*) Gauss

81. Gauss's hypothesis is

(*a*) Competitive exclusion

(*b*) Competitive inclusion

(*c*) Competitive binding

(*d*) None of the above

82. Guilds help us to identify

(*a*) Species' functional role in community

(*b*) Reduce the number of components in a community

(*c*) Identify ecological equivalents

(*d*) Both (*a*) and (*b*)

83. The organisms in different geographical regions but occupying same niche are caned

(*a*) Ecological equivalents

(*b*) Ecological polyvalents

(*c*) Ecological multivalents

(*d*) None of the above

84. The amount of water vapor present in the atmosphere is

(*a*) Humidity (*b*) Moisture

(*c*) Temperature (*d*) None of the above

85. Climate factors are all except

(*a*) Temperature (*b*) Light

(*c*) Wind (*d*) Soil

86. Examples of decomposers are

(*a*) Plants
(*b*) Fungi
(*c*) Bacteria
(*d*) Both (*b*) and (*c*)

87. The animals that consume energy as food are called

(*a*) Decomposers
(*b*) Consumers
(*c*) Producers
(*d*) Destroyers

88. Examples of producers are

(*a*) Algae
(*b*) Fungi
(*c*) Animals
(*d*) Plants

89. The biotic component of the environment are

(*a*) Plants
(*b*) Animals
(*c*) Sand
(*d*) Soil

90. The environment which has been modified by human activity is called

(*a*) Anthropogenic environment
(*b*) Man made environment
(*c*) Natural environment
(*d*) Both (*a*) and (*b*)

Answers

1	(*b*)	24	(*a*)	47	(*a*)	69	(*d*)
2	(*a*)	25	(*a*)	48	(*b*)	70	(*a*)
3	(*a*)	26	(*a*)	49	(*b*)	71	(*b*)
4	(*b*)	27	(*c*)	50	(*c*)	72	(*a*)
5	(*d*)	28	(*b*)	51	(*d*)	73	(*b*)
6	(*e*)	29	(*d*)	52	(*d*)	74	(*d*)
7	(*a*)	30	(*b*)	53	(*b*)	75	(*c*)
8	(*d*)	31	(*b*)	54	(*a*)	76	(*c*)
9	(*a*)	32	(*b*)	55	(*a*)	77	(*b*)
10	(*b*)	33	(*d*)	56	(*d*)	78	(*a*)
11	(*a*)	34	(*a*)	57	(*b*)	79	(*b*)
12	(*b*)	35	(*b*)	58	(*a*)	80	(*c*)
13	(*a*)	36	(*a*)	59	(*d*)	81	(*a*)
14	(*b*)	37	(*b*)	60	(*d*)	82	(*d*)
15	(*a*)	38	(*b*)	61	(*d*)	83	(*a*)
16	(*b*)	39	(*a*)	62	(*b*)	84	(*a*)
17	(*b*)	40	(*b*)	63	(*a*)	85	(*d*)
18	(*c*)	41	(*c*)	64	(*d*)	86	(*d*)
19	(*c*)	42	(*b*)	65	(*b*)	87	(*b*)
20	(*a*)	43	(*c*)	66	(*b*)	88	(*d*)
21	(*c*)	44	(*c*)	67	(*a*)	89	(*b*)
22	(*d*)	45	(*b*)	68	(*b*)	90	(*d*)
23	(*c*)	46	(*b*)				

Chapter 2

Natural Resources

1. **Resources means**
   (*a*) Source of supply
   (*b*) Source of support
   (*c*) Both (*a*) and (*b*)
   (*d*) None of the above

2. **Natural resources are**
   (*a*) Land
   (*b*) Soil
   (*c*) Water
   (*d*) Forests
   (*e*) All of the above

3. **An example of inexhaustible resource is**
   (*a*) Water
   (*b*) Minerals
   (*c*) Solar energy
   (*d*) Wild life

4. **A renewable resource is**
   (*a*) Water
   (*b*) Natural vegetation
   (*c*) Soil fertility
   (*d*) All of the above

5. **Hydropower is an example of____resource**
   (*a*) Exhaustible
   (*b*) Inexhaustible
   (*c*) Non-useful
   (*d*) None of the above

6. **Excessive plantation of trees to save the diminishing forest cover is known as**

(*a*) Bio-plantation (*b*) Deforestation

(*c*) Afforestation (*d*) All of the above

7. **The materials origination from photosynthesis**

(*a*) Biogas (*b*) Bio-waste

(*c*) Bio-mass (*d*) None of the above

8. **Common energy source in Indian villages is**

(*a*) Electricity (*b*) Coal

(*c*) Wood and animal dung (*d*) Sun

9. **Fossil fuels and metallic minerals are**

(*a*) Renewable resources

(*b*) Inexhaustible resources

(*c*) Non-renewable resources

(*d*) None of the these

10. **Forests and wildlife are**

(*a*) Renewable resources

(*b*) Non-renewable resources

(*c*) Inexhaustible

(*d*) None of the above

11. **The practice of farming with increased organic input to the soil is**

(*a*) Taungya cultivation (*b*) Organic farming

(*c*) Urban forestry (*c*) None of the above

12. **Conversion of grassland/forest into desert is**

(*a*) Afforestation (*b*) Desertification

(*c*) Mulching (*d*) Social forestry

13. **The best soil for healthy and vigorous growth of most plants is**

(*a*) Clay (*b*) Sandy

(*c*) Loam (*d*) Clayey loam

14. The practice of growing agricultural crops between rows of planted trees.

(*a*) Taungya cultivation (*b*) Shift irrigation
(*c*) Zoom cultivation (*d*) None of the above

15. Soil conservation is a process in which

(*a*) Soil is aerated
(*b*) Soil erosion is allowed
(*c*) Sterile soil is made fertile
(*d*) Soil is protected

16. The resources which lack the ability of recycling

(*a*) Renewable source
(*b*) Non-renewable source
(*c*) Inexhaustible resources
(*d*) None of the above

17. Components of life supporting system are

(*a*) Soil (*b*) Water
(*c*) Minerals (*d*) All of the above

18. The renewable product of wildlife ecosystem is

(*a*) Maintaining food chain (*b*) Yield timber
(*c*) Yield food and fiber (*d*) All of the above

19. A substance present in nature but not utilized is

(*a*) Resources (*b*) Not a resource
(*c*) Cannot predict (*d*) Energy

20. The resources present in unlimited quantity in nature are

(*a*) Renewable resources (*b*) Exhaustive resources
(*c*) Inexhaustible resources (*d*) None of the above

21. Abiotic resources are

(*a*) Renewable (*b*) Non–renewable
(*c*) Exhaustible (*d*) Both (*b*) and (*c*)

22. Biological species are

(*a*) Non-renewable (*b*) Cannot predict

(*c*) Renewable (*d*) Useful

23. Earth's total forest cover is

(*a*) 2/3 (*b*) 1/3

(*c*) 3/4 (*d*) 1/4

24. Decline in forests is due to

(*a*) Unplanned felling (*b*) Lack of foresight

(*c*) Increase in population (*d*) All of the above

25. Preparation forest area available in India is

(*a*) 0.06 hectare (*b*) 0.64 hectare

(*c*) 1 hectare (*d*) Both (*a*) and (*b*)

26. The forest area for plains according to National Forest policy (1988) is

(*a*) 33% (*b*) 34%

(*c*) 35% (*d*) 36%

27. At present, the forest cover in India is

(*a*) 20% (*b*) 19.4%

(*c*) 21% (*d*) 22%

28. Deforestation is more prominent in

(*a*) Tropical region (*b*) Temperate region

(*c*) Arctic region (*d*) Subtropical region

29. Jhum cultivation can be seen in____of country

(*a*) North Eastern region (*b*) Southern region

(*c*) Northern region (*d*) Western region

30. Felling and burning forests followed by cultivation of crops for few years and abandoning cultivation to allow forest's re-growth

(*a*) Shifting cultivation (*b*) Jhum Cultivation

(*c*) Taungya cultivation (*d*) Both (*a*) and (*b*)

31. The current rate of deforestation across the globe is

(*a*) 20 million hectares per year

(*b*) 10 million hectares per year

(*c*) 15 million hectares per year

(*d*) Not predictable

32. Green house gas is

(*a*) CO_2 (*b*) $S0_2$

(*c*) O_2 (*d*) Nitrogen

33. If deforestation increases the soil fertility

(*a*) Increases (*b*) Decreases

(*d*) Increases and decreases (*d*) Difficult to predict

34. Deforestation generally decreases

(*a*) Rainfall (*b*) Soil erosion

(*c*) Draught (*d*) Global warming

35. Soil erosion can be prevented by

(*a*) Overgrazing (*b*) Afforestation

(*c*) Removal of vegetation (*d*) Deforestation

36. Examples of Biogas

(*a*) Methane (*b*) CO_2

(*c*) Hydrogen (*d*) Helium

(*e*) Both (*a*) and (*b*)

37. Taungya system is an example of

(*a*) Socio forestry (*b*) Agro forestry

(*c*) Organic farming (*d*) None of the above

38. The first movement against indiscriminate felling of trees was thought by

(*a*) Amrita Devi (*b*) Sunderlal Bahuguna

(*c*) Asha Parikh (*d*) None of the above

39. The major gasses in the earth's atmosphere are

(*a*) Nitrogen (*b*) Oxygen

(*c*) CO_2 (*d*) Helium

(*e*) Both (*a*) and (*b*)

40. The composition of Nitrogen in atmosphere is

(*a*) 70% (*b*) 71%

(*c*) 72% (*d*) 72%

(*e*) 74%

41. % composition of Oxygen earth's atmosphere

(*a*) 21% (*b*) 22%

(*c*) 23% (*d*) 24%

42. Energy is transferred from earth's Surface and the atmosphere through

(*a*) Conduction (*b*) Convection

(*c*) Radiation (*d*) All of the above

43. Weather phenomena is due to

(*a*) Ocean currents (*b*). Wind

(*c*) Ice (*d*) All of the above

44. The layer of atmosphere in which we live is

(*a*) Stratosphere (*b*) Mesosphere

(*c*) Troposhere (*d*) None of the above

45. Weather occurs in

(*a*) Mesosphere (*b*) Thermosphere

(*c*) Troposhere (*d*) Stratosphere

46. The deflection of wind by the earth's rotation is known as

(*a*) Coriolis effect (*b*). Radiation

(*c*) Convection (*d*) None of the above

47. Troposhere extends from the earth's surface to an height of

(*a*) 12 km (*b*) 7 miles

(*c*) 15 km (*d*) Both (*a*) and (*b*)

48. Temperature no longer varies with height, region known as

(*a*) Troposphere (*b*) Tropein

(*c*) Stratosphere (*d*) Thermosphere

49. The leader of chipko movement was

(*a*) Sunderlal Bahuguna (*b*) Amrita Devi

(*c*) Baba Amte (*d*) Baba Ramdev

50. The first movement against indiscriminate felling of trees in India took place in the year

(*a*) 1731 (*b*) 1732

(*c*) 1733 (*d*) 1734

51. Amrita Devi belonged to

(*a*) Kejarili Village (*b*) Jodhpur

(*c*) Bikaner (*d*) Jabalpur

52. Amrita Devi was a

(*a*) Bishnoi woman (*b*) Tribal woman

(*c*) Punjabi (*d*) Rajasthani

53. Chipko movement started at

(*a*) Himalayan region (*b*) Tihri Garhwal region

(*c*) Indus Valley (*d*) Silent valley

54. Chipko movement started in the year

(*a*) 1970 (*b*) 1975

(*c*) 1972 (*d*) 1980

55. Tihri Grehwal region is in the state of

(*a*) Madhya Pradesh (*b*) Uttarakhand

(*c*) Rajasthan (*d*) Gujarat

56. Appiko movement took place in

(*a*) A.P. (*b*) Karnataka

(*c*) U.P. (*d*) Madhya Pradesh

57. Ulisu (to conserve), belesu (to grow) and balasu (rational use) are the objectives of

(*a*) Chipko movement (*b*) Appiko Movement

(*c*) Bhudan movement (*d*) None of the above

58. Van Mahotsava is related to

(*a*) Trees Plantation (*b*) Worship of trees

(*c*) Celebration for trees (*d*) None of the above

59. Intensive plantation of trees on available land for fulfilling commercial demand is known as

(*a*) Commercial forestry (*b*) Production forestry

(*c*) Rural forestry (*d*) Both (*a*) and (*b*)

60. Which is not related to the remaining

(*a*) Urban forestry (*b*) Agro forestry

(*c*) Social forestry (*d*) Rural forestry

61. Using the same land for farming and forestry is

(*a*) Agro forestry (*b*) Social forestry

(*c*) Urban forestry (*d*) None

62. Raising trees for firewood, fodder etc. is

(*a*) Agro forestry (*b*) Social forestry

(*c*) Urban forestry (*d*) Conservative forestry

63. Raising ornamental and fruit trees in urban areas along roads is

(*a*) Urban forestry (*b*) Urbanization

(*c*) Urban beauty programme (*d*) Urban coating

64. Taungya and Jhum are the methods of

(*a*) Social forestry (*b*) Urban forestry

(*c*) Socioeconomic forestry (*d*) All of the above

65. Growing agricultural crops between the rows of planted trees is

(*a*) Agro Forestry (*b*) Urban Forestry

(*c*) Social forestry (*d*) All of the above

66. The portion of water present in oceans is

(*a*) 70% (*b*) 75%

(*c*) 97.5% (*d*) 99%

67. The residence time of water in the air is

(*a*) 20 days (*b*) 10 days

(*c*) 15 days (*d*) 12 days

68. Water covers____ of earth's surface

(*a*) 1/4 (*b*) 2/4

(*c*) 3/4 (*d*) None of the above

69. The 'Temples of Modern India' are

(*a*) Tirupathi Temple (*b*) Dams and Projects

(*c*) Rivers (*d*) Cinema Theatres

70. Weathering results in

(*a*) Formation of soil (*b*) Formation of water

(*c*) Formation of teeth (*d*) Mountains

71. Mine spoil is

(*a*) Spoiling the beauty of mines

(*b*) Destruction of land by mining

(*c*) Creating mines

(*d*) None

72. Power and manure are provided by

(*a*) Nuclear plants (*b*) Thermal plants

(*c*) Biogas plants (*d*) Hydroelectric plants

73. Harnessing of nuclear energy often causes

(*a*) Air pollution (*b*) Water pollution

(*c*) Thermal pollution (*d*) Noise pollution

74. Biogas is a mixture of

(*a*) Methane (*b*) Ethane

(*c*) CO_2 (*d*) Both (*a*) and (*c*)

75. Liquid fuels obtained from Biomass are

(*a*) Methanol (*b*) Ethanol

(*c*) Kerosene (*d*) Both (*a*) and (*b*)

76. 'Agrostological' refers to cultivation of

(*a*) Fields (*b*) Grasses

(*c*) Grapes (*d*) Mangoes

77. Organic farming is possible by using

(*a*) Pesticides (*b*) Mercurials

(*c*) Biofertilisers (*d*) None of the above

78. The benefits of organic farming are

(*a*) Increased soil fertility (*b*) Less toxic residues

(*c*) Soil erosion prevented (*d*) All the above

79. Crop rotation involves

(*a*) Different crops successively (*b*) Decreases soil loss

(*c*) Decreases soil fertility (*d*) Both (*a*) and (*b*)

80. Contour ploughing is done in areas of

(*a*) High temp (*b*) Less rainfall

(*c*) High rainfall (*d*) All of the above

81. Contour ploughing is providing____in the field

(*a*) Alternate furrows and ridges

(*b*) Digging

(*c*) Water pouring

(*d*) All of the above

82. Benefits of mulching are

(*a*) Retains soil moisture

(*b*) Decreases runoff

(*c*) Increases organic matters in the soil

(*d*) All of the above

83. **Mulching is**
   (*a*) Soil untilled
   (*b*) Soiled covered with crop residue
   (c) Putting paper
   (*d*) Both (*a*) and (*b*)

84. **Growing of grass to check soil erosion is known as**
   (*a*) Agrobiology (*b*) Agrostology
   (*c*) Agronomy (*d*) All of the above

85. **Planting of trees or wind breaks in deserts is**
   (*a*) Terrace farming (*b*) Afforestation
   (*c*) Agro forestry (*d*) Both (*b*) and (*c*)

86. **The conservation of eroded soil involves**
   (*a*) Soil stabilization (b) Afforestation
   (*c*) Restoration of soil fertility (*d*) Both (*a*) and (*c*)

87. **Restoration of soil fertility is a____process**
   (*a*) Fast (*b*) Slow
   (*c*) Transitory (*d*) None of the above

88. **Growing drought resistant grasses is known as**
   (*a*) Lay farming (*b*)Jhum farming
   (*c*) Soil farming (*d*) Strip cropping

89. **Seeding the bare ground with plants that can survive harsh conditions is known as**
   (*a*) Contour ploughing (*b*) Mulching
   (*c*) Soil stability (*d*) All of the above

90. **A slope is divided into a number of small flat fields. This is known as**
   (*a*) Terrace farming (*b*) Strip cropping
   (*c*) Organic farming (*d*) None of the above

91. **Different techniques of strip cropping are**
   (*a*) Contour strip cropping (*b*) Field strip cropping
   (*c*) Wind strip cropping (*d*) All of the above

92. Planting of crops in rows or strips to check flow of water is

(*a*) Mulching (*b*) Strip cropping

(*c*) Serial cropping (*d*) All of the above

93. Soil erosion is due to

(*a*) Wind (*b*) Water

(*c*) Temperature (*d*) Ice

(*e*) All (*a*), (*b*) and (*c*)

94. Benefits of soil conservation are

(*a*) Increases soil particle size

(*b*) More water to soil

(*c*) Slow down of water movement

(*d*) All of the above

95. The previous residue crop inclusion into the soil by ploughing is known as

(*a*) Conservational tillage (*b*) Terrace tillage

(*c*) Strip cropping (*d*) Organic fanning

96. Uranium is popularly used for

(*a*) Nuclear bomb manufacture (*b*) Tinting glass

(*c*) Power generation (*d*) Both (*a*) and (*c*)

97. Dams were called the temples of modern India by

(*a*) Jawaharlal Nehru (*b*) Gandhi ji

(*c*) Tilak (*d*) Benazir Bhutto

98. Aswan Dam is across the river____

(*a*) Godavari (*b*) Nile

(*c*) Narmada (*d*) Tapti

99. Excess Nitrate in drinking water may lead to

(*a*) Blue Baby Syndrome

(*b*) Methaemoglobinemia

(*c*) Acidosis

(*d*) Both (*a*) and (*b*)

100. Derelict land is the one destroyed by

(*a*) Mining (*b*) Agriculture

(*c*) Plantation (*d*) None of the above

101. Minerals can be conserved by

(*a*) Recycling (*b*) Reuse

(*c*) Decreased consumption (*d*) Substitution

(*e*) All of the above

102. The basic resources of the earth mentioned in our ancient literature are

(*a*) Khaiti (Soil) (*b*) Pawan (energy)

(*c*) Gagan (space) (*d*) Samira (air)

(*e*) All of the above

103. The main cause of desertification of grasslands is

(*a*) Overgrazing (*b*) Droughts

(*c*) Rainfall (*d*) Both (*a*) and (*b*)

104. Desertification means

(*a*) Conversion of grasslands to deserts

(*b*) Conversion of forests to deserts

(*c*) Conversion of deserts to forests

(*d*) Both (*a*) and (*b*)

Answers

1	(*c*)	23	(*b*)	45	(*c*)	67	(*b*)
2	(*e*)	24	(*d*)	46	(*a*)	68	(*c*)
3	(*c*)	25	(*a*)	47	(*d*)	69	(*b*)
4	(*d*)	26	(*a*)	48	(*a*)	70	(*a*)
5	(*b*)	27	(*b*)	49	(*a*)	71	(*b*)
6	(*c*)	28	(*a*)	50	(*a*)	72	(*c*)
7	(*c*)	29	(*a*)	51	(*a*)	73	(*c*)
8	(*c*)	30	(*d*)	52	(*a*)	74	(*d*)
9	(*c*)	31	(*b*)	53	(*b*)	75	(*d*)
10	(*a*)	32	(*a*)	54	(*c*)	76	(*b*)
11	(*b*)	33	(*b*)	55	(*b*)	77	(*c*)
12	(*b*)	34	(*a*)	56	(*b*)	78	(*d*)
13	(*c*)	35	(*b*)	57	(*b*)	79	(*d*)
14	(*a*)	36	(*e*)	58	(*a*)	80	(*b*)
15	(*d*)	37	(*b*)	59	(*d*)	81	(*a*)
16	(*b*)	38	(*a*)	60	(*d*)	82	(*d*)
17	(*d*)	39	(*e*)	61	(*a*)	83	(*d*)
18	(*a*)	40	(*a*)	62	(*b*)	84	(*b*)
19	(*b*)	41	(*a*)	63	(*a*)	85	(*b*)
20	(*c*)	42	(*d*)	64	(*d*)	86	(*d*)
21	(*d*)	43	(*a*)	65	(*a*)	87	(*b*)
22	(*a*)	44	(*c*)	66	(*c*)	88	(*a*)

89	**(*c*)**	**93**	**(*e*)**	**97**	**(*a*)**	**101**	**(*e*)**
90	**(*a*)**	**94**	**(*d*)**	**98**	**(*b*)**	**102**	**(*e*)**
91	**(*d*)**	**95**	**(*a*)**	**99**	**(*d*)**	**103**	**(*d*)**
92	**(*b*)**	**96**	**(*d*)**	**100**	**(*a*)**	**104**	**(*d*)**

Chapter 3

Ecosystem

1. **The term 'Ecosystem' was introduced by**
   (*a*) Arthur Hailey (*b*) Arthur Tansley
   (*c*) Arthur Folley (*d*) None of the above
2. **The structural and functional unit of biosphere is**
   (*a*) Mass (*b*) Ecosystem
   (*c*) Environment (*d*) All of the above
3. **The Biotic community + Physical environment constitutes**
   (*a*) Ecosystem (*b*) Ecopond
   (*c*) Ecology (*d*) All of the above
4. **The relationship between biotic and abiotic environment is called**
   (*a*) Biogenesis (*b*) Symbiosis
   (*c*) Holocoenosis (*d*) None of the above
5. **The material and energy requirement of a biotic community fulfilled by**
   (*a*) Atmosphere (*b*) Environment
   (*c*) Trees (*d*) Sun

6. An example of a microecosystem is

(*a*) Drops of water in a pond (*b*) Ocean

(*c*) Land (*d*) Forest

7. Example of a temporary ecosystem is

(*a*) Rainfed pond (*b*) Crop field

(*c*) Grassland (*d*) Lake

8. Anthropogenic ecosystems are due to

(*a*) Human activity (*b*) Animals

(*c*) Environment (*d*) Sun

9. Example of Leutic ecosystem is

(*a*) River (*b*) Pond

(*c*) Spring (*d*) Ocean

10. Food is synthesized from inorganic raw materials by

(*a*) Heterotrophs (*b*) Autotrophs

(*c*) Homotrophs (*d*) All of the above

11. Example of Terrestrial ecosystem is

(*a*) Ocean (*b*) Forest Both

(*c*) Desert (*d*) Both (*b*) and (*c*)

12. 'Lotic' means

(*a*) Running water (*b*) Standing water

(*c*) Flowing water (*d*) Both (*a*) and (*c*)

13. Autotrophs are also called

(*a*) Producers (*b*) Decomposers

(*c*) Destructors (*d*) None

14. Biotic components are

(*a*) Consumers (*b*) Decomposers

(*c*) Autotrophs (*d*) All of the above

15. Non living components constitute

(*a*) Abiotic components (*b*) Biotic components

(*c*) Environment (*d*) All of the above

16. Producers are also called

(*a*) Convertors
(*b*) Transducers
(*c*) Consumers
(*d*) Both (*a*) and (*b*)

17. Autotrophs have the characteristics of

(*a*) Chlorophyll
(*b*) Self nourishing
(*c*) Convert raw materials into food by photosynthesis
(*d*) All of the above

18. Autotrophs prepare food by

(*a*) Hydrolysis
(*b*) Photosynthesis
(*c*) Biofermentation
(*d*) All of the above

19. An ecosystem consists of

(*a*) Green plants and animals
(*b*) Green plants and decomposers
(*c*) Producers and consumers
(*d*) Green plants, animals, decomposers and abiotic environment

20. Which one of the following is an abiotic component of the ecosystem

(*a*) Bacteria
(*b*) Plants
(*c*) Humus
(*d*) Fungi

21. Most stable ecosystem is

(*a*) Ocean
(*b*) Forest
(*c*) Desert
(*d*) Mountains

22. Herbivores are____consumers

(*a*) Secondary
(*b*) Primary
(*c*) Tertiary
(*d*) None

23. In shallow waters the dominant producers are

(*a*) Macrophytes
(*b*) Microphytes
(*c*) Zoophytes
(*d*) None

24. In aquatic systems the dominant producers are

(*a*) Phytoplankton (*b*) Macrophytes
(*c*) Zooplankton (*d*) All of the above

25. During photosynthesis, solar energy is converted into

(*a*) Chemical energy (*b*) Physical energy
(*c*) Bioenergy (*d*) Glucose

26. Heterotrophs get their food from

(*a*) Consumers (*b*) Producers
(*c*) Environment (*d*) All of the above

27. First order consumer are

(*a*) Deer (*b*) Rabbit
(*c*) Grasshopper (*d*) All of the above

28. Phagotrophs are

(*a*) Consumers (*b*) Decomposers
(*c*) Autotrophs (*d*) Producers

29. Consumers are usually

(*a*) Homotropic (*b*) Heterotrophic
(*c*) Autotrophic (*d*) None of the above

30. Micro consumers are

(*a*) Decomposers (*b*) Reducers
(*c*) Producers (*d*) Both (*a*) and (*b*)

31. Key industry animals are

(*a*) Primary consumers (*b*) Tertiary consumers
(*c*) Secondary consumers (*d*) None of the above

32. Animals that feed directly on plants are

(*a*) Carnivores (*b*) Decomposers
(*c*) Herbivores (*d*) All of the above

33. Examples of Herbivores of aquatic system are

(*a*) Protozoans (*b*) Molluses
(*c*) Rabbit (*d*) Both (*a*) and (*b*)

34. The term 'key industry animals' was coined by

(*a*) Elton (*b*) Priestley

(*c*) Baba Amte (*d*) Menaka Gandhi

35. Examples of top carnivores are

(*a*) Lion (*b*) Tiger

(*c*) Frog (*d*) Both (*a*) and (*b*)

36. The animals which feed upon other animals are

(*a*) Herbivores (*b*) Autotrophs

(*c*) Carnivores (*d*) None of the above

37. The carnivores which feed upon herbivores are called

(*a*) Primary Carnivores (*b*) Tertiary consumers

(*c*) Second order consumers (*d*) Both (*a*) and (*b*)

38. Good examples of primary carnivores are

(*a*) Frog (*b*) Bird

(*c*) Fox (*d*) All of the above

39. Third order consumers are

(*a*) Owl (*b*) Fox

(*c*) Tiger (*d*) Both (*a*) and (*c*)

40. The animals which feed upon the primary carnivores are

(*a*) Secondary carnivores (*b*) Top Carnivores

(*c*) Third order consumers (*d*) Both (*a*) and (*b*)

41. Examples of saprophytes are

(*a*) Algae (*b*) Bacteria

(*c*) Fungi (*d*) Both (*b*) and (*c*)

42. Saprophytes obtain their food from the dead bodies of

(*a*) Producers (*b*) Consumers

(*c*) Carnivores (*d*) Both (*a*) and (*b*)

43. Examples of micro consumers are

(*a*) Bacteria (*b*) Fungi

(*c*) Animals (*d*) Both (*a*) and (*b*)

44. Extracellular digestion is done by____in saprophytes

(*a*) Digestive enzymes (*b*) Hormones

(*c*) Acids (*d*) Alkalies

45. ____obtain food directly from hosts

(*a*) Parasites (*b*) Scavengers

(*c*) Carnivores (*d*) Both (*a*) and (*c*)

46. ____are the animals which feed on the dead bodies of other organisms

(*a*) Scavengers (*b*) Detritivores

(*c*) Carnivores (*d*) Both (*a*) and (*b*)

47. The term 'biosphere' refers to

(*a*) Earth (*b*) Sky

(*c*) Space (*d*) None of the above

48. The amount of abiotic material present in an ecosystem is known as

(*a*) Sitting stage (*b*) Standing stage

(*c*) Walking stage (*d*) None

49. All are abiotic components except

(*a*) Climate (*b*) Inorganic substances

(*c*) Organic substances (*d*) Animals

50. The cycling of gases in the biotic system is called

(*a*) Chemical cycle (*b*) Atmospheric cycle

(*c*) Biogeochemical cycle (*d*) None

51. Organic substances include all except

(*a*) Carbohydrates (*b*) Proteins

(*c*) Lipids (*d*) Nitrogen

52. Examples of climate are

(*a*) Temperature (*b*) Light

(*c*) Wind (*d*) All of the above

53. A large variety of species composition occurs in

(*a*) Desert (*b*) Forest

(*c*) Aquatic system (*d*) None of the above

54. Biomass of organisms for unit area is known as

(*a*) Standing state (*b*) Standing crop

(*c*) Standing vegetation (*d*) All of the above

55. The biomass of a species is expressed in terms of

(*a*) Fresh weight (*b*) Dry weight

(*c*) Kg (*d*) Both (*a*) and (*b*)

56. The organisms in an ecosystem form one or more layers. This is known as

(*a*) Composition (*b*) Stratification

(*c*) Trophism (*d*) None

57. _____feed upon the organism of all trophic levels

(*a*) Decomposers (*b*) Parasites

(*c*) Consumers (*d*) None of the above

58. The amount of nutrients present in the soil at any given time is called

(*a*) Standing state (*b*) Standing crop

(*c*) Transitionary state (*d*) None of the above

59. The rate of biomass production is called

(*a*) Productivity (*b*) Production

(*c*) Biodiversity (*d*) All of the above

60. Primary productivity is expressed in

(*a*) $gm^{-2}year^{-1}$ (*b*) $KCalm^{-2}year^{-1}$

(*c*) K Cal (*d*) Both (*a*) and (*b*)

61. The rate of total production of organic matter by the producer per unit area and time is

(*a*) Gross primary productivity (*b*) Productivity

(*c*) Net primary productivity (*d*) Energy flow

62. **The rate at which organic matter is stored after the basic metabolic processes is**
   (*a*) Primary productivity
   (*b*) Gross Primary productivity
   (*c*) Net primary productivity
   (*d*) None of the above

63. **The process by which complex organic materials are broken down is known as**
   (*a*) Energy flow (*b*) Decomposition
   (*c*) Hydrolysis (*d*) Symbiosis

64. **The main site for decomposition in the ecosystem is**
   (*a*) Lower layer of soil (*b*) Upper layers of soil
   (*c*) Earth crust (*d*) All of the above

65. **The productivity that is left after meeting the cost of respiration and maintenance of producers is**
   (*a*) Net Primary Productivity
   (*b*) Gross Primary Productivity
   (*c*) Net Secondary Productivity
   (*d*) None

66. **The high level of net primary productivity is shown by**
   (*a*) Deserts (*b*) Tropical rain forests
   (*c*) Deciduous forests (*d*) All of the above

67. **The rate of increase in the biomass of consumers per unit area and time is**
   (*a*) Primary Productivity
   (*b*) Secondary productivity
   (c) Net Primary Productivity
   (c) None of the above

68. **The herbivores consume about____of assimilated energy in respiration**
   (*a*) 60% (*b*) 30%
   (*c*) 25% (*d*) 100%

69. The net primary productivity of deserts is

(*a*) 1 ton $hectare^{-1}$ $year^{-1}$
(*b*) 2 ton $hectare^{-1}$ $year^{-1}$
(*c*) 20 ton $hectare^{-1}$ $year^{-1}$
(*d*) None of the above

70. Inorganic raw materials are reutilized by plants due to

(*a*) Hydrolysis
(*b*) Oxidation
(*c*) Decomposition
(*d*) All of the above

71. Energy for the ecosystem comes initially from

(*a*) Solar system
(*b*) Sky
(*c*) Moon
(*d*) All of the above

72. In an ecosystem, producers convert solar radiation into____

(*a*) Physical form
(*b*) Water form
(*c*) Chemical form
(*d*) All of the above

73. The flow of energy in the ecosystem is governed by the principles

(*a*) Energy is transmitted from one component to another
(*b*) Energy can neither be created nor destroyed
(*c*) Transformation of energy is accompanied by dispersion
(*d*) All of the above

74. In an ecosystem dispersion of energy is followed by

(*a*) Gain of heat
(*b*) Loss of heat
(*c*) Gain and loss of heat
(*d*) Cannot predict

75. The average trophic levels in the food chain of an ecosystem are

(*a*) 3-5
(*b*) 6-7
(*c*) 7-8
(*d*) 1-2

76. The transfer of energy from one trophic level to the next trophic level is called

(*a*) Food habitat
(*b*) Food chain
(*c*) Food nexus
(*d*) All of the above

77. All are constituents of aquatic food chain except

(*a*) Phytoplankton (*b*) Wolf

(*c*) Zooplankton (*d*) Small fish

78. The consumers that are not preyed upon by other animals are

(*a*) Top carnivores (*b*) Primary carnivores

(*c*) Secondary. carnivores (*d*) All of the above

79. All are the constituents of terrestrial food chain except

(*a*) Grasshopper (*b*) Rabbit

(*c*) Snake (*d*) Crocodile

80. The ecosystem exhibiting detritus food chain largely depends on

(*a*) Solar energy (*b*) Organic matter

(*c*) Plants (*d*) All of the above

81. A network of food chains in an ecosystem is known as

(*a*) Food world (*b*) Food web

(*c*) Food universe (*d*) All of the above

82. Ecological pyramids were developed by

(*a*) Charles Elton (*b*) James Elton

(*c*) Baud Elton (*d*) None of the above

83. Charles Elton was a____ecologist

(*a*) American (*b*) British

(*c*) Canadian (*d*) Dutch

84. The graphical representation of ecological parameters is called

(*a*) Ecological pyramid (*b*) Ecological hill

(*c*) Ecological mountain (*d*) All of the above

85. In an ecological pyramid top carnivores occupy____position

(*a*) Lower (*b*) Top

(*c*) Middle (*d*) None of the above

86. Except one, remaining are the important parameters of trophic levels in a food chain. Identify the wrong one

(*a*) Number of individuals (*b*) Amount of biomass
(*c*) Amount of energy (*d*) Pyramid location

87. Ecological pyramids are also called as

(*a*) Eltonian pyramids (*b*) Charles pyramids
(*c*) Hierarchial pyramids (*d*) All of the above

88. In an Eltonian pyramid,____occupy the base

(*a*) Top carnivores (*b*) Producers
(*c*) Herbivores (*d*) None of the above

89. A graphic representation of numerical relationship between different trophic levels of the food chain is known as

(*a*) Pyramid of biomass (*b*) Pyramid of numbers
(*c*) Pyramid of energy (*d*) All of the above

90. The amount of living organic matter present in a particular environment is called

(*a*) Biogas (*b*) Biogenetics
(*c*) Biomass (*d*) Biome

91. In a terrestrial ecosystem the maximum Biomes occurs in

(*a*) Producers (*b*) Consumers
(*c*) Herbivores (*d*) All of the above

92. ____supports a larger population

(*a*) Non-vegetarian diet (*b*) Vegetarian diet
(*c*) Green vegetables (*d*) None of the above

93. Biomass is more in____

(*a*) Primary consumers (*b*) Phytoplankton
(*c*) Zooplankton (*d*) None of the above

94. In an aquatic habitat the pyramid of biomass is

(*a*) Inverted (*b*) Spindle shaped
(*c*) Long (*d*) Both (*a*) and (*b*)

95. The energy content is expressed in

(*a*) Kcal/m^2/yr (*b*) KJ/cm^2/yr
(*c*) KK/m^2/yr (*d*) Both (*a*) and (*b*)

96. The details of entrapment of energy by different consumers in an ecosystem was elucidated by

(*a*) John Elton (*b*) Odum
(*c*) Griffith (*d*) None of the above

97. The major ecosystems with their groups of climax plants and associated animals are called

(*a*) Biosphere (*b*) Biome
(*c*) Biomass (*d*) None of the above

98. In an ecosystem of pond the secondary carnivores (fishes) have an energy content of

(*a*) 2100 KJ/m^2/yr (*b*) 126 KJ/m^2/yr
(*c*) 7980 KJ/m^2/yr (*d*) None of the above

99. In India, forests occupy____% of total land area

(*a*) 19 (*b*) 35
(*c*) 70 (*d*) 10

100. The major forest biomes of India are

(*a*) Tropical rain forests (*b*) Coniferous forests
(*c*) Desert forests (*d*) Both (*a*) and (*b*)

101. Oak trees are dominant species in

(*a*) Tropical rain forests
(*b*) Temperate broad leaf forests
(*c*) Tropical deciduous forests
(*d*) Temperate forests

102. Ocean cover about____of the earth's surface

(*a*) 1/4 (*b*) 2/3
(*c*) 3/4 (*d*) None of the above

103. The marine environment is characterized by high concentration of

(a) Salts (b) Mineral ions

(c) Fish (d) Both (a) and (b)

104. A semi-closed coastal body of water having a tree connection with sea

(a) Pond (b) Estuary

(c) River (d) Lagoon

105. The change of plant and animals communities in an orderly sequence in an area is

(a) Ecological succession (b) Biotic succession

(c) Biome (d) Both (a) and (b)

106. The process of change is called

(a) Reaction (b) Action

(c) Mimic (d) Interaction

107. The first community which inhabits a bare area is called

(a) Transitional community (b) Climax community

(c) Pioneer community (d) All of the above

108. Successions beginning in dry habitat is called

(a) Psammosere (b) Xerarch

(c) Hydrarch (d) None of the above

109. The last and stable community in an area is called

(a) Pioneer community (b) Climax community

(c) Transitional community (d) Seral community

110. Sequence of successional stages on a bare rock are called

(a) Lithosere (b) Psammosere

(c) Xerosere (d) Hydrosere

111. The entire series of communities of biotic succession from pioneer to climax community is known as

(a) Seral (b) Sere

(c) Biosis (d) None of the above

112. The biotic succession that occurs on a substratum devoid of earlier life is called

(*a*) Secondary succession (*b*) Primary succession
(*c*) Tertiary succession (*d*) None of the above

113. The sequence of successional stages of a primary succession is called

(*a*) Psammosere (*b*) Prisere
(*c*) Subsere (*d*) Lithosere

114. Dead plant parts and animal remains are known as

(*a*) Detritus (*b*) Decay
(*c*) Decomposers (*d*) Dungeon

115. Rooted plants found in shallow waters

(*a*) Zoophytes (*b*) Macrophytes
(*c*) Zooplankton (*d*) Shrubs

116. Tropical forests occur in India in

(*a*) Jammu and Kashmir (*b*) Rajasthan
(*c*) Kerala and Assam (*d*) Nowhere

117. Every food chain starts with a

(*a*) Consumer (*b*) Producer
(*c*) Decomposer (*d*) User

118. Which ecological pyramid is always straight

(*a*) Pyramid of biomass
(*b*) Pyramid of numbers
(*c*) Pyramid of energy
(*d*) Pyramid of numbers and biomass

119. Name the pioneer plants of xerarch

(*a*) Lichens (*b*) Phytoplanktons
(*c*) Mass (*d*) None of the above

120. Increase in fauna and decrease in flora would be harmful due to increase in

(*a*) Diseases (*b*) CO_2

(*c*) Radio active pollution (*d*) O_2

121. Which one of the following is an abiotic component of the ecosystem

(*a*) Bacteria (*b*) Plants

(*c*) Humus (*d*) Fungi

122. Autotrophic planktons are called

(*a*) Phytoplankton (*b*) Zooplankton

(*c*) Fungi (*d*) None of the above

123. A forest community will develop only if the climax is____

(*a*) Dry (*b*) Moist

(*c*) Cool (*d*) Hot

124. Individuals of a species affect each other's life in various ways. This is called

(*a*) Action (*b*) Co-action

(*c*) Reaction (*d*) Support

125. Dipterocarpus can be seen in____

(*a*) Tropical rain forests (*b*) Coniferous forests

(*c*) Deserts (*d*) Grasslands

126. Which one is the correct food chain

(*a*) Phytoplankton → Zooplankton → Fish

(*a*) Zooplankton → Phytoplankton → Fish

(*c*) Grass → Fish → Zooplankton

(*d*) Zooplankton → Protozoans → Fish

127. In a grass → deer → tiger foodchain, the biomass of grass is 1 ton. The tiger biomass will be

(*a*) 100 kg (*b*) 10 kg

(*c*) 150 kg (*d*) 1 kg

128. The food chain in which microorganisms breakdown dead producers is called

(*a*) Consumer food chain (*b*) Predator food chain

(*c*) Parasite food chain (*d*) Detritus food chain

129. Cynodon species are found in

(*a*) Coniferous forests (*b*) Grass lands

(*c*) Deserts (*d*) Tropical rain forests

Answers

1	(*b*)	23	(*a*)	45	(*a*)	67	(*b*)
2	(*b*)	24	(*a*)	46	(*d*)	68	(*b*)
3	(*a*)	25	(*a*)	47	(*a*)	69	(*a*)
4	(*c*)	26	(*b*)	48	(*b*)	70	(*c*)
5	(*b*)	27	(*d*)	49	(*d*)	71	(*a*)
6	(*a*)	28	(*a*)	50	(*c*)	72	(*c*)
7	(*a*)	29	(*b*)	51	(*d*)	73	(*d*)
8	(*a*)	30	(*d*)	52	(*d*)	74	(*b*)
9	(*b*)	31	(*a*)	53	(*b*)	75	(*a*)
10	(*b*)	32	(*c*)	54	(*b*)	76	(*b*)
11	(*d*)	33	(*d*)	55	(*d*)	77	(*b*)
12	(*d*)	34	(*a*)	56	(*b*)	78	(*a*)
13	(*a*)	35	(*d*)	57	(*b*)	79	(*d*)
14	(*d*)	36	(*c*)	58	(*a*)	80	(*b*)
15	(*a*)	37	(*d*)	59	(*a*)	81	(*b*)
16	(*d*)	38	(*d*)	60	(*d*)	82	(*a*)
17	(*d*)	39	(*d*)	61	(*a*)	83	(*b*)
18	(*b*)	40	(*d*)	62	(*c*)	84	(*a*)
19	(*d*)	41	(*d*)	63	(*b*)	85	(*b*)
20	(*c*)	42	(*d*)	64	(*b*)	86	(*d*)
21	(*a*)	43	(*d*)	65	(*a*)	87	(*a*)
22	(*b*)	44	(*a*)	66	(*b*)	88	(*b*)

89	(*b*)	100	(*d*)	111	(*b*)	121	(*c*)
90	(*c*)	101	(*b*)	112	(*b*)	122	(*a*)
91	(*a*)	102	(*b*)	113	(*b*)	123	(*b*)
92	(*b*)	103	(*d*)	114	(*a*)	124	(*b*)
93	(*a*)	104	(*b*)	115	(*b*)	125	(*a*)
94	(*d*)	105	(*d*)	116	(*c*)	126	(*a*)
95	(*d*)	106	(*a*)	117	(*b*)	127	(*b*)
96	(*b*)	107	(*c*)	118	(*c*)	128	(*d*)
97	(*b*)	108	(*b*)	119	(*a*)	129	(*b*)
98	(*b*)	109	(*b*)	120	(*b*)		
99	(*a*)	110	(*a*)				

Chapter 4

Biodiversity and its Conservation

1. **Variety and variability among living organisms is known as**
   (*a*) Biotechnology (*b*) Biodiversity
   (*c*) Bioequivalence (*d*) All of the above

2. **The totality of genes, species and ecosystems of a region is**
   (*a*) Biodiversity (*b*) Bioconservation
   (*c*) Both (*a*) and (*b*) (*d*) Biology

3. **The genetic variation in the same genes is known as**
   (*a*) Belle (*b*) Alleles
   (*c*) Celleles (*d*) All of the above

4. **The number of genes in man are**
   (*a*) 60,000 (*b*) 35,000 to 45,000
   (*c*) 45,000 to 50,000 (*d*) None of the above

5. **The genetic variation existing within a species is called**
   (*a*) Biodiversity (*b*) Genetic diversity
   (*c*) Genome diversity (*d*) All of the above

6. **Chilka lake is located in**
   (*a*) Andhra Pradesh (*b*) Orissa
   (*c*) Karnataka (*d*) Tamil Nadu

7. **Number of highway plants recorded all over the world is**
   (*a*) 2,90,000 (*b*) 3,00,000
   (*c*) 2,70,000 (*d*) 10,00,000

8. **Total no. of mammals known to science are**
   (*a*) 4,650 (*b*) 4,800
   (*c*) 4,900 (*d*) 5,000

9. **Kolleru lake is in**
   (*a*) Tamil Nadu (*b*) Karnataka
   (*c*) Andhra Pradesh (*d*) Madhya Pradesh

10. **The evolution of new species is known as**
    (*a*) Specialization (*b*) Speciation
    (*c*) Smog (*d*) None of the above

11. **The species diversity of a region is measured based on**
    (*a*) Species richness (*b*) Equitability
    (*c*) Evenness (*d*) All of the above

12. **Distinct units of diversity are**
    (*a*) Species (*b*) Animals
    (*c*) Bacteria (*d*) None of the above

13. **Diversity within the community**
    (*a*) Alpha diversity (*b*) Beta diversity
    (*c*) Gradient diversity (*d*) None of the above

14. **The species determining the ability of large number of other species to persist in the community**
    (*a*) Milestone species (*b*) Keystone species
    (*c*) Redstone species (*d*) None of the above

15. Diversity of the habitat over the total landscape is

(*a*) Gamma diversity (*b*) Beta diversity

(*c*) Alpha diversity (*d*) None of the above

16. The most biodiversity rich zones are

(*a*) Western ghats (*b*) North east regions

(*c*) Himalayas (*d*) Both (*a*) and (*b*)

17. The species which are confined to a particular country are called

(*a*) Endemic species (*b*) Epidemic species

(*c*) Extinct species (*d*) All of the above

18. The diversity between communities is

(*a*) Alpha diversity (*b*) Beta diversity

(*c*) Gamma diversity (*d*) All of the above

19. Prominent species which are endemic to Western Ghats

(*a*) Protozoa (*b*) Amphibians

(*c*) Crustacean (*d*) None of the above

20. Biodiversity increases when we move from

(*a*) High to low latitude (*b*) From poles to equator

(*c*) From equator to poles (*d*) Both (*a*) and (*b*)

21. The two factors that reduce diversity are

(*a*) Drop in temperature (*b*) Seasonal variability

(*c*) Taxa related problems (*d*) Both (a) and (b)

22. The mean number of vascular species in a tropical rain forest are

(*a*) 150-160 (*b*) 118-236

(*c*) 100-150 (*d*) None of the above

23. About 42% of the Indian landmass is occupied by

(*a*) Western Ghats (*b*) Deccan peninsula

(*c*) Indian ocean (*d*) Maharashtra

24. 85% of the world's food is produced by

(*a*) Plants (*b*) Animals

(*c*) Both (*a*) and (*b*) (*d*) Bacteria

25. 2/3 of the world's food sustaining the human population is from the following

(*a*) Wheat (*b*) Com

(*c*) Rice (*d*) All of the above

26. The average life of a crop variety is

(*a*) 15-30 years (*b*) 5-15 years

(*c*) 10-50 years (*d*) None of the above

27. Plant based synthetic products are known as

(*a*) Bio-chemicals (*b*) Botano-chemicals

(*c*) Phyto-chemicals (*d*) All of the above

28. Taxol is obtained from

(*a*) Taxus buccata (*b*) Taxus brevifolia

(*c*) Both (*a*) and (*b*) (*d*) None of the above

29. Quinine is used in the treatment of

(*a*) Jaundice (*b*) Malaria

(*c*) Syphilis (*d*) Grey baby syndrome

30. A free floating exotic water weed in Indian rivers and lakes is

(*a*) Water hyacinth (*b*) Nile perch

(*c*) Dragonfly (*d*) All of the above

31. The botanical name of Tulsi is

(*a*) Ficus religiosa (*b*) Ocimum sanctum

(*d*) Prosopis cineraria (*d*) None

32. New species entering a geographical region are called

(*a*) Exotic species (*b*) Alien species

(*c*) Both (*a*) and (*b*) (*d*) Foreign species

33. The effect of exotic species is maximum on ____

(*a*) Mainland (*b*) Island

(*c*) Territorial (*d*) All of the above

34. Red list of threatened species is published by

(*a*) IUPAC (*b*) IUCN

(*c*) IPEC (*d*) None of the above

35. IUCN stands for

(*a*) International Union for Conservation of Nature and Natural Resources

(*b*) International Union for Caring Nature

(*c*) International Union for Controlling Nature

(*d*) Both (*b*) and (*c*)

36. The tropical forests loss of species per year is approximately

(*a*) 1,000 (*b*) 14,000–40,000

(*c*) 80,000 (*d*) 1,00,000

37. If all the plants of the earth die suddenly, all the animals die due to the deficiency of

(*a*) Food (*b*) Shade

(*c*) Oxygen (*d*) Shelter

38. Indian bustard has become endangered species due to

(*a*) Deforestation in Rajasthan (*b*) Spread of desert

(*c*) Hunting (*d*) Change in climate

39. In our country the % of land under forest is about

(*a*) 20% (*b*) 19%

(*c*) 25% (*d*) 30%

40. The elimination of a species is known as

(*a*) Expansion (*b*) Extension

(*c*) Extinction (*d*) None of the above

41. The totality of genes, species and ecosystem of a region is known as

(*a*) Conservation (*b*) Biodiversity

(*c*) Biosphere reserve (*d*) All of the above

42. The region of the earth which harbour a very large number of species

(*a*) Mega zone (*b*) Mega diversity zone

(*c*) Macro zone (*d*) Both (*a*) and (*c*)

43. A species introduced in a country from another country is known as

(*a*) Hot spots (*b*) Extinction

(*c*) Exotic species (*d*) Both (*a*) and (*c*)

44. New approach to conservation is the establishment of

(*a*) Sanctuaries (*b*) Biosphere reserves

(*c*) National parks (*d*) Reserve forests

45. Animals and Plants are best protected in

(*a*) Zoos (*b*) Botanical gardens

(*c*) Sanctuaries (*d*) National parks

46. The animal that has recently become extinct is

(*a*) Dinosaur (*b*) Dodo

(*c*) Wooly mammoth (*d*) Archaeopteryx

47. Which group of vertebrates comprises the highest number of endangered species

(*a*) Birds (*b*) Mammals

(*c*) Fishes (*d*) Reptiles

48. Which one is matching sanctuary

(*a*) Kaziranga → Musk deer

(*b*) Gir → Lion

(*c*) N.E. Himalayan region → Samber

(*d*) Sunderban → Rhino

49. As of September 2002, India had____protected areas

(*a*) 600 (*b*) 582

(*c*) 570 (*d*) 610

50. The total national parks in India are

(*a*) 50 (*b*) 60

(*c*) 89 (*d*) 200

51. Kaziranga National park is famous for

(*a*) Tiger (*b*) Musk deer

(*c*) Elephant (*d*) Rhinoceros

52. Animals having high protection prospects are

(*a*) Lion and Leopard

(*b*) Tiger and Bustard

(*c*) Himalayan bear and Musk deer

(*d*) Elephant and Crocodile

53. The results of deforestation are

(*a*) Floods (*b*) Drought

(*c*) Both (*a*) and (*b*) (*d*) Earthquakes

54. Earth summit was held in 1992 at

(*a*) Brazil (*b*) Rio de Janeiro

(*c*) Paris (*d*) London

55. Total biosphere reserves in India are

(*a*) 12 (*b*) 13

(*c*) 10 (*d*) 20

56. In India, the two of the global biodiversity hot spots are

(*a*) Western Ghats (*b*) Eastern Himalayas

(*c*) Nilgiris (*d*) Both (*a*) and (*b*)

(*e*) Both (*a*) and (*c*)

57. WWF stands for

(*a*) World Wildlife Fund (*b*) World Wide Football

(*c*) World Wrestling Federation (*d*) World Wide Films

58. Major cause of extinction of species

(*a*) Forest loss (*b*) Habitat loss

(*c*) Food loss (*d*) All of the above

59. The richest and the most threatened reservoirs of plants and animals

(*a*) Hot spots of biodiversity (*b*) Cold spots

(*c*) Biomass (*d*) None

60. Preservation of living tissues/parts of plants in liquid nitrogen at temperature of –196°C is

(*a*) Annealing (*b*) Cold storage

(*c*) Cryopreservation (*d*) All of the above

61. Variety of species within a region

(*a*) Genetic variation (*b*) Species diversity

(*c*) Genetic combination (*d*) All of the above

62. An exotic plant species which has posed threat to the native species is

(*a*) Lantana camara (*b*) Cinchona ledgeriana

(*c*) Tulsi (*d*) All of the above

63. The protected lake of Sikkim which has been declared sacred by people is

(*a*) Chilka (*b*) Sunderbans

(*c*) Khecheopalri (*d*) None

64. Communities more vulnerable to invasion by outside animals and plants are

(*a*) Temperate forests (*b*) Tropical evergreen

(*c*) Oceanic islands (*d*) Mangroves

65. National park concerned with Rhinoceros

(*a*) Corbett (*b*) Ranathambore

(*c*) Kaziranga (*d*) Gir

66. India contributes____% species to the global diversity

(*a*) 10% (*b*) 8%

(*c*) 20% (*d*) 40%

67. The Western Ghats are spread in

(*a*) Maharashtra (*b*) Andhra Pradesh

(*c*) Kerala (*d*) Both (*a*) and (*c*)

68. The two main centres of biological diversity in Western Ghats of India are

(*a*) Agastryamalai hills (*b*) Nandi hills

(*c*) Silent valley (*d*) Both (*a*) and (*c*)

69. The western coast of Indian peninsula is about

(*a*) 1000 km (*b*) 1600 km

(*c*) 2000 km (*d*) 2500 km

70. The semi evergreen forests in Western Ghats of India occur at a height of

(*a*) 500–1500 m (*b*) 500 m or below

(*c*) 1800–2000 m (*d*) None of the above

71. The new Amambalam resort is located in

(*a*) Himalayas (*b*) Southern Ghats

(*c*) Western Ghats (*d*) Both (*a*) and (*c*)

72. Eastern Himalayas can be found in

(*a*) Bhutan (*b*) Nagaland

(*c*) Assam (*d*) All of the above

73. The main objectives of Earth summit, 1992 were

(*a*) Prevention of tree felling, grasslands

(*b*) Prevention of Hacking of computers

(*c*) Conservation of biological diversities

(*d*) All of the above

74. The following is the project for the development of Biosphere reserves

(*a*) WWF (*b*) BSF

(*c*) BOC (*d*) All of the above

75. India is a homeland of following cultivated and wild plants

(*a*) 200,500 (*b*) 167,320

(*c*) 1000, 1500 (*d*) 2000, 4000

76. India is a centre of diversity of the following animal species

(*a*) Chicken (*b*) Water buffalo

(*c*) Camel (*d*) All of the above

77. India is a secondary centre of domestication of the following animals

(*a*) Horse, goat (*b*) Sheep, Cattle

(*c*) Yak and Donkey (*d*) All of the above

78. Example of ex situ biodiversity conservation is

(*a*) Botanical gardens, zoological gardens

(*b*) Seed banks

(*c*) Biosphere research

(*d*) Both (*a*) and (*b*)

79. National parks and wildlife sanctuaries are example of

(*a*) *Ex situ* biodiversities conservation

(*b*) *In situ* biodiversity conservation

(*c*) Ecological protection

(*d*) All of the above

80. The National parks and wildlife sanctuaries are maintained by

(*a*) Ministry of zoos (*b*) Ministry of forests

(*c*) Ministry of environment (*d*) Both (*b*) and (*c*)

81. India is a centre of diversity of the following crop plants

(*a*) Rice (*b*) Sugarcane

(*c*) Banana (*d*) All of the above

82. The following no. is the countries coming under mega diversity countries

(*a*) 10 (*b*) 12

(*c*) 15 (*d*) 16

83. **India accounts____of the land area of the world**
(*a*) 5% (*b*) 10%
(*c*) 2.4% (*d*) 20%

84. **The hot spots concept was developed by**
(*a*) Norman Vincent Peale (*b*) Norman Myers
(*c*) Norman Scot (*d*) None of the above

85. **Arboneta is a**
(*a*) House (*b*) Botanical garden
(*c*) Ship (*d*) Environment

86. **Preservation of genetic resources can be done by**
(*a*) Field gene banks (*b*) Biotechnology
(*c*) Cryogenesis (*d*) All of the above

87. **Most undisturbed forests without any human impact are**
(*a*) Residual forests (*b*) Pristine forests
(*c*) Deciduous forests (*d*) All of the above

88. **Sacred forests are popular in**
(*a*) Karnataka (*b*) Maharashtra
(*c*) Kerala (*d*) All of the above

89. **Ex situ conservation of biodiversity can be done by**
(*a*) Germ plasm banks (*b*) Gene banks
(*c*) DNA banks (*d*) All of the above

90. **Khecheopalri lake in Sikkim is a**
(*a*) Big lake (*b*) Sacred lake
(*c*) Small lake (*d*) None of the above

91. **Germplasm can be stored by**
(*a*) Refrigeration (*b*) Cold storage
(*c*) Cryopreservation (*d*) All of the above

92. **The easiest way to store germplasm of wild plants is by**
(*a*) Seed gene banks (*b*) Godowns
(*c*) Bio reactors (*d*) None

93. Cryopreservation involves storage at

(*a*) –200°C in Oxygen

(*b*) –196°C in liquid Nitrogen

(*c*) –196°C in liquid Hydrogen

(*d*) All of the above

94. The storage of materials by gradual cooling and simultaneously dehydration is for

(*a*) Vegetative propagules (*b*) Tissue culture

(*c*) Biologicals (*d*) Both (*a*) and (*b*)

95. Cryopreservation is a technique very useful for preserving

(*a*) Seeds

(*b*) Vegetatively propagated parts

(*c*) Tissue culture

(*d*) All of the above

96. Undisturbed and legally protected ecosystem is known as

(*a*) Natural zone (*b*) Core zone

(*c*) Buffer zone (*d*) Both (*a*) and (*b*)

97. The storage of materials at ultra low temperature by very rapid cooling is useful for

(*a*) Vegetative propagules (*b*) Tissue culture

(*c*) Seeds (*d*) All of the above

98. The outer most part of the Biosphere resources is

(*a*) Core zone (*b*) Transitional zone

(*c*) Buffer zone (*d*) None of the above

99. The zone of biosphere where active cooperation between reserve management and the local people takes place is

(*a*) Buffer zone (*b*) National zone

(*c*) Transitional zone (*d*) All of the above

100. The important functions of Biosphere reserves are

(*a*) Conservation (*b*) Development

(*c*) Scientific research (*d*) All of the above

101. The biosphere reserves was formulated by

(*a*) UNICEF (*b*) UNESCO

(*c*) I.L.O (*d*) IMF

102. ____deals with the conservation of ecosystem and the genetic resources

(*a*) Sanctuaries (*b*) Zoo

(*c*) Biosphere Reserves (*d*) Aquariums

103. The concept of bio reserves was first time started in the year

(*a*) 2001 (*b*) 1975

(*c*) 2006 (*d*) 1991

104. A representative example of natural biome is

(*a*) Sanctuary (*b*) Beach

(*c*) Biosphere reserve (*d*) Lake

105. Keoladeo Ghana Bird sanctuary is situated at

(*a*) Bharatpur (Rajasthan) (*b*) Jaipur

(*c*) Udaipur (*d*) Jabalpur

106. An area which is reserved for the conservation of animals only is

(*a*) Biosphere (*b*) Sanctuary

(*c*) Biome (*d*) None of the above

107. At present there are____wildlife sanctuaries in our country

(*a*) 500 (*b*) 492

(*c*) 400 (*d*) 350

108. Manas wildlife sanctuary is situated at

(*a*) Kamrup (*b*) Patna

(*c*) Bhubaneshwar (*d*) Hyderabad

109. The international norms for land surface of a protected area is

(*a*) 5% (*b*) 10%

(*c*) 15% (*d*) 20%

110. The total number of protected areas around the world are

(*a*) 37,000 (*b*) 40,000

(*c*) 42,000 (*d*) 35,000

111. The first National Park established in India was

(*a*) Gir National Park

(*b*) Jim Corbett National Park

(*c*) Kanha National Park

(*d*) None of the above

112. Forestry, grazing and cultivation are not allowed in

(*a*) Biosphere (*b*) Sanctuaries

(*c*) National parks (*d*) Wildlife sanctuaries

113. Totally there are____National parks in India

(*a*) 90 (*b*) 89

(*c*) 85 (*d*) 100

114. The first National park of USA is

(*a*) Royal park (*b*) Yellow stone

(*c*) Jim Corbett National park (*d*) None of the above

115. Jim Corbett National park is at

(*a*) Shimla (*b*) Nainital

(*c*) Mysore (*d*) Cuttack

116. First National Park in India to start Tiger reserve is at

(*a*) Nainital (*b*) Junagarh

(*c*) Malwa (*d*) Jaipur

117. The National Park famous for Asiatic Lion is

(*a*) Jim Corbett (*b*) Sasan Gir

(*c*) Kanha (*d*) Kaziranga

118. The National Park known for one horned Rhinoceros is

(*a*) Jim Corbett (*b*) Sasan Gir

(*c*) Kazirange (*d*) Kanha

119. The areas which provide protection to the biological diversity are

(*a*) Protected areas (*b*) Biosphere reserves
(*c*) Sacred forests (*d*) Sacred lakes
(*e*) All of the above

120. Total no. of protected areas in India are

(*a*) 581 (*b*) 481
(*c*) 681 (*d*) 781

121. The land surface of protected areas in India is

(*a*) 10% (*b*) 4.7%
(*c*) 6% (*d*) 8%

122. Private ownership right and habit manipulation are not allowed in

(*a*) Sanctuary (*b*) National park
(*c*) Biosphere reserve (*d*) All of the above

123. Kanha National park is in

(*a*) Assam (*b*) Uttar Pradesh
(*c*) Madhya Pradesh (*d*) Tamil Nadu

124. The critically endangered species of plant in India is

(*a*) *Barberis nilghirinsis* (*b*) *Bentinekia nicobarica*
(*c*) *Cupressus cashmariana* (*d*) All of the above

125. An example of vulnerable species of animal in India

(*a*) Red Panda (*b*) Black Buck
(*c*) Pigmy Hog (*d*) None of the above

126. The critically endangered animal species of India is

(*a*) Pigmy Hog (*b*) Red Panda
(*c*) Asiatic Lion (*d*) Black Buck

127. Periyar sanctuary is in

(*a*) Tamil Nadu (*b*) Kerala
(*c*) Andhra Pradesh (*d*) Karnataka

128. Jim Corbett National park is in

(*a*) Uttaranchal (*b*) Madhya Pradesh

(*c*) Assam (*d*) Andhra Pradesh

129. The most comprehensive inventory of the global conservation status of plants and animal species is

(*a*) 2008 IUCN Red list (*b*) 2000 IUCN Red list

(*c*) Both (*a*) and (*b*) (*d*) 200 IUCN list

130. The characteristics which make a species susceptible to extinction are

(*a*) Large body size

(*b*) Small population size

(*c*) Feeding at high tropic levels

(*d*) All of the above

131. Bengal Tiger is susceptible for extinction because

(*a*) Localised distribution

(*b*) Feeding at high tropic levels in the food chain

(*c*) Fixed migratory routes

(*d*) Small population

132. The following are the susceptible species for extinction

(*a*) Elephant (*b*) Blue whale

(*c*) Fox (*d*) Both (*a*) and (*b*)

133. ____is a catalogue of Taxa that are facing the risk of extinction

(*a*) IUCN Red list (*b*) IUPAC Red list

(*c*) IUCN Orange list (*d*) All of the above

134. Blue whale faces the danger of extinction because of its

(*a*) Small population size

(*b*) Fixed migration routes and habitat

(*c*) Localised distribution

(*d*) Both (*a*) and (*b*)

135. The species which are not endangered and vulnerable presently but at risk are called

(*a*) Precious (*b*) Rare

(*c*) Threatened (*d*) All of the above

136. When no member of the species remains alive anywhere in the world, this is

(*a*) Extinction (*b*) Distinction

(*c*) Destruction (*d*) Disappearance

137. Nile perch is a

(*a*) Peacock (*b*) Exotic predatory fish

(*c*) Scorpion (*d*) Snake

138. Extinction of large number of species due to catastrophe

(*a*) Mass Extinction (*b*) Anthropogenic

(*c*) Natural Extinction (*d*) None of the above

139. Disappearance of a species from earth due to human activities is known as

(*a*) Mass Extinction (*b*) Natural Extinction

(*c*) Anthropogenic Extinction (*d*) None of the above

140. The extinction of species in the geological past is known as

(*a*) Background Extinction (*b*) Mass Extinction

(*c*) Anthropogenic Extinction (*d*) None of the above

141. Categories of threatened with extinction

(*a*) Vulnerable (*b*) Endangered

(*c*) Critically endangered (*d*) All of the above

142. The number of species under threatened category in 2000 IUCN Red list is

(*a*) 20,000 (*b*) 11,046

(*c*) 11,001 (*d*) 11,008

143. Most of the traditional drugs are obtained from

(*a*) Plants (*b*) Animals

(*c*) Human beings (*d*) All of the above

144. Morphine is obtained from

(*a*) *Taxus baccata* (*b*) *Chincona ledgeriana*

(*c*) *Papaverum somniferum* (*d*) All of the above

145. Which is the anticancer drug among the following

(*a*) Morphine (*b*) Quinine

(*c*) Taxol (*d*) Cinchonine

146. Among the following, which are the trees worshipped by Indian

(*a*) *Ocimum sanctum* (*b*) *Ficus religiosa*

(*c*) *Papaverum somniferum* (*d*) Both (*a*) and (*b*)

147. The nutrient enrichment is known as

(*a*) Symbiosis (*b*) Eutrophication

(*c*) Neutrification (*d*) Both (*b*) and (*c*)

148. Cinchona tree is from

(*a*) India (*b*) Pakistan

(*c*) Peru (*d*) Russia

149. The largest flying bird is

(*a*) Eagle (*b*) California Condor

(*c*) Bat (*d*) Vulture

150. Continuous area of habitat is divided into two or more fragment. This is known as

(*a*) Disturbance (*b*) Exotic fragmentation

(*c*) Habitat fragmentation (*d*) None of the above

151. Khejri is known as

(*a*) *Ficus religiosa* (*b*) *Ocimum sanctum*

(*c*) *Prosopis cineraria* (*d*) Both (*a*) and (*b*)

152. The following is the endangered plant species

(*a*) *Bentickia nicobarica* (*b*) *Barberis nilghiriensis*

(*c*) *Cupressus cashmariana* (*d*) All of the above

153. Identify the name of Red Panda from the following

(*a*) *Sus salvinius* (*b*) *Aliurus fulgens*

(*c*) *Antilope cervi capra* (*d*) All of the above

154. Mudumalai wildlife Sanctuary is in

(*a*) Nilgiris (*b*) Kodai kanal

(*c*) Western ghats (*d*) Both (*a*) and (*b*)

155. Research and educational activities are conducted in____zone of biosphere

(*a*) Natural zone (*b*) Buffer zone

(*c*) Transitional zone (*d*) Both (*b*) and (*c*)

156. Total no. of zoos around the world

(*a*) 100 Approx. (*b*) 500 Approx.

(*c*) 800 Approx. (*d*) 2000 Approx.

157. Silent valley is in

(*a*) Tamil Nadu (*b*) Kerala

(*c*) Andhra Pradesh (*d*) Maharashtra

158. The following is the active centre of evolution and exhibits a rich diversity of flowering plants

(*a*) Eastern Himalayas (*b*) Vindhyas

(*c*) Satpuras (*d*) Andamans

159. Majority of the hotspots of biodiversity are in

(*a*) Tropical forests (*b*) Mediterranean zone

(*c*) Tropical zone (*d*) None of the above

160. Biodiversity convention was held in which year

(*a*) 1991 (*b*) 1992

(*c*) 1993 (*d*) 1990

161. The hotspots concept came into existence in

(*a*) 1990 (*b*) 1988

(*c*) 1991 (*d*) 1992

162. Mangolia and Betula are extensively found in

(*a*) Western Ghats (*b*) Eastern Himalayas
(*c*) Aravali (*d*) Vmdhyas

163. Rio de Janeiro is the capital of

(*a*) Pakistan (*b*) Greece
(*c*) Brazil (*d*) Venezuela

164. Common place of conservation for Biodiversity is

(*a*) Botanical gardens (*b*) Zoos
(*c*) National parks (*d*) Both (*a*) and (*b*)

165. The diversity of habitat over the total landscape or geographical area is known as

(*a*) Gamma diversity (*b*) Beta diversity
(*c*) Alpha diversity (*d*) None of the above

166. The area reserved for the welfare of wildlife is called

(*a*) National Park (*b*) Botanical garden
(*c*) Sanctuary (*d*) Forest

167. The higher diversity at community level provides the following

(*a*) Stability (*b*) Strength
(*c*) Higher Productivity (*d*) Both (*a*) and (*c*)

168. The traits determining particular characteristics of a species are called

(*a*) Genes (*b*) Chromosomes
(*c*) Nucleus (*d*) All the three

169. Good examples of benefits of biodiversity are

(*a*) Food (*b*) Medicine
(*c*) Fibres (*d*) All of the above

170. Species diversity____from lower to higher altitude on a mountain

(*a*) Decreases (*b*) Increases
(*c*) Doesn't change (*d*) Cannot be predicted

171. The % of following plants endemic to our country are

(*a*) 40 (*b*) 33

(*c*) 80 (*d*) 70

172. Origin of species is the book written by

(*a*) Shakespeare (*b*) Newton

(*c*) Charles Darwin (*d*) Louis Pasteur

173. The California condor is the name of

(*a*) Animal (*b*) Plant

(*c*) Flying bird (*d*) Ocean

174. Tandoba National park is at____

(*a*) Muzaffarpur (*b*) Chandrapur

(*c*) Sholapur (*d*) Nagpur

175. Similipal Biosphere reserve is in

(*a*) Orissa (*b*) Andhra Pradesh

(*c*) Karnataka (*d*) Maharashtra

176. Spot the unrelated item

(*a*) Earth Summit (*b*) Biosphere

(*c*) Biodiversity (*d*) Cracking of petrol

177. Spot the unrelated item

(*a*) World Conservation Union

(*b*) World Wide Fund for Nature

(*c*) IUCN

(*d*) Protection of Nature Union

178. Good examples of the terrestrial biodiversity hot spots are

(*a*) Tropical Andes (*b*) Madagascar

(*c*) Central Chile (*d*) All of the above

179. Giant Panda is susceptible for extinction because of

(*a*) Small population size (*b*) Large body size

(*c*) Low reproductive rate (*d*) Both (*a*) and (*c*)

180. A Taxon not critically endangered but facing a high risk of extinction in median term future is known as

(*a*) Extinct (*b*) Vulnerable·

(*c*) Endangered (*d*) Lower risk

181. For every 1000 m increase in altitude, the temperature drops by____

(*a*) 7°C (*b*) 7.5°C

(*c*) 6.5°C (*d*) 8°C

182. What happens to biodiversity if we move from poles to equator

(*a*) Decreases

(*b*) Increases

(*c*) But decreases then increases

(*d*) Difficult to predict

183. ____region has severe climate with short growing period for plants

(*a*) Tropical (*b*) Temperate

(*c*) Mediterrenean (*d*) None of the above

184. What happens to biodiversity when we move from high to low latitude

(*a*) Increases (*b*) Decreases

(*c*) Doesn't change (*d*) Difficult to predict

185. Reduction in diversity takes place in the following conditions

(*a*) Drop in temperature (*b*) Seasonal variability

(*c*) Cold (*d*) Both (*a*) and (*b*)

186. Sunderbans tiger reserve is in

(*a*) Andhra Pradesh (*b*) West Bengal

(*c*) Madhya Pradesh (*d*) Uttar Pradesh

187. Royal National Park is in

(*a*) Australia (*b*) USA

(*c*) UK (*d*) India

188. The Man and Biosphere programme was started by

(*a*) UNICEF (*b*) UNESCO
(*c*) IMF (*d*) WWF

189. Prominent animals found in Jaisalmer National Park are

(*a*) Great Indian Bustard (*b*) Lions
(*c*) Chinkara (*d*) All (*a*), (*b*) and (*c*)

190. A species with low genetic diversity results into____

(*a*) Separation (*b*) Diversion
(*c*) Uniformity (*d*) Both (*a*) and (*b*)

191. The main endemic zones in India are

(*a*) The North East
(*b*) Western Ghats
(*c*) North-West Himalayas
(*d*) All of the above

192. The subtle form of habitat degradation is

(*a*) Environmental pollution
(*b*) Synthetic compounds usage
(*c*) Spillover of oil in sea
(*d*) None of the above

193. Match the following

1.	Jim Corbet National Park	(*a*)	Chinkara
2.	Gir National Park	(*b*)	One Horned Rhinoceros
3.	Kaziranga National Park	(*c*)	Asiatic Lion
4.	Jaisalmer Desert National Park	(*d*)	Tiger

194. Match the following

1.	Gir National Park	(*a*)	West Bengal
2.	Kaziranga National Park	(*b*)	Gujarat
3.	Kanha National Park	(c)	Assam
4.	Sunderbans	(*d*)	Madhya Pradesh

195. Match the following

1.	Critically Endangered	(*a*)	Black buck
2.	Endangered	(*b*)	Pigmy hog
3.	Vulnerable	(*c*)	Red Panda
		(*d*)	Goat

196. Match the following

1.	Vulnerable	(*a*)	Cupressus
2.	Endangered	(*b*)	*Ficus bengalensis*
3.	Critically Endangered	(c)	*Beutinckia nicobarica*
		(*d*)	*Barberis nilghiriensis*

197. Match the following

1.	Chilka lake	(*a*)	Kerala
2.	Periyar Sanctuary	(*b*)	Orissa
3.	Manas wild life Sanctuary	(*c*)	Chandigarh
		(*d*)	Assam

198. Match the following

1.	Primitive Genus	(*a*)	Magnolia
2.	Exotic Specius	(*b*)	Lantana Camara
3.	Anti cancer drug	(*c*)	Yewtree
		(*d*)	Humans

199. Match the following

1.	Biodiversity	(*a*)	Legally protected
2.	Sanctuary	(*b*)	Special category of Protected Areas
3.	Biosphere Reserve	(*c*)	Conservation of animals
4.	Natural zone	(*d*)	Totality of Genes and Species

200. Match the fresh water lakes and their location

1.	Wular	(*a*)	Kashmir
2.	Chilka	(*b*)	Andhra Pradesh
3.	Kolleru	(*c*)	Orissa
4.	Manasbal	(*d*)	Srinagar

201. Match the following

1.	Genetic Diversity	(*a*)	Variations in Biological communities
2.	Species Diversity	(*b*)	Variety of Species without a region
3.	Community Diversity	(c)	Genetic Variation within a species

202. Match the following

1.	Alpha Diversity	(a)	Diversity over the total land scape
2.	Beta Diversity	(*b*)	Diversity within community
3.	Gamma Diversity	(*c*)	Diversity between communities

203. Match the following

1.	Natural Extinction	(*a*)	Due to Human activities
2.	Mass Extinction	(*b*)	Due to change in environment
3.	Anthropogenic Extinction	(*c*)	Catastrophe

204. Match the following

1.	Core zone	(*a*)	Outermost part of Biosphere
2.	Buffer zone	(*b*)	Surrounds the core area
3.	Transitional zone	(*c*)	Undistributed and legally protected system

205. Match the following

1.	Cryopreservation	(*a*)	Confined to a country
2.	Specification	(*b*)	Record of threatened species
3.	Red data Book	(c)	Evolution of new species
4.	Endemic	(*d*)	Storage at extremely low temperature

206. Match the following

1.	Wild life sanctuaries	(*a*)	581
2.	Biosphere Reserves	(*b*)	89
3.	Protected Areas	(c)	492
4.	National Parks	(*d*)	13

207. Match the following

1.	Morphine	(*a*)	*Taxus baccata*
2.	Pipal	(*b*)	*Cincharaled giriana*
3.	Quinine	(*c*)	*Ficus religiosa*
4.	Taxol	(*d*)	*Papavareum somniferum*

208. Match the following

1.	Large body	(*a*)	Woodland caribou
2.	Small Population	(*b*)	Blue whale
3.	Fixed Migratory	(*c*)	Lion
4.	Localised Distribution	(*d*)	Giant Panda

Answers

1	(*b*)	23	(*b*)	45	(*c*)	67	(*d*)
2	(*a*)	24	(*a*)	46	(*c*)	68	(*d*)
3	(*b*)	25	(*d*)	47	(*b*)	69	(*b*)
4	(*b*)	26	(*b*)	48	(*b*)	70	(*a*)
5	(*b*)	27	(*b*)	49	(*b*)	71	(*c*)
6	(*b*)	28	(*c*)	50	(*c*)	72	(*d*)
7	(*c*)	29	(*b*)	51	(*d*)	73	(*c*)
8	(*a*)	30	(*a*)	52	(*c*)	74	(*a*)
9	(*c*)	31	(*b*)	53	(*c*)	75	(*b*)
10	(*b*)	32	(*c*)	54	(*b*)	76	(*d*)
11	(*d*)	33	(*b*)	55	(*b*)	77	(*d*)
12	(*a*)	34	(*b*)	56	(*d*)	78	(*d*)
13	(*a*)	35	(*a*)	57	(*a*)	79	(*b*)
14	(*b*)	36	(*b*)	58	(*b*)	80	(*d*)
15	(*a*)	37	(*c*)	59	(*a*)	81	(*d*)
16	(*d*)	38	(*c*)	60	(*c*)	82	(*b*)
17	(*a*)	39	(*b*)	61	(*b*)	83	(*c*)
18	(*b*)	40	(*c*)	62	(*a*)	84	(*b*)
19	(*b*)	41	(*b*)	63	(*c*)	85	(*b*)
20	(*d*)	42	(*b*)	64	(*c*)	86	(*a*)
21	(*d*)	43	(*c*)	65	(*c*)	87	(*b*)
22	(*b*)	44	(*b*)	66	(*b*)	88	(*d*)

89	(*d*)	116	(*a*)	143	(*a*)	169	(*d*)
90	(*b*)	117	(*b*)	144	(*c*)	170	(*a*)
91	(*c*)	118	(*c*)	145	(*c*)	171	(*b*)
92	(*a*)	119	(*e*)	146	(*d*)	172	(*c*)
93	(*b*)	120	(*a*)	147	(*b*)	173	(*c*)
94	(*d*)	121	(*b*)	148	(*c*)	174	(*b*)
95	(*d*)	122	(*b*)	149	(*b*)	175	(*a*)
96	(*d*)	123	(*c*)	150	(*c*)	176	(*d*)
97	(*c*)	124	(*a*)	151	(*c*)	177	(*d*)
98	(*b*)	125	(*b*)	152	(*a*)	178	(*d*)
99	(*c*)	126	(*a*)	153	(*b*)	179	(*d*)
100	(*d*)	127	(*b*)	154	(*a*)	180	(*b*)
101	(*b*)	128	(*a*)	155	(*b*)	181	(*c*)
102	(*c*)	129	(*b*)	156	(*c*)	182	(*b*)
103	(*b*)	130	(*e*)	157	(*b*)	183	(*b*)
104	(*c*)	131	(*b*)	158	(*a*)	184	(*a*)
105	(*a*)	132	(*d*)	159	(*a*)	185	(*d*)
106	(*b*)	133	(*a*)	160	(*b*)	186	(*b*)
107	(*b*)	134	(*b*)	161	(*b*)	187	(*a*)
108	(*a*)	135	(*b*)	162	(*b*)	188	(*b*)
109	(*b*)	136	(*a*)	163	(*c*)	189	(*d*)
110	(*a*)	137	(*b*)	164	(*d*)	190	(*c*)
111	(*b*)	138	(*a*)	165	(*a*)	191	(*d*)
112	(*c*)	139	(*c*)	166	(*a*)	192	(*a*)
113	(*b*)	140	(*a*)	167	(*b*)		
114	(*b*)	141	(*d*)	168	(*s*)		
115	(*b*)	142	(*b*)				

Match the Following

193	**1 (*d*), 2 (*c*), 3 (*b*), 4 (*a*)**	**201**	**1 (*c*), 2 (*b*), 3 (*a*)**
194	**1 (*b*), 2 (*c*), 3 (*d*), 4 (*a*)**	**202**	**1 (*b*), 2 (*c*), 3 (*a*)**
195	**1 (*b*), 2 (*c*), 3 (*a*)**	**203**	**1 (*b*), 2 (*c*), 3 (*a*)**
196	**1 (*a*), 2 (*c*), 3 (*d*)**	**204**	**1 (*c*), 2 (*b*), 3 (*a*)**
197	**1 (*b*), 2 (*a*), 3 (*d*)**	**205**	**1 (*d*), 2 (*c*), 3 (*b*), 4 (*a*)**
198	**1 (*d*), 2 (*b*), 3 (*a*)**	**206**	**1 (*c*), 2 (*d*), 3 (*a*), 4 (*b*)**
199	**1 (*d*), 2 (*c*), 3 (*b*), 4 (*a*)**	**207**	**1 (*d*), 2 (*c*), 3 (*b*), 4 (*a*)**
200	**1 (*d*), 2 (*c*), 3 (*b*), 4 (*a*)**	**208**	**1 (*c*), 2 (*d*), 3 (*b*), 4 (*a*)**

Chapter 5

Environmental Pollution

1. **Pollution is related to**
   (*a*) Water (*b*) Air
   (*c*) Land (*d*) All of the above
2. **Meaning of pollution is**
   (*a*) Excessive addition of harmful materials
   (*b*) Life is made unfit
   (*c*) Both (*a*) and (*b*)
   (*d*) Depletion
3. **Natural pollution is caused by**
   (*a*) Volcanoes (*b*) U.Y. Rays
   (*c*) Soil erosion (*d*) All of the above
4. **Good examples of primary pollutants**
   (*a*) DOT (*b*) Hydro carbons
   (*c*) Plastic (*d*) All of the above
5. **Secondary pollutants are due to**
   (*a*) Interaction among primary pollutants
   (*b*) Addition of primary pollutants

(*c*) Release of primary pollutants

(*d*) All of the above

6. **Examples of qualitative pollutants**

(*a*) Plastic (*b*) DDT

(*c*) Herbicides (*d*) Both (*b*) and (*c*)

7. **The atmosphere of big cities is polluted most by**

(*a*) Household waste (*b*) Radio active

(*c*) Automobile (*d*) Pesticide

8. **Acid rain is caused by increase in the atmospheric concentration of**

(*a*) Ozone and Dust (*b*) SO_2 and NO_2

(*c*) Carbon monoxide (*d*) Nitrous oxide

9. **Gas leaked in Bhopal tragedy was**

(*a*) Potassium Isothiocynate (*b*) Ethyl Isocyte

(*c*) Sodium Isothiocynate (*d*) Methyl Isocyanate

10. **Anthropogenic pollution is due to**

(*a*) Human beings (*b*) Animals

(*c*) Bacteria (*d*) None of the above

11. **Example of Biodegradable pollutant is**

(*a*) Plastic (*b*) Sewage Poisons

(*c*) Glass

12. **Carbon dioxide and nitrogen oxide are examples of**

(*a*) Qualitative pollutants

(*b*) Quantitative pollutants

(*c*) Natural pollutants

(*d*) None of the above

13. **Examples of suspended pollutants**

(*a*) Soot (*b*) Asbestos fibres

(*c*) Pesticides (*d*) All of the above

14. Attack of Asthma in certain persons may be due to

(*a*) Eating of some seasonal vegetation

(*b*) Inhalation of some air borne pollen

(*c*) Exposure to cold temperature

(*d*) Low concentration of CO_2 due to increased rate of photosynthesis

15. Radioactive fall out is due to

(*a*) Nuclear device testing (*b*) Waste recove:y

(*c*) Ore preparation (*d*) Fuel burning

16. Petroleum consists mainly of

(*a*) Hydrocarbons (*b*) Sulphur

(*c*) Nitrogen (*d*) All of the above

17. The larger suspended particles with more than 1 μm in diameter are called

(*a*) Dust (*b*) Mist

(*c*) Smoke (*c*) Both (*a*) and (*b*)

18. The particle size of setteable pollutants is

(*a*) > 10 ìm (*b*) < 10 ìm

(*c*) > 100 ìm (*d*) < 100 ìm

19. Many pollutants become highly diluted above this level____of earth's surface

(*a*) 400 m (*b*) 600 m

(*c*) 800 m (*d*) None of the above

20. The vapor of compounds whose boiling point is below 200°C is called

(*a*) Particulate pollutant (*b*) Suspended pollutant

(*c*) Gaseous pollutant (*d*) All of the above

21. When fossil fuels are burnt they produce a mixture of oxides of

(*a*) Carbon (*b*) Nitrogen

(*c*) Sulphur (*d*) All of the above

22. Nearly 50% of all carbon monoxide emission originates from

(*a*) Industries (*b*) Houses

(*c*) Automobiles (*d*) Lakes

23. ____is present in Cigarette smoke

(*a*) Sulphur (*b*) Oxygen

(*c*) Carbon monoxide (*d*) Iodine

24. Carbon monoxide is short lived in atmosphere and gets converted to

(*a*) SO_2 (*b*) CO_2

(*c*) H_2 (*d*) NO_2

25. The smell of carbon monoxide is

(*a*) Pungent (*b*) Irritating

(*c*) Odourless (*d*) None of the above

26. Oxidation of sulphur in the fossil fuel produces

(*a*) SO_2 (*b*) SO_3

(*c*) O_2 (*d*) Both (*a*) and (*b*)

27. Acid rain consists of

(*a*) Sulphurous acid (*b*) Sulphuric acid

(*c*) Hydrochloric acid (*d*) Both (*a*) and (*b*)

28. Acid precipitation involves precipitation of____as rain or snow

(*a*) Sulphuric acid (*b*) Nitric acid

(*c*) Hydrochloric acid (*d*) All of the above

29. The brown air seen in traffic congested city air consists of

(*a*) Nitrogen oxide (*b*) SO_2

(*c*) Nitrogen dioxide (*d*) Both (*a*) and (*c*)

30. Oxides of nitrogen present in petroleum on oxidation produce

(*a*) Nitrogen dioxide (*b*) Nitrogen oxide

(*c*) Carbon dioxide (*d*) Both (*a*) and (*b*)

31. The most abundant hydrocarbon in nature is

(*a*) Ethane (*b*) Methane

(*c*) Propane (*d*) All of the above

32. Compounds of hydrogen and carbon are called

(*a*) Volatile organic carbons (*b*) Hydrocarbons

(*c*) Hydrogen carbon (*d*) Both (*a*) and (*b*)

33. The antiknock agents in petroleum are

(*a*) SO_2 (*b*) Tetraethyl lead

(*c*) Tetramethyl lead (*d*) Both (*a*) and (*c*)

34. The largest source of air pollution in cities is

(*a*) Trains (*b*) Automobiles

(*c*) Houses (*d*) Steamers

35. The substance which causes cancer is called

(*a*) Teratogenic (*b*) Carcinogenic

(*c*) Hydrogenic (*d*) None of the above

36. Good examples of carcinogenic substances are

(*a*) Benzene (*b*) Formaldehyde

(*c*) Plastic (*d*) Both (*a*) and (*b*)

37. The major gaseous pollutant is

(*a*) CO (*b*) CO_2

(*c*) NO_2 (*d*) C

38. Automobile exhausts and tobacco smoke contain

(*a*) Benzene (*b*) Benzo pyrene

(*c*) Benzocaine (*d*) All of the above

39. Hampers Haemoglobin formation in the human body

(*a*) Carbon (*b*) Lead

(*c*) Arsenic (*d*) Sulphur

40. ____are responsible for the ozone layer depletion

(*a*) Chloro flouro Carbon (*b*) Chloro lead

(*c*) Carbon Tetra Chloride (*d*) All of the above

41. The widely used propellants in aerosols are

(*a*) Chlorofluromethane (*b*) Chloroflourine

(*c*) CO_2 (*d*) All of the above

42. The natural pollutants present in air are

(*a*) Pollen spores (*b*) Microbes

(*c*) CO_2 (*d*) Both (*a*) and (*b*)

43. Burning of plastic produces

(*a*) SO_2 (*b*) CO_2

(*c*) Polychlorinated biphenyls (*d*) All of the above

44. Examples of secondary air pollutants are

(*a*) Smog (*b*) Acid rain

(*c*) Methane (*d*) Both (*a*) and (*b*)

45. Secondary air pollutants are produced by a

(*a*) Oxidation process (*b*) Reduction

(*c*) Photochemical reaction (*d*) Hydrolysis

46. The pH of the acid rain is

(*a*) 3 (*b*) 5.6

(*c*) 7 (*d*) 4

47. Rain water is

(*a*) Slightly acidic (*b*) Basic

(*c*) Neutral (*d*) Difficult to predict

48. PAN stands for

(*a*) Permanent Account No. (*b*) Peroxy Acetyl Nitrate

(*c*) Permitted After Night (*d*) None of the above

49. The acidic water received through rain, for and snow is called

(*a*) Acid rain (*b*) Wet deposition

(*c*) Dry deposition (*d*) None of the above

50. About half of the acidity in the atmosphere is transferred to earth by

(*a*) Wet deposition (*b*) Dry deposition

(*c*) Hydrolysis (*d*) Both (*a*) and (*b*)

51. Particles more than 2 ìm are

(*a*) Coughed out (*b*) Swallowed

(*c*) Digested (*d*) Both (*a*) and (*b*)

52. Pneumoconiosis is a disease produced by

(*a*) Cotton dust (*b*) Zinc dust

(*c*) Raw saw dust (*d*) All of the above

53. Common plants which produce allergic pollen are

(*a*) *Ricinus communis* (*b*) *Cynodon dactylon*

(*c*) *Chemopodium album* (*d*) All of the above

54. The earth is protected from harmful UV rays by

(*a*) Hydrogen (*b*) Ozone

(*c*) Oxygen (*d*) All of the above

55. Dust, Scot and Smog cause several

(*a*) Heart diseases (*b*) Lung diseases

(*c*) Kidney diseases (*d*) All of the above

56. In human beings, hydrogen fluoride causes following disorders

(*a*) Fluorosis (*b*) Mottling of teeth

(*c*) Respiratory diseases (*d*) All of the above

57. Carbon dioxide content of the atmosphere is on rise due to

(*a*) Destruction of forests

(*b*) Excessive burning of fossil fuels

(*c*) Moisture

(*d*) Both (*a*) and (*b*)

58. Green house effect is

(*a*) Houses become green due to heat

(*b*) Rise in global temperature

(*c*) Agricultural growth

(*d*) None of the above

59. ____are used to separate at particulate matter from industrial emission

(*a*) Cyclonic separators

(*b*) Trajectory separators

(*c*) Both (*a*) and (*b*)

(*d*) None of the above

60. The most effective device to remove particulate pollutants is

(*a*) Electrostatic precipitator (*b*) Separator

(*c*) Arresters (*d*) All of the above

61. Gaseous pollutants are removed by means of

(*a*) Filters (*b*) Scrubbers

(*c*) Separators (*d*) None of the above

62. Greenhouse effect may cause

(*a*) Melting of glaciers

(*b*) Flooding of low lying areas

(*c*) Submersion of islands

(*d*) All of the above

63. Electrostatic precipitators find large applications in

(*a*) Chimneys

(*b*) Thermal power plants

(*c*) Houses (*d*) All of the above

64. Water pollution by organic wastes is measured in terms

(*a*) cm (*b*) BOD

(*c*) COD (*d*) DOD

65. BOD stands for

(*a*) Biological Ox Demand

(*b*) Biological Oxygen Demand

(*c*) Biological Oven Demand

(*d*) None of the above

66. BOD is expressed

(*a*) mg of O_2/lit of water (*b*) mg of CO_2/lit of water

(*c*) Grams of O_2/lit of water (*d*) None of the above

67. The flow of pollutants through sewerage system is known as

(*a*) Diffuse source (*b*) Non-point source

(*c*) Point source (*d*) All of the above

68. Biological Oxygen Demand (BOD) is measured for

(*a*) 5 days at 20°C in the dark

(*b*) 4 days at 20°C in the light

(*c*) 5 days at 21°C in the dark

(*d*) None of the above

69. A sample of water with weak organic waste has BOD

(*a*) Above 1500 gm/litre (*b*) Below 1500 gm/litre

(*c*) Above 2000 gm/litre (*d*) Cannot predict

70. A high BOD is an indication of

(*a*) Less microbial pollution

(*b*) Intense level microbial pollution

(*c*) No microbial pollution at all

(*d*) None of the above

71. BOD is not a reliable method because it measures only

(*a*) Physical waste (*b*) Organic waste

(*c*) Sulphur (*d*) None of the above

72. The process of decomposition of organic matter present in water by microorganisms using oxygen is known as

(*a*) Nitrification (*b*) Putrescibility

(*c*) Symbiosis (*d*) Osmosis

73. Deoxygenation of water results in

(*a*) Foul smelling gases

(*b*) Change in colour of water

(*c*) Good taste

(*d*) All of the above

74. The superiority of chemical oxygen demand is that

(*a*) It measures water pollution of both organic and inorganic wastes

(*b*) Simple method

(*c*) Not costly

(*d*) All of the above

75. Biochemical oxygen demand means

(*a*) Amount of dissolved oxygen

(*b*) Oxygen needed by bacteria

(*c*) Needed by organic wastes

(*d*) Both (*a*) and (*b*)

76. What happens when oxygen is not present in water

(*a*) Water becomes blue

(*b*) Water gives bad smell

(*c*) Aquatic life in water is destroyed

(*d*) Both (*b*) and (*c*)

77. Oxygen demand of microorganisms in waste pollutants can be measured by

(*a*) Biochemical Oxygen Demand

(*b*) Chemical Oxygen Demand

(*c*) Oxygen Physical Demand

(*d*) Both (*a*) and (*b*)

78. Match the following

1.	Primary pollutant	(*a*)	Volcanic eruptions
2.	Qualitative pollutant	(*b*)	DDT
3.	Anthropogenic pollutant	(*c*)	Fungicides
4.	Natural pollutant	(*d*)	Man made pollutant

79. Match the following

1.	Degradable pollutant	(*a*)	CO_2
2.	Non-degradable pollutant	(*b*)	Plastic
3.	Primary pollutant	(*c*)	Sewage
4.	Quantitative pollutant	(*d*)	Poisons

80. Match the following

1.	Dust	(*a*)	Liquid
2.	Mist	(*b*)	Solid
3.	Setteable pollutants	(*c*)	Particles larger than 10 μm in diameter
4.	Suspended pollutants	(*d*)	Particles less than 10 μm in diameter

81. Match the following

1.	Crop spraying	(*a*)	Sulphur and Nitrogen oxides
2.	Waste recovery	(*b*)	Radio active fallout
3.	Nuclear device testing	(*c*)	Organophosphates
4.	Fuel burning	(*d*)	Smoke, soot etc

82. Match the following

1.	Oxides of carbon	(*a*)	Hydrogen and Carbon
2.	Oxides of sulphur	(*b*)	Carbon monoxide
3.	Oxide of nitrogen	(*c*)	Nitrogen dioxide
4.	Hydrocarbons	(*d*)	Sulphur dioxide
		(*e*)	Carbon tetrachloride
		(*f*)	Nitrogen mustard

83. Match the following

1. Carbon monoxide	(*a*) Coal and Petroleum
2. Hydro Carbons	(*b*) Fossil fuel
3. Sulphur Dioxide	(*c*) Organic matter decomposition
4. Nitrogen Dioxide	(*d*) Automobile exhaust

84. Match the following

1. Gaseous pollutants	(*a*) Carcinogenic
2. Particulate lead	(*b*) Chloroflouro carbons
3. Benzopyrene	(c) Haemoglobin destruction
4. Aerosols	(*d*) Lung cancer

85. Match the following

1. Dust and scot	(*a*) Hay fever
2. Fly ash	(*b*) Pneumoconiosis
3. Cotton dust	(c) Respiratory disorder
4. Pollens	(*d*) Headache, dizziness

86. Match the following

1. Scrubbers	(*a*) Automobile exhaust
2. Filters	(*b*) Gaseous
3. Adsorption	(c) Fine particulate matter
4. Catalytic	(*d*) Gaseous pollutants

87 Burning of coal produces

(*a*) Mineral ash (*b*) Fly ash

(*c*) Smoke (*d*) All of the above

88. An accidental discharge of petroleum in oceans and estuaries is known as

(*a*) Organic spill (*b*) Oil spill

(*c*) Waste spill (*d*) None of the above

89. The measure of oxygen equivalent of the requirement for oxidation of total organic matter

(*a*) Chemical Oxygen Demand

(*b*) Biological Oxygen Demand

(*c*) Microbial Demand

(*d*) None of the above

90. Water having dissolved oxygen content less than 8 mg/litre

(*a*) Good water (*b*) . Potable water

(*c*) Contaminated water (*d*) Both (*b*) and (*c*)

91. The factors which determine the amount of dissolved oxygen present in water are

(*a*) Surface turbulence

(*b*) Photosynthetic activity

(*c*) Decomposition of organic matter

(*d*) All of the above

92. Due to inorganic compounds and decomposition of organic wastes there is a profuse growth of blue green algae on water surface. This is known as

(*a*) Algal bloom (*b*) Algal gloom

(*c*) Algae profusion (*d*) Both (*b*) and (*c*)

93. Presence of excess nitrate in drinking water leads to

(*a*) Blue baby syndrome (*b*) Grey baby syndrome

(*c*) Methaemoglobinemia (*d*) Both (*a*) and (*c*)

94. Excess fluoride in drinking water causes

(*a*) Teeth deformity (*b*) Hardening of bones

(*c*) Stiff and painful joints (*d*) All of the above

95. Chronic exposure to Arsenic causes

(*a*) Grey baby syndrome (*b*) Hey Baby Syndrome

(*c*) Black foot disease (*d*) All of the above

96. Minamata disease is due to taking contaminated fish with

(*a*) Arsenic (*b*) Mercury

(*c*) Lead (*d*) Both (*a*) and (*c*)

97. Match the following

1.	Minamata disease	(*a*)	Excess nitrates in drinking water
2.	Black foot disease	(*b*)	Excess fluorides in drinking
3.	Flourosis	(c)	Chronic exposure to Arsenic
4.	Blue baby syndrome	(*d*)	Mercury indigestion

98. Water contaminated with cadmium can cause

(*a*) Itai Itai disease (*b*) Ouch-Ouch disease

(*c*) Red colour disease (*d*) Both (*a*) and (*b*)

99. A painful disease of bones and joints is

(*a*) Ouch-Ouch disease (*b*) Black foot disease

(*c*) Blue baby syndrome (*d*) Mina mata disease

100. Maximum level of fluoride which the human body can tolerate is

(*a*) 1.5 parts per million (mg I litre of water)

(*b*) 2.5 parts per million

(*c*) 3.5 parts per million

(*d*) None of the above

101. High concentration of fluorides can be found in____states in India

(*a*) 10 (*b*) 13

(*c*) 15 (*d*) 20

102. Identify the plant which produces Allergic pollen

(*a*) *Sorghum vulgure* (*b*) *Ricinus communis*

(*c*) *Papaverum somniferum* (*d*) Both (*a*) and (*b*)

103. Depletion of ozone layer exposes the earth to

(*a*) Hot sun (*b*) UV rays

(*c*) Cosmic rays (*d*) Cold air

104. Increased ultraviolet exposure of human beings may lead to

(*a*) Heart problem (*b*) Skin cancer

(*c*) Mutations (*d*) Both (*b*) and (*c*)

105. The ultraviolet radiations in the stratosphere are absorbed by

(*a*) Ozone (*b*) Oxygen

(*c*) Sulphur dioxide (*d*) Argon

106. Nitrogen oxide and hydrocarbons released by automobiles interact to form

(*a*) Carbon monoxide (*b*) Sulphur dioxide

(*c*) PAN (*d*) Aerosols

107. Which of the following groups of plants can be used as indicators of SO_2 pollution of air

(*a*) Ferns (*b*) Epiphytic lichens

(*c*) Liver worts (*d*) Horn worts

108. Dysentery is spread due to

(*a*) Food adulteration (*b*) Humid weather

(*c*) Water pollution (*d*) Air pollution

109. Maximum deposition of DDT will occur in

(*a*) Phytoplankton (*b*) Crab

(*c*) Eel (*d*) Seagull

110. Fluoride pollution mainly affects

(*a*) Kidney (*b*) Brain

(*c*) Heart (*d*) Teeth

111. Increasing skin cancer and high mutation rate are the result

(*a*) Ozone depletion (*b*) Acid rain

(*c*) CO_2 (*d*) CO pollution

112. Taj Mahal at Agra may be damaged by

(*a*) Sulphur dioxide (*b*) Chlorine
(*c*) Hydrogen (*d*) Oxygen.

113. Sound becomes hazardous noise pollution at decibels

(*a*) Above 30 (*b*) Above 80
(*c*) Above 100 (*d*) Above 120

114. Biochemical Oxygen demand measures

(*a*) Industrial pollution
(*b*) Air pollution
(*c*) Polluting capacity of effluents
(*d*) Dissolved O_2 needed by microbes to decompose organic waste

115. Automobiles are the largest source of____pollution in cities

(*a*) Water (*b*) Air
(*c*) Noise (*d*) Both (*b*) and (*c*)

116. Increase in concentration of soluble salts in soil is called

(*a*) Chlorination (*b*) Bromination
(*c*) Salination (*d*) Both (*a*) and (*b*)

117. The total land mass of India is____million hectares

(*a*) 305 (*b*) 350
(*c*) 400 (*d*) None of the above

118. Minamata disease is caused due to biomagnification of

(*a*) Arsenic (*b*) Lead
(*c*) Mercury (*d*) Sulphur

119. Peroxy acetyl Nitrate (PAN) is formed by the interaction of oxides of nitrogen and____in presence of sunlight

(*a*) Sulphur (*b*) Hydrocarbons
(*c*) Oxygen (*d*) Nitrogen

120. Units for measurement of sound are

(*a*) Houston (*b*) Decibel
(*c*) Joules (*d*) Ergs

121. Examples of secondary air pollutants are

(*a*) Sulphur dioxide (*b*) Nitrogen dioxide

(*c*) Hydrogen peroxide (*d*) Both (*a*) and (*b*)

122. Expand DDT

(*a*) Delhi Dangerous Thymol

(*b*) Didentrite Thymol

(*c*) Both (*a*) and (*b*)

(*d*) Dichloro Diphenyl Trichloroethane

123. Acid rain is

(*a*) Showers of H_2SO_4 (*b*) Showers of HNO_3

(*c*) Showers of Benzoic acid (*d*) Both (*a*) and (*b*)

124. Converting solid waste into compost with the help of earthworms is known as

(*a*) Vermiculture (*b*) Earthworm farming

(*c*) Worming (*d*) Both (*a*) and (*b*)

125. A byproduct of paper industry is

(*a*) Bregoli (*b*) Braille

(*c*) Rangoli (*d*) Both (*a*) and (*b*)

126. COD stands for

(*a*) Corps of detective (*b*) Cash on Demand

(*c*) Chemical Oxygen Demand (*d*) Both (*a*) and (*b*)

127. Inorganic pollutants of paper and pulp industry are

(*a*) Sulphides (*b*) Blenching liquors

(*c*) Cellulose fibre (*d*) Both (*a*) and (*b*)

128. Pesticides are absorbed by lower organisms and are passed to higher Qrganisms including man through food chain. This is called

(*a*) Biological Magnification (*b*) Biomagnification

(*c*) Both (*a*) and (*b*) (*d*) Bio illumination

129. Examples of marine pollution are

(*a*) Offshore oil (*b*) Oil refineries

(*c*) Capsized oil tanks (*d*) All of the above

130. India has____million hectares of saline land

(*a*) 8 (*b*) 7

(*c*) 6 (*d*) 10

131. The condition of dryness for a prolonged periods is known as

(*a*) Dry day (*b*) Drought

(*c*) Dry period (*d*) Both (*a*) and (*c*)

132. The serious consequences of drought are

(*a*) Famine

(*b*) Starvation of human and animal population

(*c*) Change of the biotic components of natural ecosystem

(*d*) All of the above

133. The disposal of solid wastes can be done by

(*a*) Composting (*b*) Vermiculture

(*c*) Recycling (*d*) All of the above

134. Examples of urban waste are

(*a*) Garbage (*b*) Plastic

(*c*) Glass (*d*) Discarded food

(*e*) All of the above

135. Degradation of organic waste by micro organisms is known as

(*a*) Earthworm farming (*b*) Composting

(*c*) Recycling (*d*) All of the above

136. Vermiculture involves the use of following for converting solid waste into compost

(*a*) Worms (*b*) Bacteria

(*c*) Earthworms (*d*) Both (*a*) and (*b*)

137. Converting a product at the end of its service (waste) to another useful product is known as

(*a*) Vermiculture (*b*) Recycling

(*c*) Composting (*d*) Both (*a*) and (*c*)

138. Disposal of urban waste in sanitary landfill is known as

(*a*) Incineration (*b*) Land filling

(*c*) Vermiculture (*d*) All of the above

139. The process of reducing unwanted material to simple and gaseous residues is known as

(*a*) Oxidation (*b*) Incineration

(*c*) Landfilling (*d*) Both (*a*) and (*c*)

140. The process of controlled burning of the waste at high temperature is

(*a*) Ignition (*b*) Incineration

(*c*) Oxidation (*d*) Both (*a*) and (*b*)

141. The temperature is used for incineration is around

(*a*) 200°C (*b*) 400 °C

(*c*) 850°C (*d*) Both (*b*) and (*c*)

142. ____is an alternative to land filling and provides an efficient means of energy recovery

(*a*) Incineration (*b*) Ignition

(*c*) Composting (*d*) Vermiculture

143. Noise level can range from

(*a*) 0 to 120 dB (*b*) 120 to 240 dB

(*c*) 0 to 10 dB (*d*) All of the above

144. The release of unwanted sound in the atmosphere is called

(*a*) Noise pollution (*b*) Sound pollution

(*c*) Environment pollution (*d*) Both (*b*) and (*c*)

145. Decibel is symbolically represented as

(*a*) dB (*b*) db

(*c*) Db (*d*) All of the above

146. Unpleasant sound that causes discomfort is called as

(*a*) Music (*b*) Bell

(*c*) Noise (*d*) Both (*a*) and (*b*)

147. Sound travels in the form of

(*a*) Pressure waves (*b*) music

(*c*) Noise (*d*) Both (*b*) and (*c*)

148. The intensity of a sound wave is expressed in

(*a*) Decibels Icm^2 (*b*) $W m^{-2}$

(*c*) Weber per $meter^2$ (*d*) Both (*b*) and (*c*)

149. The unit of sound level decibel, a name that is given to recognize the work of

(*a*) Samantha Fox (*b*) Michael Jackson

(*c*) Alexander Graham Bell (*d*) Both (*a*) and (*c*)

150. Ionizing Radiations are

(*a*) X-rays (*b*) Cosmic rays

(*c*) Atomic radiations (*d*) All of the above

151. Short wale radiation is

(*a*) A non ionizing radiation (*b*) Ultraviolet rays

(*c*) Lower penetration power (*d*) All of the above

152. Sources of Radioactive pollution are

(*a*) Natural sources (*b*) Man made sources

(*c*) Both (*a*) and (*b*) (*d*) None of the above

153. The atomic bomb explosions for first time were done at

(*a*) Nagasaki (*b*) Hiroshima

(*c*) Kentucky (*d*) Both (*a*) and (*b*)

154. First atomic bomb explosion took place in the year

(*a*) 1940 (*b*) 1945

(*c*) 1950 (*d*) None of the above

155. The chief source of environmental radioactive pollution are

(*a*) Radio nuclides with long half life
(*b*) Radio nuclides with short half life
(*c*) Both (*a*) and (*b*)
(*d*) None of the above

156. The following is the cause of salination of soil

(*a*) Inadequate drainage
(*b*) Ground water
(*c*) Use of alkaline fertilizers
(*d*) Both (*a*) and (*c*)

157. ____crust of the earth supports land, plants and animals

(*a*) Middle
(*b*) Lower
(*c*) Upper
(*d*) Both (*a*) and (*b*)

158. Soil is

(*a*) Renewable resource
(*b*) Non-renewable resource
(*c*) Permanent resource
(*d*) Both (*a*) and (*b*)

159. Pesticides with broad spectrum activity and affecting all types of life are called

(*a*) Biocides
(*b*) Frusticides
(*c*) Germicides
(*d*) Both (*b*) and (*c*)

160. The harmful effects of chemical fertilizers are

(*a*) Decreases the population of useful bacteria
(*b*) Increases the salt contents of the soil
(*c*) Reduces the productivity of the soil
(*d*) All the above

161. The addition of solid waste in the form of discarded materials to the soil are called

(*a*) Second pollution
(*b*) Third pollution
(*c*) Waste pollution
(*d*) None of the above

162. Important steps for the control of soil pollution are

(*a*) Use of biofertilizers

(*b*) Production of biogas using waste materials

(*c*) Biological pest control

(*d*) All of the above

163. Excess use of sodium nitrate leads to

(*a*) Salination of soil (*b*) Disturbance in air

(*c*) Causes cough (*d*) Both (*b*) and (*c*)

164. UV rays injure the cells of skin producing blisters, this is known as

(*a*) Sun bums (*b*) Skin bums

(*c*) Heart bum (*d*) Both (*b*) and (*c*)

165. Radiations damaging the cells by physically breaking the macro molecules is known as

(*a*) Strontiation (*b*) Ionisation

(*c*) Radiation (*d*) Both (*a*) and (*b*)

166. Long exposure of ionization radiations leads to

(*a*) Mutations (*b*) Tumors

(*c*) Cancers (*d*) All of the above

167. The intensity of normal conservation sound ranges between

(*a*) 35 dB to 60 dB (*b*) 70 dB to 90 dB

(*c*) 10 dB to 15 dB (*d*) None of the above

168. When the ambient sound intensity is equal to the reference intensity, the noise level is

(*a*) 10 dB (*b*) 0 dB

(*c*) 15 dB (*d*) Both (*a*) and (*c*)

169. Prolonged exposure to noise of____or more may lead to impairment of hearing

(*a*) 80dB (*b*) 70dB

(*c*) 60 dB (*d*) None of the above

170. A noise above this level becomes painful and unbearable

(*a*) 180 dB (*b*) 140 dB

(*c*) 120 dB (*d*) None of the above

171. Noise level permissible for residential area is (both morning night)

(*a*) 75 dB, 70 dB (*b*) 55 dB, 45 dB

(*c*) 50 dB, 40 dB (*d*) None of the above

172. Example of industrial waste is

(*a*) Used batteries (*b*) Rrugs

(*c*) Petrochemical waste (*d*) Both (*a*) and (*c*)

173. An example of eco-friendly technique is

(*a*) Decomposition (*b*) Cycling

(*c*) Recycling of waste (*d*) Both (*a*) and (*b*)

174. Hurricanes occur in

(*a*) Caribbean sea (*b*) South eastern USA

(*c*) India (*d*) Both (*a*) and (*b*)

175. In Australia, cyclones are called as

(*a*) Hurricanes (*b*) Typhoons

(*c*) Willy Willy (*d*) John Wiley

176. Typical characteristics of cyclones are

(*a*) High tidal surges

(*b*) High intensity of rainfall

(*c*) Very low atmospheric pressure

(*d*) All of the above

177. Cyclones are called Typhoons in the following

(*a*) North Pacific Ocean

(*b*) Southern coast of chin:

(*c*) Japan

(*d*) Philippines

(*e*) All of the above

178. Match the following

1.	Cyclone	(*a*)	Caribbean sea
2.	Typhoon	(*b*)	USA
3.	Hurricane	(c)	Japan
4.	Tornado	(*d*)	Bangladesh

179. Match the following

1.	Incineration	(*a*)	Earth worms
2.	Land filling	(*b*)	Waste management
3.	Recycling	(c)	Degradation of organic waste in microorganisms
4.	Composting	(*d*)	Sanitary landfill
5.	Vermiculture	(*e*)	Controlled burning

180. The following are the drought control measures

(*a*) Afforestation (*b*) Dry farming

(*c*) Water conservation (*d*) All of the above

181. In the context of earth quakes, the point of sudden energy release is known as

(*a*) Centre (*b*) Focus

(*c*) Middle (*d*) Both (*a*) and (*c*)

182. The magnitude or intensity of energy released by an earth quake is measured by

(*a*) Richter scale (*b*) Decibel scale

(*c*) Nautical scale (*d*) Both (*b*) and (*c*)

183. The place of the origin of an earthquake is called

(*a*) Focus (*b*) Hypocentre

(*c*) Middle centre (*d*) Both (*a*) and (*b*)

184. Earthquake measurement intensity was first time devised by

(*a*) Charles Darwin (*b*) Einstein

(*c*) Charles F. Richter (*d*) Lincoln

185. **The seismic wave movement of an earthquake can be recorded by**

(*a*) Stachnometer (*b*) Seismometer

(*c*) Galvanometer (*d*) Voltmeter

186. **The place on the ground surface recording the seismic waves for the first time is called**

(*a*) Focus (*b*) Epicentre

(*c*) Original centre (*d*) Both (*a*) and (*c*)

187. **The study of earthquakes is called**

(*a*) Seismology (*b*) Graphology

(*c*) Earthology (*d*) Both (*b*) and (*c*)

188. **Earth quakes are caused due to____in any part of the crust of the earth**

(*a*) Fire (*b*) Disequilibrium

(*c*) Discontent (*d*) Both (*a*) and (*c*)

189. **Match the following**

1. Earth Quake	(*a*) mg of O_2/litre of water
2. Noise	(*b*) μm
3. BOD	(c) Decibels
4. Particular matter	(*d*) Richter scale

190. **Hay fever is caused due to**

(*a*) Deficiency of Vitamin C (*b*) Allergic reaction

(*c*) Loss of iron (*d*) Both (*a*) and (*c*)

191. **Byssinosis is caused due to**

(*a*) Smoking (*b*) Cotton dust

(*c*) Illegal sexual contacts (*d*) Kissing

192. **Sulphur dioxide on inhalation causes**

(*a*) Drying of mouth (*b*) Sore throat

(*c*) Eye irritation (*d*) All of the above

193. What happens when particles less than 2 ìm are inhaled

(*a*) Trapped in nasal hair

(*b*) Trapped in bronchial mucus

(*c*) Coughed out at or swallowed

(*d*) All of the above

194. ____combine with haemoglobin of the blood and reduce oxygen carrying capacity of the blood

(*a*) Sulphur trioxide (*b*) Nitrogen oxide

(*c*) Carbon monoxide (*d*) All of the above

195. Ozone in the lower atmosphere causes

(*a*) Chest pain (*b*) Stomach ulcer

(*c*) Coughing (*d*) Both (*a*) and (*c*)

196. Risk of cancer of lung in smokers is due to

(*a*) Benzopyrene (*b*) Benzoyl chloride

(*c*) Benzyl ammonium (*d*) Both (*b*) and (*c*)

197. Silicosis is caused in____industry

(*a*) Asbestos (*b*) Cement

(*c*) Stone grinding (*d*) Wheat

198. Grey air is due to

(*a*) Incomplete smog formation (*b*) Red colour of oxides

(*c*) Less solar radiation (*d*) Both (*a*) and (*c*)

199. Main secondary air pollutions are

(*a*) Photochemical smog (*b*) Acid rain

(*c*) Greenhouse effect (*d*) Both (*a*) and (*b*)

200. Oxides of Sulpher dioxide and nitrogen are produced during combination of

(*a*) Petroleum (*b*) Coal

(*c*) Both (*a*) and (*b*) (*d*) None of the above

201. Terrain Risk areas are identified by

(*a*) TV (*b*) Remote Sensing

(*c*) Mental prediction (*d*) Both (*a*) and (*c*)

202. Match the following

1.	Urban waste	(*a*)	Hypocentre
2.	Industrial waste	(*b*)	Waste management
3.	Recycling	(c)	Acids and alkalis
4.	Earthquake	(*d*)	Plastic glass

203. Permissible ambient noise levels for silent zone are (day and night)

(*a*) 50 dB, 40 dB (*b*) 75 dB, 70 dB

(*c*) 65 dB, 55 dB (*d*) None of the above

204. Radioactive zinc gets more accumulated in

(*a*) Fish (*b*) Oysters

(*c*) Dog (*d*) Both (*a*) and (*c*)

205. Examples of radioactive isotopes reaching human body through food chain are

(*a*) Radioactive iodine (*b*) Zinc

(*c*) Radioactive phosphorous (*d*) Both (*a*) and (*c*)

206. Spontaneous emission of protons, electrons and gamma rays is known as

(*a*) Fission (*b*) Fusion

(*c*) Radioactivity (*d*) None of the above

207. The time needed for half of the radioactive nuclide to decay is

(*a*) Half-life (*b*) Better life

(*c*) Radio potency (*d*) Both (*b*) and (*c*)

208. Half life of a radio nuclide refers to its

(*a*) Period of radioactivity (*b*) Time of emission

(*c*) Quality of emission (*d*) Difficult to predict

209. The fission material for atomic explosion is

(*a*) Uranium–235 (*b*) Plutonium–239

(*c*) Radio Iodine (*d*) Both (*a*) and (*b*)

210. The following is used as a nuclear fuel

(*a*) Uranium–235 (*b*) Plutarch

(*c*) Radio Iodine (*d*) None of the above

211. Match the following

1.	0dB	(*a*)	Noise level range
2.	0–120 dB	(*b*)	No signal
3.	120dB	(c)	Normal conversation range
4.	35 dB to 60 dB	(*d*)	Physical discomfort

212. Match the following

1.	Industry	(*a*)	50 dB
2.	Cornmercial	(*b*)	55 dB
3.	Residential	(c)	65 dB
4.	Silent zone	(*d*)	75 dB

213. Zone wise permissible ambient noise levels are prescribed by

(*a*) Central Pollution Control Board

(*b*) Pollution Control Society

(*c*) Pollution Control Federation

(*d*) None of the above

214. Oxidation ponds release CO_2 and lead to formation of

(*a*) Sludge (*b*) Bio solid

(*c*) Bio residue (*d*) Both (*a*) and (*b*)

215. Chemical oxidation of waste water can be done by

(*a*) Chlorine gas (*b*) Ozone gas

(*c*) UV radiation (*d*) All of the above

216. The addition of solid waste to soil is called as

(*a*) First pollution (*b*) Second pollution

(*c*) Third pollution (*d*) Both (*a*) and (*b*)

217. Match the following

Examples of DDT Biological magnification

1.	Phytoplankton	(*a*)	4.83 ppm
2.	Zooplankton	(*b*)	124 ppm
3.	Small fish	(*c*)	1.04 ppm
4.	Fish eating birds	(*d*)	0.123 ppm
5.	Predatory fish	(*e*)	0.025 ppm

218. Match the following

1.	DNES	(*a*)	Mumbai
2.	ITRC	(*b*)	Lucknow
3.	TERI	(*c*)	Bhuj
4.	CSSRI	(*d*)	New Delhi

219. World Environment day is

(*a*) 5th June (*b*) 6th June

(*c*) 4th June (*d*) 7th June

220. United Nations day is on

(*a*) 10th October (*b*) 24th October

(*c*) 15th October (*d*) 12th October

221. World Population day is

(*a*) 11th July (*b*) 12th July

(*c*) 15th July (*d*) None of the above

222. Expand NEERI

(*a*) National Environmental Engineering Research Institute

(*b*) National Electrical and Electronics Research Institute

(*c*) National Environmental Education and Research Institute

(*d*) None of the above

223. Anti Tobacco day is on

(*a*) 31st May (*b*) 29th May

(*c*) 30th May (*d*) 28th May

224. Match the following

1.	WHO	(*a*)	Paris
2.	IAEA	(*b*)	Vienna
3.	UNESCO	(*c*)	Geneva

Fill in the Blanks

225. NEPA stands for____

226. Ozone day is celebrated on____

227. Water Pollution by organic wastes is measured in terms of____

228. ____measures all oxygen consuming pollutant material in water

229. The larger liquid suspended particles with more than 1 ìm in diameter are called____

230. Nearly 50% of all emissions originate from____

231. Match the following

1.	Fertilizer factory	(*a*)	Radioactive fallout
2.	Waste recovery	(*b*)	Sulphur and Nitrogen oxides
3.	Nuclear device testing	(*c*)	Sulphur Oxide
4.	Fuel burning	(*d*)	Smoke, metal fumes

Fill in the Blanks

232. When fossil fuels are burnt they produce a mixture of oxides of____

233. Carbon monoxide is a____and____gas

234. The most abundant hydrocarbon in the atmosphere is____

235. ____is carcinogenic and produced in automobile exhausts and tobacco smoke.

236. Petrol contains tetraethyl lead which is a____agent.

237. During Bhopal gas tragedy there was accidental release of ____which caused several deaths.

238. Sulphur dioxide and sulphur trioxide react with water in the air to produce____

239. A very long common example in the day to day for non-degradable pollutant is

(*a*) Clothes (*b*) Bags (cotton)

(*c*) Polythene shopping bags (*d*) All of the above

240. Man made pollution is known as____pollution

(*a*) Homogenic (*b*) Heterogenic

(*c*) Anthropogenic (*d*) Both (*a*) and (*b*)

241. Indoor pollution is caused commonly

(*a*) CO_2 (*b*) Formaldehyde

(*c*) SO_2 (*d*) Both (*a*) and (*c*)

242. Burning of organic matter and fossil fuel produces

(*a*) Hydrocarbons (*b*) Chlorine

(*c*) SO_2 (*d*) Both (*b*) and (*c*)

243. Formaldehyde is a derivative of

(*a*) Benzene (*b*) Chlorine

(*c*) Oxygen (*d*) Both (*b*) and (*c*)

244. Bhopal gas tragedy took place on

(*a*) 2nd Dec. 1984 (*b*) 1 Dec. 1984

(*c*) 3rd Dec. 1984 (*d*) 5th Dec. 1984

245. Peroxy acetyl Nitrate (PAN) of smog causes the following in plants

(*a*) Damages chloroplasts

(*b*) Inhibits electron transport system

(*c*) Interferes with enzyme systems

(*d*) All of the above

246. Excessive calcification of bone and teeth is known as

(*a*) Laminosis (*b*) Cheliosis

(*c*) Fluorosis (*d*) Both (*a*) and (*b*)

247. The toxic contaminants in exhaust and fuels can be minimized by using

(*a*) Compressed Natural Gas (CNG)

(*b*) Methane

(*c*) Carbon Tetra Chloride

(*d*) Both (*b*) and (*c*)

248. Throughout the globe, the global temperature is increasing by 2 or 3 °C because of

(*a*) Lack of good air (*b*) Green house effect

(*c*) People are becoming hot (*d*) Both (*a*) and (*c*)

249. Match the following

1.	Black foot disease	(*a*)	Mercury contamination
2.	Fluorosis	(*b*)	Excess Nitrates
3.	Blue baby syndrome	(c)	Excess fluorine in water
4.	Mina Mata disease	(*d*)	Arsenic contamination

250. The compounds of lead cause

(*a*) Anaemia (*b*) Headache

(*c*) Loss of muscle power (*d*) All of the above

251. Hyperkeratosis is due to excess of_____in body

(*a*) Arsenic (*b*) Lead

(*c*) Chromium (*d*) Nickel

252. Use of one tonne of recycled paper can save

(*a*) 100 trees (*b*) 17 trees

(*c*) 50 trees (*d*) Difficult to predict

253. Only the radioactive wastes with____can be discharged into sewerage

(*a*) High radiation (*b*) Low radiation

(*c*) Medium radiation (*d*) None of the above

254. The urban solid waste can be disposed off by an anaerobic digestion, process known as

(*a*) Recycling (*b*) Vermiculture

(*c*) Composting (*d*) None of the above

255. Loss of species diversity due to nutrient enrichment of water is known as

(*a*) Photochemical smog (*b*) Eutrophication

(*c*) Wet deposition (*d*) Both (*a*) and (*c*)

256. The substance which causes pollution is known as

(*a*) Polluant (*b*) Pollant

(*c*) Pollutant (*d*) All of the above

257. Acid rain is

(*a*) Showers of acid

(*b*) Showers of blessings from god

(*c*) Tears of gods in the form of rain

(*d*) Both (*b*) and (*c*)

258. Importance of ozone layer

(*a*) Protects us from harmful UV radiation

(*b*) Gives colour and beauty to our female population

(*c*) It protects ladies from becoming ugly

(*d*) Both (*b*) and (*c*)

259. Most of the air pollutants cause harmful effects on

(*a*) Alimentary canal (*b*) Cheeks

(*c*) Nose and Lips (*d*) Respiratory system

260. Main causes of increased CO_2 concentration in atmosphere

(*a*) Cutting of trees

(*b*) Burning of Hydrocarbons

(*c*) Use of wood for fuel

(*d*) All of the above

261. The most earthquake prone countries are

(*a*) Philippines (*b*) Japan

(*c*) Indonesia (*d*) All of the above

262. Taj Mahal was recently in news because of

(*a*) Marble damage by pollutants

(*b*) Drying up of Yamuna river

(*c*) Lovers being killed

(*d*) None of the above

263. The natural disaster Management involves

(*a*) Relief measures

(*b*) Disaster prediction

(c) Disaster research

(*d*) Geographic information system and surveys

(*e*) All of the above

264. Health dangers after a natural disaster are predominantly________in character

(*a*) Environmental (*b*) Medical

(*c*) Personal (*d*) All of the above

265. Floods can be controlled by adopting suitable measures like

(*a*) Hasten the discharge of water

(*b*) Divert the flow of flood water

(*c*) Forewarn regarding floods

(*d*) All of the above

266. An earthquake is

(*a*) A mild tremor

(*b*) Wild motion

(*c*) Fissures and cracks in the ground

(*d*) All of the above

267. Cyclones are common in

(*a*) USA, China (*b*) Japan, Philippines

(*c*) India (*d*) All of the above

268. During cyclones winds blow at a speed of

(*a*) 50 km/hr (*b*) 180–400 km/hr

(*c*) Slowly (*d*) Both (*a*) and (*c*)

269. The byproducts of composting are

(*a*) Organic manure (*b*) Methane

(*c*) CO_2 (*d*) Both (*a*) and (*b*)

270. Landslides occur in

(*a*) Hilly areas (*b*) Mountains

(*c*) Rivers (*d*) Both (*a*) and (*b*)

271. Landslides occur due to

(*a*) Shock produced by earthquake

(*b*) Rain water

(*c*) Heavy pollution

(*d*) Both (*b*) and (*c*)

272. Tsunami was due to

(*a*) Heavy rains

(*b*) Heavy pollution

(*c*) The seismic waves of earthquake

(*d*) Difficult to predict

273. Radioactivity is

(*a*) Radio Mirchi FM programme

(*b*) Views activity due to radios

(*c*) Emission of protons and electrons by certain elements

(*d*) Both (*a*) and (*b*)

274. Examples of electrons are

(*a*) Alpha particles (*b*) Beta particles

(*c*) Gamma rays (*d*) None of the above

275. Gamma rays are

(*a*) Short wave electromagnetic rays

(*b*) Used for sterilization of antibiotics

(*c*) Used for sterilization of fruits

(*d*) All of the above

276. Groundwater becomes saline by

(*a*) Capillary action (*b*) Leaching

(*c*) Absorption (*d*) All of the above

277. What happens when the pH of water is below 5

(*a*) Aquatic life ends (*b*) Plants die

(*c*) People enjoy the taste (*d*) Both (*a*) and (*b*)

278. The death of lichens in an area is an indicator of

(*a*) Air pollution (*b*) Sound pollution

(*c*) Water pollution (*d*) All of the above

279. Byssinosis is also known as

(*a*) Pneumoconiosis (*b*) Lung fibrosis

(*c*) Silicosis (*d*) Both (*a*) and (*b*)

280. Tobacco causes abnormal changes in____of the smokers

(*a*) Heart

(*b*) Bronchial epithelium of lungs

(*c*) Heart muscle

(*d*) Kidney

281. Passive smoking is from

(*a*) Those who smoke

(*b*) Those who don't smoke

(*c*) Both (*a*) and (*b*)

(*d*) Difficult to answer

Answers

1	(*d*)	23	(*c*)	45	(*c*)	67	(*c*)
2	(*c*)	24	(*b*)	46	(*b*)	68	(*a*)
3	(*d*)	25	(*c*)	47	(*a*)	69	(*b*)
4	(*d*)	26	(*d*)	48	(*b*)	70	(*b*)
5	(*a*)	27	(*d*)	49	(*b*)	71	(*b*)
6	(*d*)	28	(*a*)	50	(*b*)	72	(*b*)
7	(*c*)	29	(*d*)	51	(*d*)	73	(*a*)
8	(*b*)	30	(*d*)	52	(*a*)	74	(*a*)
9	(*d*)	31	(*b*)	53	(*d*)	75	(*d*)
10	(*a*)	32	(*d*)	54	(*b*)	76	(*d*)
11	(*b*)	33	(*d*)	55	(*b*)	77	(*d*)
12	(*b*)	34	(*b*)	56	(*d*)	87	(*d*)
13	(*d*)	35	(*b*)	57	(*d*)	88	(*b*)
14	(*b*)	36	(*d*)	58	(*b*)	89	(*a*)
15	(*a*)	37	(*a*)	59	(*c*)	90	(*c*)
16	(*d*)	38	(*b*)	60	(*a*)	91	(*d*)
17	(*d*)	39	(*b*)	61	(*b*)	92	(*a*)
18	(*a*)	40	(*a*)	62	(*d*)	93	(*d*)
19	(*b*)	41	(*a*)	63	(*b*)	94	(*d*)
20	(*c*)	42	(*d*)	64	(*b*)	95	(*c*)
21	(*d*)	43	(*c*)	65	(*b*)	96	(*b*)
22	(*c*)	44	(*d*)	66	(*a*)	98	(*d*)

99	**(*a*)**	**129**	**(*d*)**	**159**	**(*a*)**	**192**	**(*d*)**
100	**(*a*)**	**130**	**(*c*)**	**160**	**(*d*)**	**193**	**(*d*)**
101	**(*b*)**	**131**	**(*b*)**	**161**	**(*b*)**	**194**	**(*d*)**
102	**(*d*)**	**132**	**(*d*)**	**162**	**(*d*)**	**195**	**(*d*)**
103	**(*b*)**	**133**	**(*d*)**	**163**	**(*a*)**	**196**	**(*a*)**
104	**(*d*)**	**134**	**(*e*)**	**164**	**(*a*)**	**197**	**(*c*)**
105	**(*a*)**	**135**	**(*b*)**	**165**	**(*b*)**	**198**	**(*d*)**
106	**(*c*)**	**136**	**(*c*)**	**166**	**(*d*)**	**199**	**(*d*)**
107	**(*b*)**	**137**	**(*b*)**	**167**	**(*a*)**	**200**	**(*c*)**
108	**(*c*)**	**138**	**(*b*)**	**168**	**(*b*)**	**201**	**(*b*)**
109	**(*d*)**	**139**	**(*b*)**	**169**	**(*a*)**	**203**	**(*a*)**
110	**(*d*)**	**140**	**(*b*)**	**170**	**(*b*)**	**204**	**(*b*)**
111	**(*a*)**	**141**	**(*c*)**	**171**	**(*b*)**	**205**	**(*d*)**
112	**(*a*)**	**142**	**(*a*)**	**172**	**(*d*)**	**206**	**(*c*)**
113	**(*b*)**	**143**	**(*a*)**	**173**	**(*c*)**	**207**	**(*a*)**
114	**(*d*)**	**144**	**(*a*)**	**174**	**(*d*)**	**208**	**(*a*)**
115	**(*d*)**	**145**	**(*a*)**	**175**	**(*c*)**	**209**	**(*d*)**
116	**(*c*)**	**146**	**(*c*)**	**176**	**(*d*)**	**210**	**(*a*)**
117	**(*a*)**	**147**	**(*a*)**	**177**	**(*e*)**	**213**	**(*a*)**
118	**(*c*)**	**148**	**(*d*)**	**180**	**(*d*)**	**214**	**(*d*)**
119	**(*b*)**	**149**	**(*c*)**	**181**	**(*b*)**	**215**	**(*d*)**
120	**(*b*)**	**150**	**(*d*)**	**182**	**(*a*)**	**216**	**(*c*)**
121	**(*d*)**	**151**	**(*d*)**	**183**	**(*d*)**	**219**	**(*a*)**
122	**(*d*)**	**152**	**(*c*)**	**184**	**(*c*)**	**220**	**(*b*)**
123	**(*d*)**	**153**	**(*d*)**	**185**	**(*b*)**	**221**	**(*a*)**
124	**(*d*)**	**154**	**(*b*)**	**186**	**(*b*)**	**222**	**(*a*)**
125	**(*a*)**	**155**	**(*a*)**	**187**	**(*a*)**	**223**	**(*a*)**
126	**(*c*)**	**156**	**(*d*)**	**188**	**(*b*)**	**239**	**(*c*)**
127	**(*d*)**	**157**	**(*c*)**	**190**	**(*b*)**	**240**	**(*c*)**
128	**(*c*)**	**158**	**(*b*)**	**191**	**(*b*)**	**241**	**(*b*)**

242	(*a*)	253	(*b*)	263	(*e*)	273	(*c*)
243	(*a*)	254	(*c*)	264	(*a*)	274	(*b*)
244	(*a*)	255	(*b*)	265	(*d*)	275	(*d*)
245	(*d*)	256	(*c*)	266	(*d*)	276	(*d*)
246	(*c*)	257	(*a*)	267	(*d*)	277	(*d*)
247	(*a*)	258	(*a*)	268	(*b*)	278	(*a*)
248	(*c*)	259	(*d*)	269	(*d*)	279	(*d*)
250	(*d*)	260	(*d*)	270	(*d*)	280	(*b*)
251	(*a*)	261	(*d*)	271	(*a*)	281	(*b*)
252	(*b*)	262	(*a*)	272	(*c*)		

Match the Following

78	1 (*b*), 2 (*c*), 3 (*d*), 4 (*a*)	179	1 (*e*), 2 (*d*), 3 (*b*), 4 (*c*), 5 (*a*)
79	1 (*c*), 2 (*d*), 3 (*a*), 4 (*b*)	189	1 (*d*), 2 (*c*), 3 (*a*), 4 (*b*)
80	1 (*b*), 2 (*a*), 3 (*c*), 4 (*d*)	202	1 (*d*), 2 (*c*), 3 (*b*), 4 (*a*)
81	1 (*c*), 2 (*a*), 3 (*b*), 4 (*d*)	211	1 (*b*), 2 (*a*), 3 (*d*), 4 (*c*)
82	1 (*b*), 2 (*d*), 3 (*c*), 4 (*a*)	212	1 (*d*), 2 (*c*), 3 (*b*), 4 (*a*)
83	1 (*d*), 2 (*c*), 3 (*b*), 4 (*a*)	217	1 (*e*), 2 (*d*), 3 (*c*), 4 (*b*), 5 (*a*)
84	1 (*d*), 2 (*c*), 3 (*a*), 4 (*b*)	218	1 (*d*), 2 (*b*), 3 (*a*), 4 (*c*)
85	1 (*c*), 2 (*d*), 3 (*b*), 4 (*a*)	224	1 (*c*), 2 (*b*), 3 (*a*)
86	1 (*d*), 2 (*c*), 3 (*b*), 4 (*a*)	231	1 (*c*), 2 (*d*), 3 (*a*), 4 (*b*)
97	1 (*d*), 2 (*c*), 3 (*b*), 4 (*a*)	249	1 (*d*), 2 (*c*), 3 (*b*), 4 (*a*)
178	1 (*d*), 2 (*c*), 3 (*b*), 4 (*a*)		

Fill in the Blanks

225. National Environmental Policy Act
226. 16th September
227. Biochemical Oxygen Demand
228. Chemical Oxygen Demand
229. Mist
230. Automobiles

232. Carbon, Nitrogen and Sulphur

233. Colourless and Odourless

234. Methane

235. Benzopyrene

236. Anti-knocking agent

237. Methyl Isocynate

238. Acid rain or Acid precipitation

Chapter 6

Social Issues and the Environment

1. **The ability to last out and keep from falling**
   (*a*) Environment (*b*) Sustainability
   (*c*) Adaptability (*d*) Both (*b*) and (*c*)

2. **Sustainability is a____concept**
   (*a*) Rotating (*b*) Static
   (*c*) Dynamic (*d*) Both (*b*) and (*c*)

3. **Important measures for management of waste are**
   (*a*) Water harvesting (*b*) Well management
   (*c*) Watershed management (*d*) Both (*a*) and (*c*)

4. **Green house gases are**
   (*a*) Methane (*b*) Carbondioxide
   (*c*) Nitrous oxide (*d*) All of the above

5. **Arthshashtra was written by**
   (*a*) Vishnu Sharma (*b*) Kautilya
   (*c*) Max Muller (*d*) Chanakya

6. World's first forest conservation and wild life management was mentioned in the book of

(*a*) Jungle Book (*b*) Mahabharata

(*c*) Ramayana (*d*) Arthshashtra

7. Radiatively active gases are

(*a*) CO_2 (*b*) Methane

(*c*) Nitrous oxide (*d*) All of the above

8. Green house gases are also known as

(*a*) Hot gases (*b*) Wild gases

(*c*) Radiatively active gases (*d*) None of the above

9. ____recognizes an enduring allegiance of human kind to mother earth

(*a*) Rig veda (*b*) Atharva veda

(*c*) Yajur veda (*d*) Sam veda

10. Basic resources of earth are

(*a*) Khsiti (*b*) Jal

(*c*) Pawan (*d*) Gagan, Samira

(*e*) All of the above

11. 'Live and Let Live, Let no one species encroach over the rights and Privileges of other species' is quoted in

(*a*) Isho Upanishad (*b*) Atharva veda

(*c*) Chandogya Upanishad (*d*) All of the above

12. The most damaging agent of ozone layer is

(*a*) CFC's (*b*) CH_4

(*c*) N_2O (*d*) UV Radiation

13. The conference that discussed the role of different countries in mitigating climate changes

(*a*) Earth Summit (*b*) Kyoto Protocol

(*c*) Paris convention (*d*) All of the above

14. The discoveries related to ozone destruction were made by

(*a*) Sherwood Rowland (*b*) Mario Molino

(*c*) Paul Crutezen (*d*) All of the above

15. The mission of National Wasteland Development Board is

(*a*) Check land degradation

(*b*) Bring wastelands into sustainable use

(*c*) Increase biomass availability

(*d*) All of the above

16. Which of the following is not a 'Greenhouse' gas?

(*a*) Oxygen (*b*) Carbon dioxide

(*c*) Chlorofluoro Carbons (*d*) Methane

17. Greenhouse effect is related to

(*a*) Green trees on house (*b*) Global warming

(*c*) Grasslands (*d*) Greenery in country

18. The greenhouse effect is due to

(*a*) Impermeability of long wavelength radiation through CO_2 of the atmosphere

(*b*) Penetrability of low wavelength radiations through O_3 layer

(*c*) Penetrability of low wavelength radiations through CO_2

(*d*) Impermeability of long wavelength radiations through O_3 layer

19. Which of the following gases contribute maximum to the 'Greenhouse effect' on earth

(*a*) Carbon Dioxide (*b*) Methane

(*c*) Chloroflouro Carbon (*d*) Freon

20. Ozone day is observed on

(*a*) January 30 (*b*) April 21

(*c*) September 16 (*d*) December 25

21. Formation of hole in ozone is maximum over

(*a*) India (*b*) Antarctica

(*c*) Europe (*d*) Africa

22. Chlorofluoro Carbon releases a chemical harmful to ozone is

(*a*) Chlorine (*b*) Fluorine

(*c*) Nitrogen Peroxide (*d*) Sulphur dioxide

23. An area characterized by water flow

(*a*) Wasteland (*b*) Watershed

(*c*) Water logged (*d*) All of the above

24. Which one of the following gases can deplete the ozone layer in the upper atmosphere

(*a*) Methane (*b*) Ammonia

(*c*) Carbon dioxide (*d*) Sulphur dioxide

25. Which important green house gas other than methane is being provided from the agricultural fields

(*a*) SO_2 (*b*) Nitrous Oxide

(*c*) SO_3 (*d*) Ammonia

26. Peeling of Ozone umbrella which protects us from UVrays is caused by

(*a*) PAN (*b*) Carbon dioxide

(*c*) CFCs (*d*) Coal Burning

27. Which gas catalytically destroy ozone layer by converting it into oxygen

(*a*) Sulphur (*b*) Nitrogen

(*c*) Chlorine (*d*) Iodine

28. Till now, how many countries have signed the Montreal Protocol?

(*a*) 100 (*b*) 125

(*c*) 175 (*d*) 180

29. Expand 'NWDB'

(*a*) National Wasteland Development Board

(*b*) National Water and Development Board

(*c*) National Wastage Development Board

(*d*) None of the above

30. National forest policy is framed in the year

(*a*) 1987 (*b*) 1988

(*c*) 1989 (*d*) None of the above

31. Match the following

1.	The environment Protection Act	(*a*)	1981
2.	The insecticides Act	(*b*)	1974
3.	The water Act	(c)	1968
4.	The air (prevention and control of pollution) Act	(*d*)	1986

32. Examples of nuclear power plant accidents are

(*a*) Three mile island nuclear power plant

(*b*) Chemobyl nuclear power plant

(*c*) Both (a) and (b)

(*d*) None of the above

33. Important initiatives regarding the mitigation of global changes are

(*a*) Montreal Protocol (*b*) Earth summit

(*c*) Kyoto protocol (*d*) All of the above

34. The land which is not in use is called

(*a*) Dry land (*b*) Waste land

(*c*) Grass land (*d*) None of the above

35. The classical literature which gives the message that resources should not be wasted but should be conserved are

(*a*) Atharva Veda (*b*) Isho Upanishad

(*c*) Artha Shastra (*d*) All of the above

36. Which Upanishad say so–'One can enjoy nature by giving up greed'

(*a*) Isho Upanishad (*b*) Atharva Veda

(*c*) Chandogya Upanishad (*d*) None of the above

37. The most abundant greenhouse gas in the atmosphere is

(*a*) Methane (*b*) CO_2

(*c*) CFCs (*d*) All of the above

38. Increase in global mean temperature of the earth is known as

(*a*) Global warming

(*b*) Greenhouse warming

(*c*) Red hot warming

(*d*) Both (*a*) and (*b*)

39. Methane is a product of incomplete decomposition caused by a group of bacteria called

(*a*) Methanogens (*b*) Methanians

(*c*) Methyl oxide (*d*) None of the above

40. Chlorofluorocarbons are synthetic gaseous compounds of

(*a*) Carbon (*b*) Nitrogen

(*c*) Halogens (*d*) Both (*a*) and (*c*)

41. The most common use of chlorofluorocarbons is in

(*a*) Refrigerants (*b*) Shock absorbers

(*c*) Catalysts (*d*) Both (*a*) and (*b*)

42. Chlorofluorocarbons are

(*a*) Non toxic (*b*) Non-flammable

(*c*) Highly stable (*d*) All of the above

43. ________is produced from garbage dumps, swamps, flooded rice fields etc

(*a*) Chlorofluoro carbons (*b*) Methane

(*c*) Carbon dioxide (*d*) All of the above

44. The main sources of chlorofluorocarbons in the atmosphere are

(*a*) Leaking air conditioner
(*b*) Leaking refrigeration units
(*c*) Evaporation of industrial solvents
(*d*) All of the above

45. River valley projects provide the following benefits

(*a*) Irrigation
(*b*) Power generation
(*c*) Water supply
(*d*) All of the above

46. The social consequences of constructing big dams are

(*a*) Displacement of millions of tribals from their homeland
(*b*) Influx into urban area
(*c*) No proper food
(*d*) Both (*a*) and (*b*)

47. The well known environmentalist is

(*a*) Sunderlal Bahuguna
(*b*) Soniya Gandhi
(*c*) Vinoba Bhave
(*d*) Morarji Desai

48. Bodhghat project is in

(*a*) Andhra Pradesh
(*b*) Uttar Pradesh
(*c*) Madhya Pradesh
(*d*) Assam

49. The famous critical watersheds in the world are

(*a*) Andes mountains
(*b*) Alps
(*c*) The Himalayas
(*d*) All of the above

50. The ingenious technique of rain water harvesting is

(*a*) Bandhs
(*b*) Johads
(*c*) Lakes
(*d*) Both (*a*) and (*b*)

51. CO_2 contributes____% of global warming.

(*a*) 30
(*b*) 50
(*c*) 60
(*d*) 80

52. The response of plants to elevate concentrations of CO_2 are

(*a*) Symbiotic effect

(*b*) Carbon dioxide fertilization effect

(*c*) Interactive CO_2 effect

(*d*) Both (*a*) and (*b*)

53. The concentration of O_3 (Ozone) is highest during

(*a*) July–October (*b*) February-April

(*c*) November–December (*d*) All of the above

54. Ozone has a steady state of concentration layer between____of stratosphere

(*a*) 30–60 km (*b*) 20–26 km

(*c*) 50–60 km (*d*) Both (*a*) and (*b*)

55. The thickness of Ozone layer is measured in

(*a*) Pascal (*b*) Dabson Units

(*c*) cm/sec (*d*) Both (*a*) and (*c*)

56. Ozone is broken into O_2 and O in the stratosphere by

(*a*) Infrared radiation (*b*) UV radiation

(*c*) Normal light (*d*) Both (*a*) and (*b*)

57. Match the following

1.	Ozone hole	(*a*)	Land not in use
2.	Ozone day	(*b*)	Antarctica
3.	Greenhouse effect	(*c*)	September 16
4.	Wasteland	(*d*)	CO_2

58. The following got Nobel Prize for chemistry in 1995

(*a*) Albert Ross (*b*) Sherwood Rowland

(*c*) Mario Molina (*d*) Both (*b*) and (*c*)

59. Kyoto protocol is regarding

(*a*) Water harvesting (*b*) Greenhouse emission

(*c*) Conservation of energy (*d*) Both (*a*) and (*b*)

60. Maximum amount of UV radiation is absorbed in

(*a*) Near the equator (*b*) Near poles

(*c*) Away from the equator (*d*) None of the above

61. The effect of global warming on crop production will be

(*a*) Increased incidence of plant disease

(*b*) Explosive growth of weeds

(*c*) Enhanced basal rate of respiration of plants

(*d*) All of the above

62. The objectives of National Wasteland Development Board are

(*a*) Arrest land degradation (*b*) Arrest deforestation

(*c*) Increase rainfall (*d*) Both (*a*) and (*b*)

63. Noise is a

(*a*) Air pollution (*b*) Water pollution

(*c*) Sound pollution (*d*) Both (*a*) and (*c*)

64. The sum total of all conditions and influences that affect the development and life of organisms

(*a*) Climate (*b*) Environment

(*c*) Biosphere (*d*) Bio reserve

65. Worst affected by water scarcity are

(*a*) USA (*b*) Africa

(*c*) West Asia (*d*) Both (*b*) and (*c*)

66. Harvesting rain water by collecting rain water in check dams is known as

(*a*) Full fledging harvesting (*b*) Artificial recharging

(*c*) Rain water harvesting (*d*) Both (*a*) and (*b*)

67. Narmada Sagar project is in

(*a*) Gujarat (*b*) Madhya Pradesh

(*c*) U.P. (*d*) Bihar

68. A glass house used for raising delicate plants is called

(*a*) Glass garden (*b*) Green house

(*c*) Plant house (*d*) Both (*a*) and (*c*)

69. Global climate is changing due to increased amount of green house gases. This is known as

(*a*) Global change (*b*) Global revolution

(*c*) Global metamorphosis (*d*) None of the above

70. Montreal Protocol was regarding

(*a*) Global warming

(*b*) Protect Stratospheric Ozone

(*c*) Conservation of water

(*d*) Saving fuel

71. Chernobyl nuclear power plant is in

(*a*) USA (*b*) USSR

(*c*) Germany (*d*) France

72. The concentration of O_3 in the stratosphere is lowest during

(*a*) February–April (*b*) July–October

(*c*) December–January (*d*) None of the above

73. In the 20th century, the global mean temperature has increased by about

(*a*) 1% (*b*) 1.5%

(*c*) 0.6% (*d*) None of the above

74. In the years to come, the rise in temperature will be more in the____regions

(*a*) Middle latitude (*b*) Lower latitude

(*c*) Higher latitude (*d*) Both (*a*) and (*c*)

75. Due to global warming, the precipitation will____at higher latitude but will____at lower latitude.

(*a*) Increase, Decrease (*b*) Decrease, Increase

(*c*) Progressive changes (*d*) Difficult to predict

76. What is the effect of global warming on the range of species distribution?

(*a*) Many species will shift slowly towards poles

(*b*) Many species will move towards higher elevations or mountain areas

(*c*) No effect

(*d*) Both (*a*) and (*b*)

77. Match the following

1.	Weather and Climate	(*a*)	Rise due to thermal expansion of ocean and melting of glaciers
2.	Sea Level	(*b*)	Decreases
3.	Range of Species distribution	(*c*)	Shift slowly to polar regions
4.	Food production	(*d*)	Temperature increases

78. The effect of global warming on food production is

(*a*) Decreases

(*b*) Increases

(*c*) First increases and then decreased

(*d*) Cannot predict

79. The food production decreases during global warming because

(*a*) Increased incidence of plant diseases and pests

(*b*) Explosive growth of weeds

(*c*) Both (*a*) and (*b*)

(*d*) None of the above

80. During global warming, the sea levels increase due to

(*a*) Thennal expansion of oceans

(*b*) Melting of glaciers

(*c*) Greenland ice sheets melt

(*d*) All of the above

81. Even a small rise in the sea level due to global warming, what happens

(*a*) Nothing happens

(*b*) People enjoy more tides at beaches

(*c*) Coastal cities and towns will be hit by storms and floods

(*d*) Low lying areas will be submerged

(*e*) Both (*b*) and (*d*)

82. Global warming effect on crop productivity is

(*a*) Increases (*b*) Decreases

(*c*) No effect (*d*) Difficult to predict

83. Match the following

Due to global warming, the effects on natural habitat and environment

1.	Decrease in food production	(*a*)	Droughts and floods increase
2.	Migration of species	(*b*)	Shortage of Pulses & wheat
3.	Increase in temperature	(*c*)	Death of animal flora

84. Global warming can be reduced by following the below methods

(*a*) Using substitutes for chlorofluorocarbons

(*b*) Increasing forest cover

(*c*) Reducing the use of fossil fuels

(*d*) All of the above

85. Examples of culturable wasteland are

(*a*) Water logged and marsh land

(*b*) Saline lands with laterite soils

(*c*) Degraded forest lands

(*d*) All of the above

86. "By 2012, all countries should reduce their green house emission to 5% level by adopting appropriate methods" this was the protocol of

(*a*) Montreal protocol (*b*) Earth summit

(*c*) Kyoto protocol (*d*) Both (*a*) and (*b*)

87. Good examples of unculturable wastelands are

(*a*) Rocky areas (*b*) Glaciers

(*c*) Mine spoils (*d*) Both (*a*) and (*b*)

88. Identify the odd one

(*a*) Steep slopes (*b*) Snow capped hills

(*c*) Glaciers (*d*) Shifting cultivation areas

89. Match the following

1.	NWDB	(*a*)	UV radiation
2.	Kyoto Protocol	(*b*)	Discovery of ozone layer destruction
3.	Sherwood Rowland	(c)	Reduction in green house
4.	Ozone depletion	(*d*)	Reclamation of wasteland

90. Identify the wrong one

(*a*) Gagan (*b*) Samira

(*c*) Pawan (*d*) Rekha

91. The hymn "What of thee I dig out, let that quickly grow ever, let me not hit thy vitals or thy heart" is from

(*a*) RigVeda (*b*) Atharva Veda

(*c*) Sam Veda (*d*) Yajur Veda

92. Good examples of water harvesting and watershed management are seen in the states of

(*a*) Rajasthan (*b*) Maharashtra

(*c*) Madhya Pradesh (*d*) All of the above

93. A watershed is an area characterized by

(*a*) Air (*b*) Water flow

(*c*) Water logging (*d*) Both (*a*) and (*c*)

94. The main idea in harvesting rain water is to check the

(*a*) Run off water (*b*) Logged water

(*c*) Bad water (*d*) Both (*b*) and (*c*)

95. Normally rainwater is slightly acidic because

(*a*) CO_2 of the air reacts with water to form weak acid

(*b*) O_2 of the air forms strong acid

(*c*) Both (*a*) and (*b*)

(*d*) None of the above

96. Sardar Sarovar project is in

(*a*) U.P. (*b*) Gujarat

(*c*) M.P. (*d*) Rajasthan

97. Tribal Panchsheel was formulated during the period of

(*a*) Mahatma Gandhi (*b*) Jawaharlal Nehru

(*c*) Lal Bahadur Shastri (*d*) All of the above

98. The following water projects have created a much hype and controversy

(*a*) Hirakud (*b*) Sardar Sarovar

(*c*) Tehri (*d*) Both (*b*) and (*c*)

99. Ancient literature in which details of water conservation and environmental conservation are found are

(*a*) Atharva veda (*b*) Rig veda

(*c*) Isho upanishad (*d*) Both (*a*) and (*c*)

100. The main source of N_2O are

(*a*) Agriculture (*b*) Biomass burning

(*c*) Industrial processes (*d*) All of the above

101. The increasing concentration of green house gasses in the atmosphere has the following effects

(*a*) Global warming

(*b*) Depletion of ozone layer

(*c*) CO_2 fertilization effect on plants

(*d*) All of the above

102. The atmospheric CO_2 concentration is expected to reach____by the end of 21st century

(*a*) 500 ppm (*b*) 540–970 ppm

(*c*) 1000 ppm (*d*) None of the above

103. The contribution of methane to global warming is about

(*a*) 60% (*b*) 20%

(*c*) 40% (*d*) 80%

104. The expansion of ppm is

(*a*) Pretty Positive mind (*b*) Parts per Million

(*c*) Peaceful and Popular men (*d*) Both (*a*) and (*b*)

105. The concentration of N_2O in the atmosphere today is____

(*a*) 270 ppm (*b*) 360 ppb

(*c*) 415 ppb (*d*) None of the above

106. Chlorofluorocarbons produce____in the presence of UV radiation

(*a*) Greenhouse effect (*b*) Active chlorine

(*c*) Active bromine (*d*) Sulphuric acid

107. The two important summits 'that took place on the principle of reducing the effect of green house gases are

(*a*) Montreal protocol (*b*) Earth summit

(*c*) Kyoto protocol (*d*) Both (*b*) and (*c*)

108. The major effect of ozone depletion is

(*a*) Allows more UV–B radiation to the earth's surface

(*b*) UV radiations are harmful to living systems

(*c*) More rains

(*d*) Both (*a*) and (*b*)

Answers

1	(*b*)	23	(*b*)	46	(*d*)	69	(*a*)
2	(*c*)	24	(*a*)	47	(*a*)	70	(*b*)
3	(*d*)	25	(*b*)	48	(*c*)	71	(*b*)
4	(*d*)	26	(*c*)	49	(*c*)	72	(*b*)
5	(*b*)	27	(*c*)	50	(*d*)	73	(*c*)
6	(*d*)	28	(*c*)	51	(*c*)	74	(*d*)
7	(*d*)	29	(*a*)	52	(*b*)	75	(*a*)
8	(*c*)	30	(*b*)	53	(*b*)	76	(*d*)
9	(*b*)	32	(*c*)	54	(*b*)	78	(*a*)
10	(*e*)	33	(*d*)	55	(*b*)	79	(*c*)
11	(*a*)	34	(*b*)	56	(*b*)	80	(*d*)
12	(*a*)	35	(*d*)	58	(*d*)	81	(*e*)
13	(*b*)	36	(*a*)	59	(*b*)	82	(*b*)
14	(*d*)	37	(*b*)	60	(*a*)	84	(*d*)
15	(*d*)	38	(*a*)	61	(*d*)	85	(*d*)
16	(*a*)	39	(*a*)	62	(*d*)	86	(*c*)
17	(*b*)	40	(*d*)	63	(*d*)	87	(*d*)
18	(*a*)	41	(*a*)	64	(*b*)	88	(*d*)
19	(*a*)	42	(*d*)	65	(*d*)	90	(*d*)
20	(*c*)	43	(*b*)	66	(*b*)	91	(*b*)
21	(*b*)	44	(*d*)	67	(*b*)	92	(*d*)
22	(*a*)	45	(*d*)	68	(*b*)	93	(*b*)

94	(*a*)	**97**		**100**	(*d*)	**106**	(*b*)
95	(*a*)	**98**	(*d*)	**104**	(*b*)	**107**	(*d*)
96	(*b*)	**99**	(*d*)	**105**	(*c*)	**108**	(*d*)

Match the Following

31	**1 (*d*), 2 (*c*), 3 (*b*), 4 (*a*)**	**83**	**1 (*b*), 2 (*c*), 3 (*a*)**
57	**1 (*b*), 2 (*c*), 3 (*d*), 4 (*a*)**	**89**	**1 (*d*), 2 (*c*), 3 (*b*), 4 (*a*)**
77	**1 (*d*), 2 (*a*), 3 (*c*), 4 (*v*)**	**101**	**(*d*), 102 (*b*), 103 (*b*)**

Chapter 7

Human Population and the Environment

1. A group of individuals of a species occupying a definite geographic area at a given time is known as

(*a*) Family (*b*) Unit

(*c*) Population (*d*) Both (*a*) and (*b*)

2. Human population over the world belongs to species

(*a*) *Homo sapiens* (*b*) *Mongolia richens*

(*c*) Both (*a*) and (*b*) (*d*) None of the above

3. Modern human beings appeared around____years ago

(*a*) 20,000 (*b*) 40,000

(*c*) 50,000 (*d*) 70,000

4. Human population of the entire world belongs to____species

(*a*) 2 (*b*) 5

(*c*) 1 (*d*) 10

5. **The natural balance of environment is disturbed because of**

(*a*) Ever growing population (*b*) Bad population
(*c*) Huge population (*d*) Both (*a*) and (*c*)

6. **The dramatic increase in population size over a relatively short period is called**

(*a*) Population explosion (*b*) Population growth
(*c*) Population demography (*d*) None of the above

7. **Currently the world population is increasing____millions per year**

(*a*) 70 (*b*) 80
(*c*) 100 (*d*) 120

8. **The rapid growth of population is due to**

(*a*) Spread of public health programmes
(*b*) Rise in food production
(*c*) Reduction in infant mortality
(*d*) All of the above

9. **By 2050, India will stand____position in the world as far as population is concerned**

(*a*) First (*b*) Second
(*c*) Third (*d*) Fourth

10. **The maximum population size that can be supported with a given space is called**

(*a*) Maximum carrying capacity
(*b*) Maximum population capacity
(*c*) Maximum people capacity
(*d*) None of the above

11. **Examples of production systems are**

(*a*) Croplands (*b*) Orchards
(*c*) Fruit trees (*d*) All of the above

12. **The maximum carrying capacity of the environment depends upon**
    (*a*) Productive system (*b*) Protective system
    (*c*) Waste Assimilative system (*d*) Both (*a*) and (*b*)

13. **Examples of protective systems are**
    (*a*) Forests (*b*) Oceans
    (*c*) Orchards (*d*) Both (*a*) and (*b*)

14. **The component that assimilates the waste produced by hum. activities is**
    (*a*) Productive system (*b*) Protective system
    (*c*) Waste assimilative system (*d*) Both (*a*) and (*b*)

15. **The components of life supportive capacity are**
    (*a*) Productive system (*b*) Protective system
    (*c*) Waste assimilative system (*d*) Both (*a*) and (*b*)

16. **Exponential growth shows a____curve**
    (*a*) J–shaped (*b*) Z–shaped
    (*c*) C–shaped (*d*) Y–shaped

17. **The increase in a population nearly in a fixed proportion of____own size during any period of time is called**
    (*a*) Incremental growth (*b*) Exponential growth
    (*c*) Rapid growth (*d*) Both (*a*) and (*b*)

18. **Human population growth rate is measured as**
    (*a*) Yearly growth rate
    (*b*) Annual average growth rate
    (*c*) Randomized growth rate
    (*d*) Both (*a*) and (*c*)

19. **The formula for average annual growth rate in % is**
    (*a*) $[P_2 - P_1/P_1 \times N] \times 100$ (*b*) $[P_1 - P_2/P_1 \times N] \times 100$
    (*c*) $[P_2 \times 100]/P_1$ (*d*) None of the above

20. The number of individuals present in a given region at a given time can be ascertained by

(*a*) Computer data (*b*) Census

(*c*) Colossus (*d*) Both (*a*) and (*c*)

21. The time required for a population to double itself is called

(*a*) Doubling time (*b*) Duplicating time

(*c*) Matching time (*d*) None of the above

22. Important indicators of the pace of population growth are

(*a*) Annual average growth rate

(*b*) Stock Exchange Index

(*c*) Doubling Time

(*d*) Both (*a*) and (*c*)

23. The number of babies produced per thousand individuals is

(*a*) Baby rate (*b*) Birth rate

(*c*) Baby count (*d*) Both (*a*) and (*c*)

24. Determinant of the current growth of population is

(*a*) Sex drive (*b*) Sexual capacity

(*c*) Fertility (*d*) Both (*a*) and (*b*)

25. Death rate per thousand individuals is called

(*a*) Death count (*b*) Mortality

(*c*) Kick bucket rate (*d*) Both (*a*) and (*b*)

26. If birth and death rates are equal, a zero population growth rate would result which is called

(*a*) Zero baby syndrome

(*b*) Demographic transition

(*c*) Heaven's bless phenomena

(*d*) Both (*a*) and (*c*)

27. Migration of individuals into a country is known as

(*a*) Emigration (*b*) Immigration

(*c*) Transgression (*d*) Both (*a*) and (*c*)

28. Number of live births in the middle of year are called

(*a*) Mortality (*b*) Crude birth rate

(*c*) Crude death rate (*d*) None of the above

29. In most countries, the mortality rate has come down because

(*a*) Improved sanitation (*b*) Modern medicine

(*c*) Increased body size (*d*) Both (*a*) and (*b*)

30. A decrease in death rate would result in

(*a*) More competition for attractive girls

(*b*) Increased population growth rate

(*c*) Birth of more intelligent people

(*d*) None of the above

31. Number of deaths in the middle of a given year is known as

(*a*) Crude death rate (*b*) Real death rate

(*c*) Predictive death rate (*d*) Both (*b*) and (*c*)

32. The difference between the number of births and that of deaths is the rate of

(*a*) Yearly increase (*b*) Natural increase

(*c*) Notional increase (*d*) Both (*a*) and (*c*)

33. The current population growth rate of India is____per year

(*a*) 5% (*b*) 6%

(*c*) 1.6% (*d*) 10%

34. Movement of people into or out of a place or country is known as

(*a*) Migration (*b*) Gyration

(*c*) Jubilation (*d*) Both (*b*) and (*c*)

35. The age–sex structure of a developed world gives a____like structure

(*a*) Pyramid (*b*) Reverse pyramid

(*c*) Polygonal (*d*) Rhombic

36. 15% of the world's population controls about____percent of the resources

(*a*) 100 (*b*) 85

(*c*) 70 (*d*) 60

37. Urban people characteristics are

(*a*) Produce little food

(*b*) Consume more natural resources

(*c*) Generate more waste products per capita

(*d*) All of the above

38. Nearly 80% of the diseases in developing countries are

(*a*) Air borne (*b*) Water borne

(*c*) Material borne (*d*) Both (*a*) and (*c*)

39. Life expectancy of human beings today is (average)

(*a*) 100 years (*b*) 90 years

(*c*) 66 years (*d*) 40 years

40. In Africa, the life expectancy of people is (average)

(*a*) 60 (*b*) 70

(*c*) 40 (*d*) 100

41. The life expectancy in Africa is less because of

(*a*) More prevalence of AIDS

(*b*) Genetic abnormality

(*c*) Cholera and Vomiting

(*d*) Because of color of people

42. About 95% of AIDS patients are from

(*a*) Developing countries

(*b*) Underdeveloped countries

(*c*) USA& UK

(*d*) Both (*a*) and (*b*)

43. **____countries accounts for 13% of the world's population but 69% of the cases of HIV infection**

(*a*) Third worlds countries (*b*) African

(*c*) North American (*d*) None

44. **By the end of year 2000, about____million people in India were infected with AIDS**

(*a*) 10 (*b*) 5

(*c*) 3.7 (*d*) 20

45. **Examples of sexually transmitted diseases are**

(*a*) Syphilis (*b*) Gonorrhoea

(*c*) Lung cancer (*d*) Both (*a*) and (*b*)

46. **High incidence of AIDS cases are in the following states**

(*a*) Maharashtra (*b*) Andhra Pradesh

(*c*) Tamil Nadu (*d*) Karnataka

(*e*) All of the above

47. **About 89% of reported AIDS cases in India were from the age group**

(*a*) 18-40 years (*b*) 40-60 years

(*c*) 13-18 years (*d*) Both (*b*) and (*c*)

48. **Family planning programme in India was started in**

(*a*) 1960 (*b*) 1951

(*c*) 1962 (*d*) 1947

49. **A zero population growth rate due to equal birth and death rates is known as**

(*a*) Doubling time (*b*) Emigration

(*c*) Demographic transition (*d*) None of the above

50. **The number of babies produced per thousand individuals is known as**

(*a*) Natality (*b*) Birth rate

(*c*) Nationality (*d*) Both (*a*) and (*b*)

51. The system which provides food and fibre etc is known as

(*a*) Productive system (*b*) Reproductive system

(*c*) Manufacturing system (*d*) Both (*b*) and (*c*)

52. Expand TFR

(*a*) Total Female Rate

(*b*) Therapeutic Female Reproduction

(*c*) Total Fertility Rate

(*d*) Both (*a*) and (*b*)

53. The average number of children that would be born to a woman during her lifetime is

(*a*) Reproductive capacity (*b*) Total fertility rate

(*c*) Total Reproductive rate (*d*) Both (*a*) and (*c*)

54. The dramatic increase in population size over a relatively short period is

(*a*) Population Explosion (*b*) Population growth

(*c*) Population revolution (*d*) Both (*b*) and (*c*)

55. Population explosion has occurred in the last

(*a*) 500 years (*b*) 300 years

(*c*) 150 years (*d*) 700 years

56. Today, the world's number one problem is

(*a*) Pollution (*b*) Population explosion

(*c*) Nuclear proliferation (*d*) Natural calamities

57. Study of trends in human population growth and prediction of future growth is called

(*a*) Demography (*b*) Biography

(*c*) Kalography (*d*) Psychology

58. The force which acts against the achievement of highest possible level of population growth is known as

(*a*) Population pressure (*b*) Saturation level

(*c*) Environmental resistance (*d*) Carrying capacity

59. One of the critical mechanism by which the environment controls population of species is

(*a*) Removal of excreta
(*b*) Spread of disease
(*c*) Check on death rate
(*d*) Supply of food

60. If the rate of addition of new members increases with respect to individual lost of the same population then the graph obtained is

(*a*) Declined growth
(*b*) Exponential growth
(*c*) Zero growth
(*d*) None of the above

61. Full form of AIDS is____

(*a*) Acquired Intelligence Disposal Syndrome
(*b*) Acquired Immuno Deficiency Syndrome
(*c*) Artificial Intelligence Development System
(*d*) Both (*a*) and (*c*)

62. Match the following

1. Productive systems	(*a*) No. of babies/ thousand individuals
2. Protective systems	(*b*) Ability of reproduction
3. Fertility	(c) Forests, Oceans
4. Birth rate	(*d*) Croplands, Orchards

63. Match the following

1. Mortality	(*a*) Movement
2. Migration	(*b*) Death
3. STDs	(*c*) 1951
4. Family Planning Programme	(*d*) Syphilis

64. Human Rights are also known as

(*a*) Fundamental Rights
(*b*) Natural Rights
(*c*) Basic Rights
(*d*) All of the above

65. The commission on Human Rights was established by

(*a*) USA

(*b*) UK

(*c*) United Nations Economic and Social Council

(*d*) Both (*a*) and (*b*)

66. The commission on Human Rights was established in the year

(*a*) 1950 (*b*) 1946

(*c*) 1948 (*d*) 1960

67. Human Rights occupy a significance place in

(*a*) UN Charter (*b*) USA Charter

(*c*) British Parliament (*d*) Both (*b*) and (*c*)

68. By remote sensing technique, it is possible to predict well in advance the following environmental issues_____

(*a*) Droughts (*b*) Floods

(*c*) Volcanic eruptions (*d*) All of the above

69. The human population at the beginning of 20th century is

(*a*) 1.2 billion (*b*) 1.6 billion

(*c*) 2 billion (*d*) None of the above

70. Expand NACO

(*a*) The National AIDS Control Organisation

(*b*) National Association for Communicable Diseases

(*c*) National Association for Control of Organisms

(*d*) None of the above

71. Match the following

1.	Life expectance	(*a*)	Veneral warts
2.	AIDS	(*b*)	2000
3.	National Population Policy	(*c*)	Africa
4.	STD	(*d*)	66 years

72. Europe's population will be expected to decline to____% by 2050

(*a*) 12 (*b*) 11

(*c*) 7 (*d*) 10

73. Proportion of individuals of different ages within the population is known as

(*a*) Age structure (*b*) Age form

(*c*) Age classification (*d*) Both (*b*) and (*c*)

74. Active reproduction age in case of females is

(*a*) 15–44 years (*b*) 21–30 years

(*c*) 10–30 years (*d*) None of the above

75. In the age–sex structure of the developed world, we get a steeper pyramid which implies

(*a*) Less population (*b*) More population

(*c*) Stable population (*d*) None

76. Having more children leads to

(*a*) Better standard of living

(*b*) Lower standards of living

(*c*) Depends on the situation

(*d*) Cannot predict

77. More developed countries have

(*a*) More fertility rates (*b*) Lower fertility rates

(*c*) Cannot be predicted (*d*) None of the above

78. Birth rate

(*a*) Cannot be negative

(*b*) Do not indicate the current fertility pattern

(*c*) No. of babies produced per thousand individuals

(*d*) All of the above

79. **When population grows exponentially, utilization of resources and generation of waste grows**
(*a*) Exponentially (*b*) Drastically
(*c*) Not proportionally (*d*) Both (*b*) and (*c*)

80. **The number of children a couple must produce to replace themselves is known as**
(*a*) Progeny level (*b*) Replacement level
(*c*) Genetic level (*d*) None of the above

81. **Some children will die before reaching reproductive age. Hence, the replacement value is slightly higher than**
(*a*) 3.0 (*b*) 2.0
(*c*) 4.0 (*d*) None of the above

82. **Replacement level in developing countries is**
(*a*) 2 (*b*) 3
(*c*) 2.7 (*d*) 4

83. **Replacement level is more in case of developing countries because**
(*a*) Higher death rate at immature age
(*b*) Big population
(*c*) Less money
(*d*) Both (*b*) and (*c*)

84. **The replacement level in developed countries like USA is**
(*a*) 3 (*b*) 2.5
(*c*) 2.1 (*d*) 2.0

85. **India had a peak population growth rate of 2.1 % during**
(*a*) 1975–80 (*b*) 1999–2002
(*c*) 1965–70 (*d*) 2001–2005

86. **Population growth rate depends on**
(*a*) Birth rate (*b*) Death rate
(*c*) Migration (*d*) All of the above
(*e*) Age-sex ratio

87. The incidence of STD (Sexually Transmitted Disease) is highest among people

(*a*) Who kiss more (*b*) Sex Workers

(*c*) Multisexual partners (*d*) Both (*b*) and (*c*)

88. HIV infected women pass HIV to their children

(*a*) During delivery

(*b*) Through breast feeding

(*c*) Eating food in the same vessel

(*d*) Both (*a*) and (*b*)

89. Expand HIV

(*a*) Human Immunodeficiency Virus

(*b*) Herpes Influx Virus

(*c*) Hepatic Insufficiency Virus

(*d*) None of the above

Answers

1	(*d*)	23	(*b*)	45	(*d*)	69	(*b*)
2	(*a*)	24	(*c*)	46	(*e*)	70	(*a*)
3	(*c*)	25	(*b*)	47	(*a*)	72	(*c*)
4	(*c*)	26	(*b*)	48	(*b*)	73	(*a*)
5	(*d*)	27	(*b*)	49	(*c*)	74	(*a*)
6	(*a*)	28	(*b*)	50	(*d*)	75	(*c*)
7	(*b*)	29	(*d*)	51	(*a*)	76	(*b*)
8	(*d*)	30	(*b*)	52	(*d*)	77	(*b*)
9	(*a*)	31	(*a*)	53	(*b*)	78	(*d*)
10	(*a*)	32	(*b*)	54	(*a*)	79	(*a*)
11	(*d*)	33	(*c*)	55	(*c*)	80	(*b*)
12	(*d*)	34	(*a*)	56	(*b*)	81	(*b*)
13	(*d*)	35	(*a*)	57	(*a*)	82	(*c*)
14	(*c*)	36	(*b*)	58	(*c*)	83	(*a*)
15	(*d*)	37	(*d*)	59	(*d*)	84	(*c*)
16	(*a*)	38	(*b*)	60	(*b*)	85	(*c*)
17	(*b*)	39	(*c*)	61	(*b*)	86	(*e*)
18	(*b*)	40	(*c*)	64	(*d*)	87	(*d*)
19	(*a*)	41	(*a*)	65	(*c*)	88	(*d*)
20	(*b*)	42	(*d*)	66	(*b*)	89	(*a*)
21	(*a*)	43	(*b*)	67	(*a*)		
22	(*d*)	44	(*c*)	68	(*d*)		

Match the Following

62 1 (*d*), 2 (*c*), 3 (*b*), 4 (*a*) 63 1 (*b*), 2 (*a*), 3 (*d*), 4 (*c*)

71 1 (*d*), 2 (*c*), 3 (*b*), 4 (*a*)

Practice Exercises–I

1. **The response of plants to elevate concentrations of CO_2 is known as____**

2. **Collecting and storage of rain water for future use is known as_____**

3. **Spot the wrong one**

   (*a*) Chlorofluorocarbons (*b*) Carbon dioxide

   (*c*) Methane (*d*) Oxygen

4. **Identify the odd one**

   (*a*) Nandadevi (*b*) Nilgiris

   (*c*) Periyar (*d*) Pachmarhi

5. **Identify the odd one**

   (*a*) Keoladeo (*b*) Mudumalai

   (*c*) Chilka (*d*) Simlipal

6. **Identify the odd one**

   (*a*) *Su salvinius* (*b*) *Aliurus fulgens*

   (*c*) *Barberis nilghiriensis* (*d*) *Antilope cervicapra*

7. **Identify the odd one**
   (*a*) Strontium
   (*b*) Cesium
   (*c*) Arsenic
   (*d*) Iodium

8. **Identify the odd one**
   (*a*) X-rays
   (*b*) Cosmic Rays
   (*c*) Atomic rays
   (*d*) Infra Red rays

9. **Find the odd one**
   (*a*) Kautilya
   (*b*) Earth Summit
   (*c*) Kyoot protocol
   (*d*) Montreal treaty

10. **Find the odd one**
    (*a*) Producers
    (*b*) Consumers
    (*c*) Decomposers
    (*d*) Abiotic components

11. **Identify the odd one**
    (*a*) Lichen stage
    (*b*) Moss stage
    (*c*) Herb stage
    (*d*) Parent stage

12. **Find the odd one**
    (*a*) Reaction
    (*b*) Co-action
    (*c*) Stabilisation
    (*d*) Drama

13. **Identify the odd one**
    (*a*) Phytoplankton
    (*b*) Zooplankton
    (*c*) Grasslands
    (*d*) Fish

14. **Identify the odd one**
    (*a*) Photochemical smog
    (*b*) Brown air
    (*c*) White air
    (*d*) Grey air

15. **Identify the odd one**
    (*a*) Coal
    (*b*) Sulphur dioxide
    (*c*) Petroleum
    (*d*) Gas

16. **Identify the odd one**
    (*a*) Sunderlal Bahuguna
    (*b*) Amrita Devi
    (*c*) Salim Ali
    (*d*) Indira Gandhi

17. Most Biotic resources are____

18. The practice of felling and burning of forests followed by cultivation of crop for few years is called____

19. The first movement against tree felling was fought by a Bishnoi women of Keharili Village named____

20. In Karnataka, Chipko movement is known as____

21. Our big dams were called "The temples of modern India" by our late Prime Minister____

22. Chipko movement in Tihri Garhwal region was initiated by ____

23. The practice of farming with increased organic input to the soil is known as____

24. The land that has been destroyed due to mining is known as____

25. Conversion of grass lands/forests into deserts is known as____

26. The practice of growing agricultural crops between rows of planted trees is known as____

27. Find the odd one

(*a*) Coal (*b*) Forests

(*c*) Water (*d*) Wildlife

28. Find out the odd one

(*a*) Over grazing

(*b*) Removal of vegetation

(*c*) Afforestation

(*d*) Deforestation

29. Raising of plant species having high calorific value and growth rate is known as ____

30. The book list containing a record of threatened species is____

31. The most rich biodiversity zones of India are____

32. Match the following

1.	Red data book	(*a*)	19%
2.	Land Under forests	(*b*)	Data of threatened species
3.	Fluorosis	(*c*)	15-44 years
4.	Reproductive age	(*d*)	1.5 parts per million

33. A zero population growth when birth and death rates are equal is known as____

34. The life expectancy of people in Zambia is ____

35. UNESCO stands for____

36. World Wildlife week is celebrated during ____

37. The two very much complex interdependent mutually reactive entities are

38. The various principles, that govern the relationship between the organisms and environment is called____

39. The term 'Ecology' was first time introduced by____

40. Lithosphere, Hydrosphere and the Atmosphere will constitute the____

41. The basic unit of ecological hierarchy is known as____

42. Competition, mutualism and predation are the interactions occurring between____

43. The assemblage of population of plants, animals, bacteria and fungi that live in and is known as____

44. A unit of land with a natural boundary having a mosaic of patches is known as____

45. A large regional unit characterized by a major vegetation type and associated fauna found in a specific climate zone is known as____

46. ____extends to about 22.5 km in thickness from ocean to mountain tops.

47. The solid earth crust is known as____

48. The surroundings in which the organisms live is known as____

49. Short term properties of atmosphere is known as____

50. The climate is largely determined by the________and of an area

51. Because of their small size, decomposers are known as____

52. The minerals and atmospheric gases enter into biotic system and after the death and decay of organisms return to the soil and atmosphere. This is known as____

53. The rate of increase in the biomass of consumers per unit and time is called____

54. The larger carnivores prey upon the primary carnivores are called____

55. All types of ecosystems possess two types of food chain, they are____

56. Ecological pyramids were first devised by a British ecologist____

57. A semi-closed coastal body of water that has a free connection with sea is known as____

58. The sequence of succession stages on a bare rock are called as____

59. The last and stable community in the area is called as____

60. The first arrivals in a bare area are called____

61. The sequence of successional stages on sand are known as ____

62. The sequence of successional stages of secondary succession is called____

63. A byproduct of paper industry resembling raw dust is known as____

64. The acidic water received through rain, for and snow is known as____

65. Plants which are non-green, grow on other plants and obtain food from the host are known as____

66. **An area which is strictly reserved for the welfare of wild life and activities like forestry, grazing and cultivation are not allowed is known as____**

67. **The first National Park established in India was _____**

68. **A species is considered____when no member of the species remains alive.**

69. **The most undisturbed forests without any human impact are known as_____**

70. **The major drawback of incineration is_____**

71. **The major disappearance of species from the face of earth due to human activities is known as_____**

72. **____is the most comprehensive inventory of the global conservation status of plant and animal species.**

73. **Growing ornamental and fruit trees in urban areas alongside roads is known as_____**

74. **Using the same land for farming and forestry is known as_____**

75. **About 3/4 of the earth's surface is occupied by_____**

76. **On an average, the residence time of water in the air is only about_____**

77. **Kaveri river sharing water dispute is between_____**

78. **The dam across the river Nile in Egypt is_____**

79. **The biomass can be used to obtain fuels like**

80. **Biogas is a mixture of two gases namely**

81. **The study that deals with the origin, formation and geographic distribution of the soil is known as_____**

82. **____converts direct solar energy into electricity.**

83. **All materials originating from photosynthesis are known as_____**

84. **The energy produced from the kinetic energy of water falling from a height is known as_____**

85. Breaking down of rock into small particles is known as____

86. Maturation of soil through development of humus is known as____

87. The removal of top soil from its place by various physical agencies is called____

88. Men folk opposing the large scale felling of forests trees, the movement was known as____

89. Chipko movement was started in____

90. The main objectives of Appiko movement were____

91. A special programme of tree plantation is held every year in the country known as____

92. The measures and practices taken to protect the soil against loss is known as____

93. Water logging and application of non-degradable chemicals leads to loss of____

94. Planting of crops in rows or strips to check flow of water is called

95. The best soil for healthy and vigorous growth of most plants is____

96. Both power and manure are provided by____

97. The successful establishment of a species after reaching to new area, as a result of adjustment with the conditions prevailing there is known as____

98. Successive replacement of communities in an area over a period of time is known as____

99. Autotrophic plankton are called____

100. A primary succession on bare rock starts with____

101. Tropical forests occur in India in____

102. The rate of biomass or organic matter production by the organisms is known as____

103. The food chain in which microorganisms breakdown dead producers is called____

104. Anthropogenic or man made ecosystems basically agricultural fields is known as____

105. The pyramid of____is always upright

106. New approach to conservation of environment is the establishment of____

107. The area reserved for welfare of wildlife is called____

108. The totality of genes, species and the ecosystems of a region are known as____

109. The whole sequence of communities that replaces one another in the given area is called____

110. Formation of different layers or strata of different species in a community is known as____

111. A population is able to adapt to its environment and respond to natural selection due to its____

112. The evenness in the number of individuals of a species is known as____

113. Nile perch is an exotic predatory____

114. Water hyacinth is a free floating exotic____

115. The extinction of species in the geological past is called____

116. Chilka lake bird sanctuary is in____

117. The two main centres of biological diversity in Western Ghats are____

118. Animals having high protection prospects are____

119. The main natural sources of air pollution are____

120. Burning of plastic producers____

121. The widely used aerosol propellant and refrigerant is____

122. ____dissolves silver and lead paints.

Answers

1 Carbon dioxide fertilization effect

2 Rain water harvesting

3 (*d*) (Oxygen)

4 (*c*) (Periyar)

5 (*d*) (Simlipal)

6 (*c*) (*Barberis nilghiriensis*)

7 (*c*) (Arsenic)

8 (*d*) (Infra Red rays)

9 (*a*) (Kautilya)

10 (*d*) (Abiotic component)

11 (*d*) (Parent stage)

12 (*d*) (Drama)

13 (*c*) (Grasslands)

14 (*c*) (White air)

15 (*b*) (Sulphur dioxide)

16 (*d*) (Indira Gandhi)

17 Renewable

18 Jhum (Shifting cultivation)

19 Amrita Devi

20 Appiko movement

21 Pandit Jawaharlal Nehru

22 Sunderlal Bahuguna

23 Organic Farming

24 Mine spoil

25 Desertification

26 Taungya cultivation

27 (*a*) (Coal)

28 (*c*) (Afforestation)

29 Energy plantation

30 Red data book

31 Western Ghats and North East

32 1 (*b*), 2 (*a*), 3 (*d*), 4 (*c*)

33 Demographic Transition

34 37 years

35 United Nations Educational, Scientific and Cultural Organization

36 1st–7th October

37 Organism and Environment

38 Ecology

39 Ernst Haeckel

40 Physical Environment

41 Organism

42 Different populations in an area

43 Biological Community

44 Landscape

45 Biome

46 Biosphere

47 Lithosphere

48 Environment

49 Weather

50 Temperature and Rainfall

51 Micro Consumers

52 **Biogeochemical cycle**
53 **Secondary Productivity**
54 **Third order consumers or secondary carnivores**
55 **Grazing food chain and detritus food chain**
56 **Charles Elton**
57 **Estuary**
58 **Lithosphere**
59 **Climax community**
60 **Pioneers or Pioneer colonizers**
61 **Psammosere**
62 **Subsere**
63 **Bregoli**
64 **Wet deposition**
65 **Parasites**
66 **National park**
67 **Jim Corbett National Park, Nainital**
68 **Extinct**
69 **Pristine forests**
70 **That it releases a number of toxic chemicals in the atmosphere**
71 **Anthropogenic extinction**
72 **2000 IUCN Red list**
73 **Urban forestry**
74 **Agro Forestry**
75 **Oceans**
76 **10 days**
77 **Karnataka and Tamil Nadu**
78 **Aswan Dam**
79 **Methanol and Ethanol**
80 **Methane and Carbon dioxide**

81 Pedology
82 Photo Voltaic cell
83 Biomass
84 Hydro power
85 Weathering
86 Pedogenesis
87 Soil Erosion
88 Chipko movement
89 Tihri Garhwal region of Uttar Pradesh
90 Ulisu, Belesu and Balasu
91 Van Mahotsava
92 Soil conservation
93 Soil fertility
94 Strip cropping
95 Loamy soil
96 Biogas plants
97 Ecesis
98 Succession
99 Phytoplankton
100 Lichens
101 Kerala and Assam
102 Productivity
103 Detritus food chain
104 Agroecosystem
105 Energy
106 Biosphere reserves
107 National park
108 Biodiversity
109 Sere
110 Stratification

111 Genetic diversity

112 Equitability

113 Fish

114 Water weed

115 Background extinction

116 Balagaon (Orissa)

117 Agastryamalai hills, Silent valley and the new Ammambalam reserve

118 Himalayan bear and musk deer

119 Volcanic eruptions and solar flames

120 Polychlorinated biphenyls

121 Chlorofluoro Methane

122 Hydrogen Sulphide

Practice Exercises–II

1. Match the following

1.	Herbivores	(*a*)	Lion, Tiger
2.	Carnivores	(*b*)	Micro consumers
3.	Decomposers	(*c*)	Eat other animals
4.	Top Carnivores	(*d*)	Feed on plants

2. Match the following

1.	Forests	(*a*)	Food Chain
2.	Wildlife	(*b*)	Milk, Meat and Wool
3.	Grassland	(*c*)	Timber

3. Match the following

1.	Weather	(*a*)	Grinnel
2.	Climate	(*b*)	The place
3.	Habitat	(*c*)	Temperature
4.	Niche	(*d*)	Arthur Tansely

4. Match the following

1.	Ecology	(*a*)	Khejri trees
2.	Bishnoi women	(*b*)	Haekel

3. Ecosystem (c) Charles Elton
4. Ecological (d) Arthur Tansely

5. Match the following

1. Pyramid of energy	(a) Norman Myers
2. Hot spots	(b) Odum
3. Minamata disease	(c) Richter scale
4. Earth quake intensity	(d) Contaminated fish with mercury

6. Match the following

1. Incineration	(a) Australia
2. Willy Willy	(b) Ponds
3. Johada	(c) Green house gas
4. Methane	(d) 85D *DC*

7. Match the following

1. Alpha diversity	(a) Vulnerable but at risk
2. Taxol	(b) Within the community
3. Rare	(c) Anti Cancer drug
4. Anthropogenic	(d) Man made

8. Match the following

1. Benzo pyrene	(a) Gujarat
2. Fluorosis	(b) Tobacco
3. Sardar Sarovar	(c) Teeth deformity
4. Chlorofluorocarbons	(d) Ozone depletion

9. Match the following

1. Xerophytes	(a) Syphilis
2. STD	(b) Himachal Pradesh
3. Shikari Devi sanctuary	(c) Plants in dry soil
4. Floating plants	(d) Pistia

10. Match the following

1.	Macrophytes	(*a*)	Bharatpur
2.	Detritus	(*b*)	Routed plants found in shallow waters
3.	Keoladeo Sanctuary	(*c*)	Irritation of eyes
4.	Paroxy Acetyl Nitrate	(*d*)	Dead plant parts and animals

11. Match the following

1.	Extinction	(*a*)	40 dB (night)
2.	Specification	(*b*)	Elimination
3.	Greenhouse gas	(*c*)	Evolution of new species
4.	Silent zone	(*d*)	Methane

12. Match the following

1.	Montreal protocol	(*a*)	2000
2.	National population policy	(*b*)	66
3.	Hot spots	(*c*)	1987
4.	Life expectancy	(*d*)	25

13. Match the following

1.	Itai Itai	(*a*)	Flourine excess
2.	Blue baby syndrome	(*b*)	Black foot disease
3.	Arsenic	(*c*)	Cadmium
4.	Flourosis	(*d*)	Excess nitrate

14. Match the following

1.	Decomposers	(*a*)	Organic substances
2.	Abiotic component	(*b*)	Diversity of habitat over total landscape
3.	Gamma diversity	(*c*)	Hot spots
4.	Eastern Himalayas	(*d*)	Actinomycetes

15. Match the following

1.	Indian Bustard	(*a*)	Earth worm farming
2.	Vermiculture	(*b*)	CO_2
3.	Most abundant greenhouse gas	(*c*)	Reproducing capacity
4.	Fertility	(*d*)	Endangered species

16. Match the following

1.	Morphine	(*a*)	*Ocimum Sanctum*
2.	Quinine	(*b*)	*Taxus Baccata*
3.	Taxol	(c)	*Papaverum somniferum*
4.	Tulsi	(*d*)	*Cinchona ledgeriana*

17. Match the following

1.	Sultanpur lake bird sanctuary	(*a*)	Assam
2.	Kaziranga National park	(*b*)	1500 gm/litre
3.	BOD	(*c*)	Weed
4.	Water hyacinth	(*d*)	Gurgaon

18. Match the following

1.	Endangered	(*a*)	Conservational method
2.	Transitional zone	(*b*)	Red Panda
3.	Germ bank	(*c*)	75dB
4.	Industrial zone	(*d*)	Outer part of biosphere reserve

19. Match the following

1.	Western Ghats	(*a*)	Alexander Graham Bell
2.	Global warming	(*b*)	Air
3.	Samira	(c)	Ozone layer depletion
4.	Decibel	(*d*)	Hot spots

20. Match the following

1.	Kyoto protocol	(*a*)	Ozone layer destruction
2.	Sherwood Rawland	(*b*)	Living mater at different trophic levels
3.	Standing crop	(*c*)	Excessive plantation of trees
4.	Afforestation	(*d*)	Green house emission level

21. Match the following

1.	World environment day	(*a*)	16th September
2.	Earth day	(*b*)	5th June
3.	Anti tobacco day	(*c*)	31st May
4.	Ozone day	(*d*)	22nd April

22. Match the following

1.	Organism	(*a*)	Local population
2.	Biome	(*b*)	Distinct unit of life
3.	Climate	(c)	Regional unit–desert
4.	Ecotype	(*d*)	Seasons

23. Match the following

1.	Niche	(*a*)	Gause's Hypothesis
2.	Competitive Exclusion	(*b*)	Crowded cities
3.	Urbanization	(*c*)	Crust of earth
4.	Lithosphere	(*d*)	Grinnel

24. Match the following

1.	Tropical rain forests	(*a*)	Pine and deodar
2.	Deserts	(*b*)	Diptero carpus
3.	Coniferous	(*c*)	Cacti
4.	Grassland	(*d*)	Cynodon

25. Match the following

1.	13,000 genes	(*a*)	Magnolia
2.	Anti cancer drug	(*b*)	Humans
3.	Exotic species	(*c*)	Yew free
4.	Primitive genes	(*d*)	Lantana Camara

26. Match the following

1.	Kaziranga National Park	(*a*)	Tiger
2.	Corbett National Park	(*b*)	Lion
3.	Gir National Park	(*c*)	Rhinoceros
4.	Bandipur	(*d*)	Sambhar

27. Match the following

1.	Biosphere	(*a*)	Grass, Sea weed
2.	Endemic	(*b*)	Threatened species
3.	Red book	(*c*)	Confined to a place
4.	Producers	(*d*)	Ecosphere

28. Match the following

1.	Fusion material	(*a*)	Uranium-235
2.	Fission material	(*b*)	Lithium
3.	Earth quake	(*c*)	Madhya Pradesh
4.	Bodhghat project.	(*d*)	Seismology

29. Match the following

1.	Royal park	(*a*)	Elton
2.	Key industry animals	(*b*)	Consumers
3.	Standing stage	(*c*)	Abiotic material
4.	Phargotrophs	(*d*)	Australia

30. Match the following

1.	Producers	(*a*)	Malaria
2.	Quinine	(*b*)	Transducers
3.	Penicillin	(*c*)	*Ficus religiosa*
4.	Pipal	(*d*)	Antibiotic

31. Match the following

1.	Solar energy	(*a*)	Methane and CO_2
2.	Minerals	(*b*)	Agro forestry
3.	Taungya system	(*c*)	Mining
4.	Derelict land	(*d*)	Inexhaustible
5.	Biogas	(*e*)	Exhaustible

32. Match the following

1.	Ecology	(*a*)	Bacteria and Fungi
2.	Niche	(*b*)	Earth crust
3.	Edaphic	(*c*)	Soil
4.	Lithosphere	(*d*)	Haeckel
5.	Decomposers	(*e*)	Grinnel

33. Match the following

1.	Earth summit	(*a*)	Ozone hole
2.	Kyoto protocol	(*b*)	1988
3.	Antarctica	(*c*)	Rio de Janeiro
4.	Kautilya	(*d*)	Japan
5.	National forest policy	(*e*)	World's first forest conservation programme

34. Match the following

1.	Plankton	(*a*)	Dicanthium
2.	Floating	(*b*)	Cephalanthus
3.	Woodland	(*c*)	Lemma
4.	Sedge or marsh	(*d*)	Cyanobacteria

35. Match the following

1.	Birmotibagh wild life sanctuary	(*a*)	Assam
2.	Dachigam sanctuary	(*b*)	West Bengal
3.	Jaladapura sanctuary	(c)	Patiala
4.	Manas wild life sanctuary	(*d*)	Srinagar

36. Match the following

1.	Rice	(*a*)	*Prosopis cineraria*
2.	Khejri	(*b*)	*Oryza sativa*
3.	Pipal	(*c*)	*Ficus sanctum*
4.	Tulsi	(*d*)	*Ficus religiosa*

37. Match the following

1.	Exotic species	(*a*)	Flying bird
2.	Drought	(*b*)	Cadmium
3.	Inorganic pollutant	(*c*)	Alien species
4.	California condor	(*d*)	Dryness

38. Match the following

1.	Organic Farming	(*a*)	Cover with crop residue and litter
2.	Crop rotation	(*b*)	Increased organic input to soil
3.	Contour ploughing	(c)	Growing different crops
4.	Mulching	(*d*)	Alternate furrows and ridges

Fill in the Blanks

39. The term 'Ecology' was coined by____

40. The study of reciprocal relationship between organisms and their environment is called____

41. ____can be visualized as a thin life supporting belt on and around earth

42. The microscopic, photosynthetic organisms that float on water surface are called____

43. The ecological factors related to soil and substratum are called____factors.

44. The solid or dry crust of the earth that forms the contents and other land masses is called____

45. The water mantle of the earth is known as____

46. The ability of some animals to blend with the surrounding is known as____

47. The study of soil and its related aspects is known as____

48. ____are groups of species exploiting a common resource in a similar fasion

49. ____showed that closely related birds could occur together due to difference in their nesting and feeding habits

50. Bisons of Asia, Kangaroos of Australia and Zebra of Africa are____

51. Physical space occupied by an organism is ____

52. The place occupied by an organism, population or community is called____

53. The term 'Niche' was first introduced by in the year____

54. The first atomic bomb explosion was done in____and second in____

55. Prolonged exposure to noise of____or more may lead to impaired hearing

56. Degradation of organic wastes by microorganisms in presence of oxygen is known as____

57. The process of controlled burning of the waste at high temperature of 850°C is known as____

58. The magnitude or intensity of energy released by an earthquake is measured by____

59. Tropical cyclones in Caribbean sea are called as____

60. Property of certain elements to emit protons or electrons is known as____

61. Ozone is formed in the atmosphere by____

62. Tobacco smoke contains a hydrocarbon called____

63. Petroleum used in automobiles contains____as a antiknock agent

64. ____is the most abundant hydrocarbon in the atmosphere
65. Increase in concentration of soluble salts in soil is called____
66. Minamata disease is caused due to bio-magnification of____
67. The total land mass of India is____million hectares.
68. Deposition of acids (sulphuric acid/nitric acid) on the earth from atmosphere is known as____
69. The unit of measurement of sound is____
70. Excess of nitrates in drinking water causes____in humans.
71. The radioactive gases are also called____
72. The thinning of stratospheric zone layer during spring time is called____
73. The management of rainfall with resultant runoff is called____
74. The type of vegetation found in temperature forests is____
75. Deep channels worn by water are called____
76. Soil on river bank is poor in nutrients because of____
77. A land for grazing is known as____
78. The three components of our environment are____
79. The difference between the number of births and that of death is known as____
80. The replacement levels in developed and developing countries are____
81. The total fertility rate in developed countries is close to____
82. Replacement level in developing countries is always higher than 2.0 because____
83. The number of children a couple must produce to replace themselves is called____
84. The movement of individuals into a place or country is known as____

85. The time required for a population to double itself is known as____

86. The number of babies produced per thousand individuals is known as____

87. Biochemical Oxygen Demand measures____

88. ____can serve as indicators to detect pollution of air by SO_2.

89. Sound becomes hazardous noise pollution at decibels____

90. Taj Mahal at Agra is damaged by____

91. Preservation of living tissues/parts of plants in liquid nitrogen at a temperature of –196°C known as____

92. An exotic plant species which has posed threat to the native species is____

93. The total number of National Parks in India are____

94. Derivative chemicals synthesized from plant products are called____

95. The diversity of communities along the environmental gradient is____

96. Oak species are predominantly found in____forests

97. The pioneer plant of xerarch is________

96. The primary source for all living organisms is____

99. The ecological pyramids were devised by ____

100. The plant species that invade the bare habitat initially are known as____

101. Graphic representation of ecological parameters at each trophic level in an ecosystem is____

102. Interconnected matrix of food chain in an ecosystem is known as____

103. A protected area where multiple use of the land is permitted by dividing it into certain zone is known as____

104. The richest and most threatened reservoirs of plant and animal life on earth is known as____

105. **Excess fluoride in drinking water causes____**

106. **The ability of the reproductively active individuals to produce babies is known as____**

107. **If the individuals of a species remain alive only in captivity or other human controlled conditions, the species is said to be____**

108. **The increased concentration of greenhouse gases in the atmosphere leads to____**

109. **A glass house used for raising delicate plants is called____**

110. **____is a product of incomplete decomposition caused by a group of bacteria called ethanogens.**

Answers

Match the Following

1	1 (*d*), 2 (*c*), 3 (*b*), 4 (*a*)	20	1 (*d*), 2 (*a*), 3 (*b*), 4 (*c*)
2	1 (*c*), 2 (*a*), 3 (*b*)	21	1 (*b*), 2 (*d*), 3 (*c*), 4 (*a*)
3	1 (*c*), 2 (*d*), 3 (*b*), 4 (*a*)	22	1 (*b*), 2 (*c*), 3 (*d*), 4 (*a*)
4	1 (*b*), 2 (*a*), 3 (*d*), 4 (*c*)	23	1 (*d*), 2 (*a*), 3 (*b*), 4 (*c*)
5	1 (*b*), 2 (*a*), 3 (*d*), 4 (*c*)	24	1 (*b*), 2 (*c*), 3 (*a*), 4 (*d*)
6	1 (*d*), 2 (*a*), 3 (*b*), 4 (*c*)	25	1 (*b*), 2 (*c*), 3 (*d*), 4 (*a*)
7	1 (*b*), 2 (*c*), 3 (*a*), 4 (*d*)	26	1 (*c*), 2 (*a*), 3 (*b*), 4 (*d*)
8	1 (*b*), 2 (*c*), 3 (*a*), 4 (*d*)	27	1 (*d*), 2 (*c*), 3 (*b*), 4 (*a*)
9	1 (*c*), 2 (*a*), 3 (*b*), 4 (*d*)	28	1 (*b*), 2 (*a*), 3 (*d*), 4 (*c*)
10	1 (*b*), 2 (*d*), 3 (*a*), 4 (*c*)	29	1 (*d*), 2 (*a*), 3 (*c*), 4 (*b*)
11	1 (*b*), 2 (*c*), 3 (*d*), 4 (*a*)	30	1 (*b*), 2 (*a*), 3 (*d*), 4 (*c*)
12	1 (*c*), 2 (*a*), 3 (*d*), 4 (*b*)	31	1 (*d*), 2 (*e*), 3 (*b*), 4 (*c*), 5 (*a*)
13	1 (*c*), 2 (*d*), 3 (*b*), 4 (*a*)	32	1 (*d*), 2 (*e*), 3 (*c*), 4 (*b*), 5 (*a*)
14	1 (*d*), 2 (*a*), 3 (*b*), 4 (*c*)	33	1 (*c*), 2 (*d*), 3 (*a*), 4 (*e*), 5 (*b*)
15	1 (*d*), 2 (*a*), 3 (*b*), 4 (*c*)	34	1 (*d*), 2 (*c*), 3 (*a*), 4 (*b*)
16	1 (*c*), 2 (*d*), 3 (*b*), 4 (*a*)	35	1 (*c*), 2 (*d*), 3 (*b*), 4 (*a*)
17	1 (*d*), 2 (*a*), 3 (*b*), 4 (*c*)	36	1 (*b*), 2 (*a*), 3 (*d*), 4 (*c*)
18	1 (*b*), 2 (*d*), 3 (*a*), 4 (*c*)	37	1 (*c*), 2 (*d*), 3 (*b*), 4 (*a*)
19	1 (*d*), 2 (*c*), 3 (*b*), 4 (*a*)	38	1 (*b*), 2 (*c*), 3 (*d*), 4 (*a*)

Fill in the Blanks

39 Ernst Haeckel
40 Ecology
41 Biosphere
42 Phytoplankton
43 Edaphic
44 Lithosphere
45 Hydrosphere
46 Camouflage
47 Pedalogy
48 Guilds
49 Lack
50 Ecologically Equivalents
51 Spacial
52 Habitat
53 Grinnel, 1917
54 Nagasaki, Hiroshima
55 80 dB
56 Composting
57 Incineration
58 Richter scale
59 Hurricanes
60 Radioactivity
61 Photochemical reaction
62 Benzo pyrene
63 Tetra ethyl lead
64 Methane
65 Salination
66 Mercury
67 305

68 Acid rain

69 Decibel

70 Methamoglobinemia–Blue baby syndrome

71 Greenhouse gases

72 Ozone hole

73 Watershed management

74 Conifers mixed with broad leaves trees

75 Gullies

76 Leaching

77 Pasture

78 Productive system, protective system and waste assimilative system

79 Rate of natural increase

80 2.1 and 2.7

81 Replacement level

82 Some children die before reaching reproductive age

83 Replacement level

84 Immigration

85 Doubling time

86 Natalilty

87 Dissolved O_2 needed by microbes to decompose organic waste

88 Lichens

89 Above 80

90 Sulphur Dioxide–SO_2

91 Cryopreservation

92 Lanata Camara

93 89

94 Botano chemicals

95 Beta diversity

96 Temperate broad leaf

97 Lichens

98 Sun

99 Charles Elton

100 Pioneer species

101 Ecological Pyramids

102 Food web

103 Biosphere reserve

104 Hotspots

105 Fluorosis

106 Fertility

107 Extinct in the wild

108 Global warming

109 Greenhouse

110 Methane

Part II

Glossary

2,4-D – An herbicide used to kill weeds and considered to be a carcinogen and a mutagen.

20/30/10 standard: 20 mg/l Biochemical Oxygen Demand (BOD), 30mg/l Suspended Solids (SS), 10 units of *E. coli* the water quality standard for greywater use in toilets, laundry and surface irrigation.

3R's: are Reduce, Reuse, Recycle.

4R's: are Reduce, Reuse, Recycle, Replenish/Rethink

5R's: (sustainability) Reduce, Remanufacture, Reuse, Recycle, Recover.

7Q10: Seven-day, consecutive low flow with a ten year return frequency; the lowest stream flow for seven consecutive days that would be expected to occur once in ten years. This minimum 7Q10 is used in the water quality permit arena to designate a minimum flow to which permit conditions apply. However for dissolved gas standards, the maximum 7Q10 is used, which is the average peak annual flow for seven consecutive days that has a recurrence interval of ten years.

A

Abandoned Well: Any well (drinking water, oil and gas etc.) which is not used for a long period of time, is not maintained properly, and/or is not properly sealed when its useful life is over.

Abatement: Decreasing the degree or intensity of, or eliminating, pollution.

Abatement debris: Waste generated from remediation activities.

Abiocoen: refers to the non-living components of environment.

Abiota: The soil, water and air of an ecosystem or environment *i.e.*, nonliving component is known as Abiota.

Abiotic: Non-living chemical and physical factors of the environment. Non-living; usually applied to the physical characteristics of biological systems, such as moisture, nutrients, soils, solar radiation, etc.

Abiotic environment: all physical and non-living chemical factors such as soil, water and atmosphere which influence living organisms.

Abiotic substance: any fundamental chemical element or compound in the environment.

Ablation: A geological term for the process of wearing away of solid ice from a glacier. It includes surface, internal and basal melting, sublimation, and calving of icebergs.

Aborigines: The native inhabitants or original of a region or country.

Abrasion: General grinding or wearing of rock material by impact of objects in transport. Eolian wind-accomplished erosion of rock or soil masses by impact of saltating grains.

Absolute: In physics and chemistry, absolute generally refers to something viewed apart from modifying influences or without comparison with other objects; actual; real; - opposed to relative and comparative. In pollution context – a chemical substance that is relatively free of impurities.

Absolute age (geology): The geologic age of a fossil, or a geologic event or structure expressed in units of time, usually years. A variety of dating methods are used by geologists to achieve this. The determination of the absolute age of rocks, minerals and fossils, in years before the present, is the basis for the field of geochronology. The measurement of the decay of radioactive isotopes, especially uranium, strontium, rubidium, argon and carbon, has allowed geologists to more precisely determine the age of rock formations. Tree rings and seasonal sedimentary deposits called verves can be counted to determine absolute age. Although the term implies otherwise, "absolute" ages typically have some amount of potential error and are inexact.

Absolute dating: It is the process of determining a specific date for an archaeological or palaeontological site or artifact. Absolute dating is usually based on the physical or chemical properties of the materials of artifacts, buildings, or other items that have been modified by humans. One of the most widely used and well-known absolute dating techniques is carbon-14 (or radiocarbon) dating, which is used to date organic remains.

Absolute humidity: The total moisture present in a given amount of air, measured in units of mass per unit volume. Expressed as grams of water vapor per cubic meter of air.

Absolute pressure: The total pressure within a vessel or pipe or etc., not offset by external atmospheric pressure. Units are millimeters of mercury (mm Hg) and pounds per square inch (Absolute).

Absorbed dose: In exposure assessment, the amount of a substance that penetrates an exposed organism's absorption barriers (*e.g.*, skin, lung tissue, gastrointestinal tract) through physical or biological processes. Absorbed Dose is synonymous with 'Internal Dose'.

Absorption: The passage of one substance into or through another, either physically or chemically. It also can be defined as the uptake of water or other fluids or dissolved chemicals by a cell or an organism by capillary, osmotic solvent or chemical action (as tree roots absorb dissolved nutrients in soil).

Absorption barrier: Any of the exchange sites of the body that permit uptake of various substances at different rates (*e.g.*, skin, lung tissue, and gastrointestinal-tract wall).

Absorption pit (soak away): A hole dug in permeable ground and filled with broken stones or granular material and usually covered with earth allowing collected water to soak into the ground.

Absorptive capacity: The maximum amount of waste that can be absorbed by the environment.

Acceptable daily intake: A daily exposure level that will not cause adverse health effects.

Accident site: The location of an unexpected occurrence, failure or loss, either at a plant or along a transportation route, resulting in a release of hazardous materials.

Acclimation or acclimatization: The process of an organism adjusting or adaptation to chronic change or new conditions in its environment. Biology – adaptation.

Acetic acid: A clear liquid with a strong odor and flavour that is the prime ingredient of vinegar. The Chemical formula is CH_3COCH_3. Acetic acid is a weak acid and an important solvent used to manufacture organic chemicals and plastics.

ACGIH: American Conference of Governmental Industrial Hygienists. An organization of professional in governmental agencies or educational institutions engaged in occupational safety and health programs. ACGIH develops and publishes recommended occupational exposure limits for chemical substances and physical agents.

Acid: An acid is a substance that increases the hydrogen ion (H^+) or hydronium ions (H_3O^+) concentration in a solution and neutralizes bases and conduct electrically to form a salt plus water. Acids undergo dissociation in water with the formation of hydrogen ions. Acids are characterized by their tendency to lose a proton (or to gain an electron pair) in reactions. Acid is a corrosive solution with a pH less than 7. Acids have a sour taste and may cause severe burns. They turn litmus paper red and have pH values of 0 to 6. Acids will neutralize bases or alkaline media. Acids turn litmus paper red and can be recognized by their usually sour taste (as in vinegar, acetic acid) and low pH.

Acid aerosol: Acidic liquid or solid particles small enough to become airborne are called as Acid Aerosol. Air borne particles composed of sulfates, sulfuric acid, nitrates and or nitric acid. Moderate air concentrations have been shown to cause constriction of the respiratory passages in exercising asthmatics and much higher levels have been linked to excess mortality rates in past air pollution episodes. Dry particle diameter are typically less than 1 – 2 microns. High concentrations can irritate the lungs and have been associated with respiratory diseases like asthma.

Acid deposition: A complex chemical and atmospheric phenomenon that occurs when emissions of sulfur and nitrogen compounds and other substances are transformed by chemical processes in the atmosphere, often far from the original sources, and then deposited on earth in either wet or dry form. The wet forms, popularly called "Acid Rain," can fall to earth as rain, snow, or fog. The dry forms are acidic gases or particulates.

Acid fog: Air borne water droplets with typical diameters of 5 – 30 micrometers containing sulfuric acid and (or) nitric acid.

Acid gas: Air pollutant with a pH value of less than 5 released as a byproduct of incomplete combustion of solid wastes and fossil fuels.

Acid mine drainage: Acidic water that flows into streams from abandoned coal mines (or) pills of coal mining wastes. Acid mine drainage is the highly acidic, sediment laden discharge from exposed mines that is released into the ambient aquatic environment or the outflow of acidic water from metal mines or coal mines. The water has a low pH because of its contact with sulfur – bearing material and is harmful to aquatic organisms. Acid mine drainage is not generated by strip mining itself but by the nature of the rock where it takes place. Three conditions are responsible to form acid mine drainage and they are pyrite bearing rock, oxygen and iron oxidizing bacteria.

Acid neutralizing capacity: Measurability of a base (*e.g.*, water or soil) to resist changes in pH.

Acid rain or acid precipitation: Rain or other forms of precipitation that is unusually acidic. Acid rain is caused by emissions from the burning of fossil fuels. When fuels such as coal, oil and natural gas are burned, many substances enter the atmosphere. Sulfur dioxide and nitrogen compounds, which contribute to air pollution, travel through the air and react with each other in sunlight to form secondary pollutants such as sulfuric acid and nitric acid. When rain contains these acids, it is called "acid rain". (distilled water is neutral pH 7; normal precipitation has a pH of about 5.6 because dissolved carbon dioxide from the atmosphere produces carbonic acid; acid rain has pH of even 2.4 because of dissolved interacts with oxides of sulfur and nitrogen in the atmosphere produces sulfuric acid and nitric acid). When sulfur dioxide and nitrogen oxides are released from power plants and other sources, winds blow them far from their source. If the acid chemicals in the air are blown into areas where the weather is wet, the acids can fall to Earth in the form of rain, snow, fog or mist. In areas where the weather is dry, the acid chemicals may become incorporated into dusts or smokes. Acid rain can damage the environment, human health and property. Long term deposition of these acids is linked to adverse effects on aquatic organisms and vegetation in geographical areas with poor buffering capacity within the sediments. Acid deposition results from the deposition of air borne acidic pollutants on land and in bodies of water. These pollutants can cause damage to forests as well as lakes and streams.

Acidic: The condition of water of soil that contains a sufficient amount of acidic substances to lower the pH below 7.0.

Acidification: Raising the acidity of a fluid by adding an acid or reduction in pH, usually in soil. Acidification of soils and natural waters, by acid rain or acid wastes can result in reduced biological productivity if the pH is sufficiently reduced. Acidification can result in poorly structured or hard – setting topsoils that cannot support sufficient vegetation to prevent erosion.

Acidity: It is measured in a solution with a concentration of hydrogen ions per mole greater than 1×10^{-7} (or) a pH less than 7 is called acidity of a solution.

Acidosis: Condition of decreased alkalinity of the blood and tissues marked by sickly sweet breath, headache, nausea, vomiting, and visual disturbances; usually the result of excessive acid production.

Acquired immuno deficiency syndrome (AIDS): A disease caused by the Human Immunodeficiency Virus (HIV), which reduces the body's ability to fight other diseases.

Acrid: A chemical or substance or liquid or solid with sharp and harsh or unpleasantly bitterly pungent in taste or odor; irritating to the eyes, nose, etc.; acrid smoke from burning rubber.

Activated carbon: A highly adsorbent form of carbon used to remove odors and toxic substances from liquid or gaseous emissions. In waste treatment, it is used to remove dissolved organic matter from waste drinking water. It is also used in motor vehicle evaporative control systems.

Activated Charcoal: A material produced by burning a carbon material such as wood (or) coconut shells and then increasing the adsorptive capacity by steam treatment; which creates a very large surface area. The product is used to adsorb organic materials, either as an air (or) water pollution control techniques (or) as an air sampling device.

Activated sludge: Product that results when primary effluent is mixed with bacterialaden sludge and then agitated and promote biological treatment; speeding the breakdown of organic matter in raw sewage undergoing secondary waste treatment. The suspended solids, mostly microorganisms present in the aeration tank of a sewage treatment plant.

Activated sludge process: The activated sludge process is in aerobic (oxygen-rich) continuous flow biological method for the treatment of domestic and biodegradable industrial waste water, in which organic matter is utilized by microorganisms for life sustaining process, that is; for energy, for reproduction, digestion, movement etc. A from of secondary treatment of sewage, this process uses micro-organisms, suspended in well-aerated waste water to degrade the organic material. A sedimentation tank placed after the activated sludge/aeration tank removes the sludge, routing some of it back to the aeration tank . The rest is pumped outside the system for treatment and disposal .

Activator: A chemical added to increase the activity of a pesticide.

Active Ingredient: A product ingredient that actually does what the product is designed to do in a pesticide product . The remaining ingredients may be inert . Pesticides are regulated primarily on the basis of active ingredients .

Active solar heating: The use of solar panels to collect and concentrate the sun's energy for heating.

Acute: A brief but high level of exposure to a hazardous substance or an adverse health effect resulting from a brief exposure to a hazardous substance. In medicine, an acute disease is a disease with either of a hazardous substance. In medicine, an acute disease is a disease with either of a rapid onset or of a short course (as opposed to a chronic course) or both. Normally it is confused with severe, something can be acute but not severe.

Active effect: Adverse effect on any living organism like on a human or animal body or plants that takes place soon after exposure, which results in severe symptoms that develop rapidly and may quickly reach a crisis.

Acute exposure: A single exposure to a toxic substance which may result in severe biological harm or death. Single exposures over short time are referred to as acute exposure, and repeated exposure over a longer time as chronic exposure. Duration of exposure should be reported with toxicity data in safety data sheets.

Acute lethality: Animals death immediately or within 14 days after a single does or of exposure to a toxic substance.

Acute radiation effects: Effects that are caused by (usually short-term) exposure to radiation and that occur immediately after the exposure.

Acute radiation syndrome (ARS): Serious illness that occurs when the body receives a high dose of penetrating radiation (that is, able to reach internal organs) over a short period of time (usually minutes).

Adaptation: Adjustment to a new or changing environment. Various types of adaptation can be distinguished, including anticipatory and reactive adaptation, private and public adaptation, and autonomous and planned adaptation. A characteristic of an organism that has been favoured by natural selection (or) changes in an organism's physiological structure or function or habits that allow it to survive in new surroundings (or) Biological modification that allows species to better exist in a specific environment.

Adaptive management: A systematic process for continually improving management policies and practices by learning from the outcomes of previously employed policies and practices.

Adaptive radiation: Closely related species that look very different, as a result of having adapted to widely different ecological niches. It describes the rapid speciation of a single or a few species to fill many ecological niches. This is an evolutionary process driven by mutation and natural selection. This often occurs when a species in introduced to a new ecosystem, or when a species can survive in an environment that was unreachable before.

Additives: Chemical substances added to food, often considered to represent a threat to human health.

Add – on control device: An air pollution control apparatus or device such as carbon absorber (or) incinerator added to the exhaust system to prevent air pollution. The device usually does not reduce pollutants through a modification of the process rather, potential pollutants are removed from the exhaust gases.

Adenocarcinoma: A malignant tumor originating in glandular tissue.

Adenosis: A disease of a gland, especially one marked by the abnormal formation or enlargement of glandular tissue.

Adhesion: A union of two surfaces those are normally separate.

Administered dose: In exposure assessment, the amount of a substance given to a test subject (human or animal) to determine dose – response relationships. Since exposure to chemicals is inadvertent, this quantity is often called potential dose.

Administrative Procedures Act: A law spells out procedures and requirements related to the promulgation of regulations.

Adsorb: Collecting gas or liquid molecules on the surface of another material.

Adsorbate: The chemicals being adsorbed (or) taken up in a pollution control device used for the adsorption of a chemical substance.

Adsorbent: The material to which a pollutant is being adsorbed in a pollution control device operating by adsorption.

Adsorption: One substance taking up another at its surface or the adhesion of molecules of gas, liquid, or dissolved solids to a surface. The term also refers to a method of treating wastes in which activated carbon removes organic matter from wastewater. Removal of a pollutant from air or water by collecting the pollutant on the surface of a solid material; *e.g.*, an advanced method of treating waste in which activated carbon removes organic matter from waste-water.

Adulterant: A chemical impurity (or) substance that by law should not be in either a food (or) a pesticide. Any substance present in food (or) drugs of concentrations exceeding those allowed by the food and drug administration authority.

Adulterated: Any pesticide whose strength or purity falls below the quality stated on its label. A food, feed, or product that contains illegal pesticide residues.

Advanced treatment: A level of wastewater treatment more stringent than secondary treatment; requires an 85-per cent reduction in conventional pollutant concentration or a significant reduction in non-conventional pollutants. Sometimes called tertiary treatment.

Advisory: A non-regulatory document that communicates risk information to those who may have to make risk management decisions.

Aerated lagoon: A holding and/or treatment pond that speeds up the natural process of biological decomposition of organic waste by stimulating the growth and activity of bacteria that degrade organic waste.

Aeration: Process of mixing with air or oxygen. Aeration is the basis for secondary wastewater treatment. Air is bubbled through the water to increase amount of dissolved oxygen. A process which promotes biological degradation of organic matter in water. More oxygen promotes the action of aerobic bacteria, which break down organic waste and thereby purify the water. The process may be passive (as when waste is exposed to air), or active (as when a mixing or bubbling device introduces the air).

Aeration lagoon: The pond used to process organic wastes by ensuring that sufficient oxygen in the water to promote biological decomposition. Often the devices used agitate the contents of the pond to promote contact between the contents and the air. The presence of oxygen stimulates the activity of microorganisms capable of degrading organic materials.

Aeration tank: A chamber used to inject air into water.

Aeroallergens: Air borne particles that can cause allergic responses (such as hay fever) in susceptible persons.

Aerobic: Life or processes that require air or oxygen.

Aerobic treatment: Process by which microbes decompose complex organic compounds in the presence of oxygen and use the liberated energy for reproduction and growth. (Such process include extended aeration, trickling filtration, and rotating biological contactors).

Aerosol: A finely divided solid material or liquid particles suspended in air or other gaseous environment in the atmosphere. It is also defined as a suspension of extremely small particles or tiny droplets of liquids in air or in a gas. Gases under pressure are used to dispense a variety of substances as aerosols, from spray paints and insecticides to asthma medications. The use of chlorofluorocarbons (CFC's) as propellants in aerosol cans has been linked to the destruction of the earth's protective ozone layer.

Adduct: In biology, an adduct is a complex that forms when a chemical binds to a biological molecule, such as DNA or protein.

Affected public: The people who live and/or work near a hazardous waste site. The human population adversely impacted following exposure to a toxic pollutant in food, water, air, or soil.

Afforestation: The establishment of a forest through tree planting or seeding on land that has lacked forest cover for a very long time or has never been forested.

Aflatoxins: Aflatoxins are potent toxic, carcinogenic, mutagenic, immunosuppressive substances that can be found on poorly stored grains and nuts. They are produced as secondary metabolites by certain types of molds (*Aspergillus flavus* and *Aspergillus parasiticus*). Food contaminated with aflatoxins, can affect both humans and animals and lead to liver cancer.

After burner: An air pollution control device that incinerates organic compounds in an air stream to carbondioxide and water. In incinerator technology, a burner is located so that the combustion gases are made to pass through its flame in order to remove smoke and odors. It may be attached to or be separated from the incinerator proper. Also known as 'Vapour Incinerates'.

Age distribution or structure: The proportions of a population falling into three distinct groups pre productive, reproductive, and post productive. If the population is divided equally among the three groups, the population growth is stable. When a large proportion of the population is in the pre productive stage, the long-term trend is population growth. If a large proportion of the population is in the reproductive stage, a baby boom is experienced.

Age tank: A tank used to store a chemical solution of known concentration for feed to a chemical feeder. Also known as "Day Tank".

Agency: Extent to which humans are able to create and change the world in which they live.

Agenda 21: One of several documents emerging from the Earth Summit in Rio de Janeiro in June 1992. Major issues of environment and development were examined, including poverty, population, and human health.

Agent: Any substance, force, radiation, organism, or influence that affects the body. Effects may be beneficial or injurious (or) any physical, chemical, or biological entity that can be harmful to an organism. Synonym for "Stressors".

Agent orange: A mixture of several herbicides that is considered to be a carcinogen because it is contaminated with dioxin. A toxic herbicide and defoliant used in the Vietnam by American military to deprive the enemy of food and of hiding places in the jungles, by killing all vegetation. It contained 2, 4, 5 – trichlorophen – oxyacetic acid (2, 4, 5-T) and 2-4 dichlorophenoxyacetic acid (2, 4-D) with trace amounts of dioxin. The name comes from the orange – banded barrels in which it was marketed.

Agribusiness: Industrialized agriculture controlled by corporations.

Agricultural economy: An economic system based primarily on crop production.

Agricultural pollution: The adverse consequences associated with the operation of a farm. Runoff and leaching of pesticides and fertilizer from cultivated fields, dust and erosion problems resulting from cultivation practices, in proper disposal of chemical manure and carcasses and accumulated crop residues and other debris potentially cause environmental problems.

Agricultural revolution: A shift that took place 10,000 to 12,000 years ago and was characterized by the movement of human activity from hunting and gathering to agriculture.

Agricultural waste: Poultry and livestock manure, and residual materials in liquid or solid form generated from the production and marketing of poultry, livestock or fur – bearing animals; also includes grain, vegetable, and fruit harvest residue.

Agroecosystem: Land used for crops, pasture, and livestock; the adjacent uncultivated land that supports other vegetation and wildlife; and the associated atmosphere, the underlying soils, groundwater, and drainage networks.

Agroforestry (sustainability): An agricultural system based on the cultivation of trees with other crops. An ecologically – based farming system, that, through the integration of trees in farms, increases social, environmental and economic benefits to land users.

Air binding: Situation where air enters the filter media and harms both the filtration and backwash processes.

Air changes per hour (ACH): The movement of a volume of air in a given period of time; if a house has one air change per hour, it means that the air in the house will be replaced in an one – hour period.

Air cleaning: Indoor – air quality – control strategy to remove various airborne particulates and / or gases from the air. Most common methods are particulate filtration, electrostatic precipitation, and gas sorption.

Air curtain: A pneumatic barrier for containing oil spills. Air bubbling through a perforated pipe causes an upward water flow that slows the spread of oil. It can also be used to stop fish from entering polluted water.

Air exchange rate: The rate at which outside air replaces indoor air in a given space.

Air gap: Open vertical gap or empty space that separates drinking water supply to be protected from another water system in a treatment plant or other location. The open gap protects the drinking water from contamination by backflow or back siphonage.

Air mass: An air mass is a large body of air that has similar temperature and moisture properties throughout.

Air padding: Pumping dry air into a container to assist with the withdrawal of liquid or to force a liquefied gas such as chlorine out of the container.

Air permeability: Permeability of soil with respect to. Important to the design of soil-gas surveys. Measured in darcys or centimeters per second.

Air plenum: Any space used to convey air in a building, furnace, or structure. The space above a suspended ceiling is often used as an air plenum.

Air pollutant: Any substance in air that could, in high enough concentration, harm man, other animals, vegetation, or material. Pollutants may include almost any natural or artificial composition of airborne matter capable of being airborne. They may be in the form of solid particles, liquid droplets, gases, or in combination thereof. Generally, they fall into two main groups 1. those emitted directly from identifiable sources and 2. those produced in the air by interaction between two or more primary pollutant, or by reaction with normal atmospheric constituents, with or without photoactivation. Exclusive of pollen, fog, and dust, which are of natural origin, about 100 contaminants have been identified. Air pollutants are often grouped in categories for ease in classification; some of the categories are particulate matter, sulfur compounds, volatile organic chemicals, nitrogen compounds, oxygen compounds, halogen compounds, radioactive compounds, and odors.

Air pollutant transport: Air pollutant transport is the advection (or) horizontal convection of air pollutants from an area where emission occurs to a down wind receptor area by local (or) regional winds.

Air pollution: The presence of contaminants or pollutant substances in the air that interfere with human health or welfare, or produce other harmful environmental effects. The modification of the natural characteristics of the atmosphere by a chemical, particulate matter, or biological agent. Pollution can occur from natural changes or from human activities. The five major primary pollutants are carbon monoxide, hydrocarbons, nitrogen compounds, particulate matter, and sulfur dioxide.

Air pollution control device: Mechanism or equipment that cleans emissions generated by a source (*e.g.*, an incinerator, industrial smokestack, or an automobile exhaust system) by removing pollutants that would otherwise be released to atmosphere.

Air pollution episode: A period of abnormally high concentration of air pollutants, often due to low winds and temperature inversion, that can cause illness and death.

Air pollution hot spot: A location where emissions from specific sources may expose individuals and population groups to elevated risks of adverse health effects – including but not limited to cancer – and contribute to the cumulative health risks of emissions from other sources in the area.

Air quality criteria: The levels of pollution and lengths of exposure above which adverse health and welfare effects may occur.

Air quality standards: The level of pollutants prescribed by regulations that are not be exceeded during a given time in a defined area.

Air sparging: Injecting air or oxygen into an aquifer to strip or flush volatile contaminants as air bubbles up through the ground water and is captured by a vapor extraction system. Air sparging may be a good choice of treatment technology at sites contaminated with solvents and other volatile organic compounds (VOC).

Air stripping: A treatment system that removes or strips volatile organic compounds (VOCs) from contaminated ground water or surface water by forcing an airstream through the water and causing the compounds to evaporate.

Air/oil table: The surface between the vadose zone and ambient oil; the pressure of oil in the porous medium is equal to atmospheric pressure.

Airborne particulates: Total suspended particulate matter found in the atmosphere as solid particles or liquid droplets. Chemical composition of particulates varies widely, depending on location and time or year. Sources of airborne particulates include dust, emissions from industrial processes, combustion products from the burning of wood and coal, combustion products associated with motor vehicle or non-road engine exhausts, and reactions to gases in the atmosphere.

Airborne release: Release of any pollutant in to the air.

Alachlor: A herbicide, marketed under the trade name lasso, used mainly to control weeds in corn soybean fields.

Alar: Trade name for daminozide, a pesticide that makes apples redder; firmer, and less likely to drop off trees before growers are ready to pick them. It is also used to a lesser extent on peanuts; tart cherries, concord grapes, and other fruits.

ALARA: Acronym for "Law as Reasonably Achievable".

Albedo: Reflectance; refers to the ratio of light from the Sun that is reflected by the Earth's surface, to the light received by it. Unreflected light is converted to infrared radiation (heat), which causes atmospheric warming. Thus, surfaces with a high albedo, like snow and ice, generally contribute to cooling, whereas surfaces with a low albedo, like forests, generally contribute to warming. Changes in land use that significantly alter the characteristics of land surfaces can alter the albedo.

Aldicarb: An insecticide sold under the trade name Temik. It is mad from ethyl isocywwanate.

Alage: Simple rootless plants that grow in sunlit waters in proportion to the amount of available nutrients. A general category of photosynthesizing unicellular and multicellular organisms, primarily aquatic. They can affect water quality adversely by lowering the dissolved oxygen in the water. They are food for fish and small aquatic animals. Algae (singular alga) include many unicellular planktonic organisms, diatoms, and multicellular seaweeds.

Algal bloom: The rapid and excessive growth of algae; generally caused by high nutrient levels combined with other favourable conditions. Sudden spurts of algal growth, which can affect water quality adversely and indicate potentially hazardous changes in local water chemistry. Blooms can deoxygenate the water leading to the loss of wildlife.

Algicide: Substance or chemical used specifically to kill or control algae.

Alien species: Also called as Introduced Species.

Aliquot: A measured portion of a sample taken for analysis. One or more aliquots make up a sample.

Alkali: Any chemical substance which forms soluble soaps with fatty acids. Alkalis are also referred to as bases. Alkali may cause severe burns to the skin. Alkalis turn litmus paper blue and have pH values from 8 to 14.

Alkaline: The condition of water or soil which contains a sufficient amount of alkali substance to raise the pH above 7.0. It is the opposite of acidic; also called basic; having a high pH value and thus a low concentration of hydrogen ions and a high concentration of hydroxide ions.

Alkalinity: The capacity of bases to neutralize acids. An example is lime added to lakes to decrease acidity.

Allergen: A substance that causes an allergic reaction in individuals sensitive to it.

Allergic reaction: Abnormal physiological response to chemical stimuli by a sensitive person.

Allergic respiratory reaction: Labored breathing, coughing, or gasping caused by inhaling a particular substance.

Allergic skin reaction: Reddening, swelling and/or itching of the skin following contact with a substance to which a person has become sensitized due to pervious skin contact or natural body condition.

Alluvial: Relating to and/or sand or soil deposited by flowing water.

Alopecia: Loss of hair.

Alternative agriculture: Agriculture based on reduced use of chemical fertilizers and pesticides, increased use of crop rotation, and reduced tillage of the soil.

Alternative compliance: A policy that allows facilities to choose among methods for achieving emission-reduction or risk-reduction instead of command-and control regulations that specify standards and how to meet them. Use of a theoretical emissions bubble over facility to cap the amount of pollution emitted while allowing the company to choose where and how (within the facility) it complies.

Alternative crops: Nontraditional crops that can be grown in an area to diversify rotations and increase income.

Alternative energy: Energy produced from sources other than fossil fuels (solar, wind, hydroelectric, geothermal, and biomass).

Alternative fuels: Fuels that can replace ordinary gasoline. Alternative fuels may have particularly desirable energy efficiency and pollution reduction features. Alternative fuels include compressed natural gas, alcohols, liquefied petroleum gas (LPG), and electricity. The 1990 Clean Air Act encourages development and sale of alternative fuels.

Altruism: Seeking the good for others.

Aluminosis: A disorder of the lung resulting from excessive inhalation of aluminum containing particulate matter. In some cases, pulmonary fibrosis is present.

Aluminum: A silvery nonferrous metal found in the bauxite ore. It is used in making hard, light, corrosion-resistant materials.

Ambient: Usual or surrounding environment conditions.

Ambient air: Any unconfined portion of the atmosphere, open air, surrounding air.

Ambient air quality standards: Refer Criteria Pollutants and National Ambient Air Quality Standards.

Ambient measurement: A measurement of the concentration of a substance or pollutant within the immediate environs of an organism; taken to relate it to the amount of possible exposure.

Ambient temperature: Temperature of the surrounding air or other medium.

Ames test: It is a biological assay to assess the mutagenic potential of chemical compounds. Named after its developer, Bruce Ames.

Amprometric titration: A way of measuring concentrations of certain substances in water using an electric current that flows during a chemical reaction.

Anaerobic: A life or process that not requires air or oxygen and is not destroyed by, the absence of oxygen. Used in reference to decomposition processes that occur in the absence of oxygen.

Anaerobic decomposition: Reduction of the net energy level and change in chemical composition of organic matter caused by microorganisms in an oxygen-free environment.

Anaerobic digestion: The biological degradation of organic materials in the absence of oxygen to yield methane gas (that may be combusted to produce energy) and stabilized organic residues (that may be used as a soil additive).

Analgesia: Loss of sensitivity to pain.

Ancient forests: Forests that have never been cut and typically consist of trees 250 years of age and older.

Anderson sampler: An air sampling device that separates air borne particles into size categories using stacked collection plates.

Anesthetic: Chemical that causes a total or partial loss of sensation by a living organism. Overexposure to anesthetics can cause impaired judgment, dizziness, drowsiness, headache, unconsciousness, and even death.

Anhydride: Compound derived from other compound by removing elements composing water (hydrogen and oxygen).

Anhydrous: Containing no water.

Animal dander: Scales of dead animals skin shed in the normal course of activity. This material is a common constituent of indoor air pollution in those homes with pet dogs. One of the allergens that contribute to asthma.

Animal rights: The belief that animals have rights similar to those afforded to humans.

Anisotropy: In hydrology, the condition under which one or more hydraulic properties of an aquifer vary from a reference point.

Annular space, annulus: The space between two concentric tubes or casing, or between the casing and the borehole wall.

Anorexia: Loss of appetite.

Anosmia: Loss of the sense of smell.

Anoxia: Means "without oxygen", an extreme form of hypoxia or "low oxygen". Is the absence of oxygen caused by inhaling a breathing gas that contains no oxygen or unable to inhale any breathing gas.

Antagonism: Interference or inhibition of the effect of one chemical by the action of another.

Antarctic ozone hole: Refers to the seasonal depletion of ozone in the upper atmosphere above a large area of Antarctica.

Anthropic: Relating to the period during which humans have existed on earth.

Anthropocentric ethic: The belief that only humans have value and that the environment exists solely for the benefit of humans; nature has no rights.

Anthropogenic: Man-made, not natural (or) Based on human activities; often used to refer to environmental changes caused by human activity.

Anthroposophy: Literally "the wisdom of man". Knowledge and practical techniques based on the body of teaching given out by an Austrian spiritual philosopher, scientist, artist and educator Rudolf Steiner, who lived from 1861 to 1925. Steiner was the initiator of biodynamic gardening.

Antidote: Remedy to relieve, prevent, or counteract the effects of a poison.

Anti-microbial: An agent that kills microbes.

Appearance: Physical state of a material.

Appropriated carrying capacity: Another name for the Ecological Footprint, but often used in referring to the imported ecological capacity of goods from overseas.

Aquaculture: The cultivation of aquatic organisms under controlled conditions.

Aquatic toxicity (AQTX): Adverse effects on marine life that result from their being exposed to toxic substance.

Aqueous: Water-based solution or suspension. Frequently, a gaseous compound dissolved in water.

Aqueous solubility: The maximum concentration of a chemical that will dissolve in pure water at a reference temperature.

Aquifer: A porous layer of underground rock that holds water. A bed or layer yielding water for wells and springs etc.; an underground geological formation capable of receiving, storing and transmitting large quantities of water. An aquifer is not an underground lake, but it very much resembles, a soaked sponge. Aquifer types include confined (sealed and possibly containing "fossil" water); unconfined (capable of receiving inflow); and Artesian (an aquifer in which the hydraulic pressure will cause the water to rise above the upper confining layer).

Aquifer depletion: Depletion of water of an aquifer resulting from withdrawal that is greater than natural or artificial recharge.

Aquifer test: A test to determine hydraulic properties of an aquifer.

Aquitard: Geological formation that may contain groundwater but is not capable of transmitting significant quantities of it under normal hydraulic gradients. May function as confining bed.

Arable land: Land that can be used for growing crops or for cultivation of crops.

Architectural coatings: Coverings such as paint and roof tar that are used on exteriors of buildings.

Area source: Any source of air pollution that is released over a relatively small area but which cannot be classified as a point source. Such sources may include vehicles and other small engines, small businesses and household activities, or biogenic sources such as a forest that releases hydrocarbons.

Argyria: Local or generalized gray/blue – colored impregnation of the body tissue with silver.

Arid: A condition in which less than 10 inches of rain falls each year and the level of evaporation is greater than the level of precipitation.

Arithmetic growth: An increase in some phenomenon at a constant rate over a specified time period.

Aromatics: A type of hydrocarbon, such as benzene or toluene, with a specific type of ring structure. Aromatics are sometimes added to gasoline in order to increase octane. Some aromatics are toxic.

Arsenicals: Pesticides containing arsenic.

Artesian (Aquifer or Well): Water held under pressure in porous rock or soil confined by impermeable geological formations.

Artificial fertilizer: A chemical added to soil to enhance crop production.

Artificial recharge: Artificially adding water to an aquifer.

Asbestos abatement: Procedures to control fiber release from asbestos – containing materials in a building or to remove them entirely, including removal, encapsulation, repair, enclosure, encasement, and operations and maintenance programs.

Asbestosis: A chronic lung disease caused by inhalation of asbestos fibers. The disease makes breathing progressively more difficult and can be fatal.

A-scale sound level: A measurement of sound approximating the sensitivity of the human ear, used to note the intensity or annoyance level of sounds.

Ash: The noncombustible solid residue of incineration; the mineral content of a product remaining after complete combustion.

Ash monofill cell: A specially constructed lined landfill used only for ash/residue from incinerated garbage, from the waste-to-energy facilities/plants.

Asphyxia: is a condition of severely deficient supply of oxygen to the body that arises from being unable to breathe normally. An example of asphyxia is choking.

Asphyxian: Vapor or gas which causes unconsciousness or death by suffocation. Most simple asphyxiants are harmful to the body only when they become so concentrated that they reduce oxygen in air (normally 21%) to dangerous levels (16% or lower). Asphyxiation is a potential hazard of working in confined spaces. Some examples of asphyxiants are Carbon Dioxide, Carbon Monoxide, Argon, etc. They function as asphyxiants by reducing the blood's ability to carry oxygen.

Assay: A test for a specific chemical, microbe, or effect.

Assessment endpoint: Environmental characteristics, which, if they were found to be significantly affected, would indicate a need for remediation. In ecological risk assessment, an explicit expression of the environmental value to be protected; includes both an ecological entity and specific attributed thereof. Entity (*e.g.*, salmon are a valued ecological entity; reproduction and population maintenance – the attribute – form an assessment endpoint.)

Assimilation cycle: A natural process by which a water body uses microbes, which convert nonliving substances into protoplasm or cells, to purify itself from pollutants.

Assimilative capacity: The capacity of a natural body of water to receive wastewaters or toxic materials without deleterious effects and without damage to aquatic life or humans who consume the water.

Association of boards of certification: An international organization representing boards which certify the operators of waterworks and wastewater facilities.

ASTM (American Society for Testing and Materials): The ASTM sets standards for many services, including methods of sampling and testing of hazardous waste and media contaminated with hazardous waste.

Asymptomatic: Neither causing nor exhibiting symptoms.

Atmosphere: The air that surrounds the earth and is bound by the earth's gravitational attraction. The gas layer surrounding a planet. Excluding the water vapor, the earth's atmosphere consists (by volume) of 78 per cent nitrogen, 21 per cent oxygen,

0.9 per cent argon, 0.035 per cent carbon dioxide (this level is rising), and minute quantities of other gases (including neon, krypton, helium, and air pollutants). Water vapor may constitute up to 3 per cent, depending on the relative humidity and temperature of the atmosphere. The atmosphere plays an important role in the protection of life on Earth; it absorbs ultraviolet solar radiation and reduces temperature extremes between day and night. It has four segments: tropo, strato, meso and thermospheres.

Atmosphere (atm.): Pressure measurement. One atmosphere (atm)=14.7 lbs/sq in.

Atmosphere, tertiary: Polluted atmosphere of today containing air pollutants besides the oxidizing gases.

Atmospheric inversion: A situation in which a layer of warm air traps pollutants. Basin cities such as Denver and Los Angeles experience this problem.

Atmospheric, oxidizing: Present atmosphere with oxidizing atoms dominating in the atmospheric gases (CO_2, H_2O, C_2 and N_2).

Atmospheric, reducing: Early atmosphere with H atoms dominating in the atmospheric gases (H, H_2O, CH_4, NH_3 and HCN).

Attainment area: An area considered to have air quality as good as or better than the national ambient air quality standards as defined in the Clear Air Act. An area may be an attainment area for one pollutant and a not-attainment area for others.

Attenuation: The process by which a compound is reduced in concentration over time, through absorption, adsorption, degradation, dilution, and/or transformation. And also be decreased with distance of sight caused by attenuation of light by particulate pollution.

Attractant: A chemical or agent that lures insects or other pests by stimulating their sense of smell.

Attrition: Wearing or grinding down of a substance by friction. Dust from such processes contributes to air pollution. The action of the particle rubbing against the other in a filter media or ion exchange bed that can in time cause breakdown of the particles.

Autoignition: Temperature: Minimum temperature which a substance must be heated without application of flame or spark to cause substance to ignite. Materials should not be greater than 80% of this tem perature.

Autotroph: An organism that produces complex organic compounds from simple inorganic molecules using energy from light or inorganic chemical reactions.

B

Back pressure: A pressure that can cause water to back flow into the water supply when a user's waste water system is at a higher pressure than the public system.

Backflow/back siphonage: A reverse flow condition created by a difference in water pressures that causes water to flow back into the distribution pipes of a drinking water supply from any source other than the intended one.

Background air pollution monitoring network (BAPMON): Over 200 off-monitoring sites located in remote areas around the world, established by the world meteorological organization in 1968 to record background concentration of particulate matter, precipitation, acidity and green such as CO_2, CFC and methane.

Background level: The amount or level of any constituent that occurs naturally or is not the result of human activities in soil, water, or air.

Backwashing: Reversing the flow of water back through the filter media to remove entrapped solids.

Backyard composting: Diversion of organic food waste and yard trimmings from the municipal waste stream by composting them in one's yard through controlled decomposition of organic matter by bacteria and fungi into a humus like product. It is considered as source reduction, not recycling, because the composted materials never enter the municipal waste stream.

BACT- Best available control technology: An emission limitation based on the maximum degree of emission reduction (considering energy, environmental and economic impacts) achievable through application of production processes and available methods, systems, and techniques. BACT does not permit emissions in excess of those allowed under any applicable Clean Air Act provisions. Use of the BACT concept is allowable on a case by case basis for major new or modified emissions sources in attainment areas and applies to each regulated pollutant.

Bacteria: (Singular: bacterium) Microscopic living organisms that can aid in pollution control by metabolizing organic matter in sewage, oil spills or other pollutants. However, bacteria in soil, water or air can also cause plant, animal and human health problems.

Bactericide: A pesticide used to control or destroy bacteria, typically in the home, schools, or hospitals.

Baffle: Broadly refers to an obstruction to trap debris in drainage water. A flat board or plate, deflector, or similar device constructed or placed in flowing water or slurry systems to cause more uniform flow velocities to absorb energy and to divert, guide, or agitate liquids.

Baffle chamber: In incinerator design, a chamber designed to promote the settling of fly ash and coarse particulate matter by changing the direction and/or reducing the velocity of the gases produced by the refuse or sludge.

Bag house: An air pollution abatement device used to trap particulate by filtering gas streams through large fabric bags usually made of glass fibers. The device is similar to a vacuum cleaner bag. As the exhaust air passes through the fabric filter, particles are separated from the air stream by sticking either to the filter fabric or to the cake of particles previously collected on the inside of the filter. Bag houses vary according to filter material and the method for dislodging the collected dust. It is effective in the elimination of intermediate and large (greater than 20 microns in diameter) particles a bag house system is usually much cheaper to install and operate than a system using electrostatic precipitation to remove particulates.

Bagasse: The fibrous residue of sugarcane milling used as a fuel to produce steam in sugar mills.

Bagassosin: A lung disease linked to the inhalation of dust from sugarcane residues.

Bailer: A pipe with a valve at the lower end, used to remove slurry from the bottom or side of side of a well as it is being drilled, or to collect groundwater samples from wells or open bore holes.

Balance of nature: An idea popularized by George Perkins Marshall that all life is interrelated and is in balance. This idea has been rejected by modern biological ecology but embraced by proponents of deep ecology.

Baler: A machine that compacts wastes materials, usually into rectangular bales. Balers often are used on newspaper, plastics and corrugated cardboard.

Baling: Compacting solid waste into blocks to reduce volume and simplify handling solid waste baling system provides an environmentally friend and economically viable solution to the challenges of transporting and storing municipal solid waste. The system produces large cylindrical bales, which are completely sealed in plastic film. This provides containment for shipment and prevents aerobic and anaerobic decomposition. The bales can easily be stored onsite at waste transfer facilities prior to shipment via flatbed trailer to landfill or incineration plants.

Ballistic separator: A device that takes out noncompostable material like stones, glass, metal, and rubber, from solid waste by passing the waste over a rotor that has impellers to fling the material in the air; the lighter organic (compostable) material travels a shorter distance than the heavier (noncompostable) material. By throwing the material up and forward by a paddle, Ballistic Separator sorts the materials due to their ballistic properties. A 90% precise material separation into heavy, screening and light fraction is achieved. In addition, the material is cleaned from dirt and dust.

Band application: The application of fertilizers over, or next to, each row of plants in a field.

Banking: A system for recording qualified air emission reductions for later use in bubble, offset, or netting transactions.

Bar screen: In wastewater treatment, a device used to remove large solids.

Barrel sampler: A drive – type soil sampler with a open end or split barrel.

BASE: Substances that (usually) liberate OH anions when dissolved in water. Base react with acids to form salts and water. Base have a pH greater than 7, turn litmus paper blue. A strong base is called alkaline or caustic.

Baseline risk assessment: A baseline risk assessment is an assessment conducted before cleanup activities begin at a site to identify and evaluate the threat to human health and the environment. After remediation has been completed, the information obtained during a baseline risk assessment can be used to determine whether the cleanup levels were reached.

Baseload: The steady and reliable supply of energy through the grid. This is punctuated by bursts of higher demand known as "peak-load". Supply companies must be able to respond instantly to extreme variation in demand and supply, especially during extreme conditions. Gas generators can react quickly while coal is slow but provides the steady "baseload". Renewable energies are generally not available on demand in this way.

Basic: Alkaline; demonstrating the qualities of a base.

Basic needs: The basic items and services needed by an individual to ensure a reasonable standard of living.

BAUME: Arbitrary scale of specific gravities; used to determine specific gravities and in graduation of hydrometers.

Bean sheet: Common term for a pesticide data package record.

Bed load: Sediment particles resting on or near the channel bottom that are pushed or rolled along by the flow of water.

Bedrock: Bedrock is the rock that underlies the soil; it can be permeable or non-permeable.

BEN: EPA's computer model for analyzing a violator's economic gain from not complying with the law.

Bench mark: Concentrations of chemical pollutants that are used in calculating a hazard quotient, a measure of the likelihood that a chemical will cause an adverse effect. The selected value can be observed – adverse effect level, water quality criteria, sediment quality criteria (or) a similar standard.

Benchmark station: The location within a contaminated environment that will serve as the site from which samples will be collected to determine the likelihood that damage to individuals will occur, if they contact the contaminants.

Bench-scale tests: Laboratory testing of potential cleanup technologies.

Benefit-Cost analysis: An economic method for assessing the benefits and costs of achieving alternative health – based standards at given levels of health protection.

Benign: Not recurrent or not tending to progress. Not cancerous.

Benthic/Benthos: An organism that feeds on the sediment at the bottom of a water body such as an ocean, lake, or river.

Bentonite: A colloidal clay, expansible when moist, commonly used to provide a tight seal around a well casing.

Beryllium: A metal hazardous to human health when inhaled as an airborne pollutant. It is discharged by machine shops, ceramic and propellant plants, and foundries.

Best Available Control Measures (BACM): A term used to refer to the most effective measures (according to EPA guidance) for controlling small or dispersed particulates and other emissions from sources such as roadway dust, soot and ash from woodstoves and open burning of rush, timber, grasslands, or trash.

Best Available Control Technology (BACT): Best available control technology is a standard used in air pollution control in the prevention of significant deterioration (PSD) of air quantity in the United States. Under the Clean Air Act, a major stationary new source of air pollution, such as an industrial plant, is remained to have a permit that sets emission limitations for the quantity. As part of the permit, limitations are based on lends achievable by the use of BACT for each pollutant.

Best available retrofit technology: The level of air pollution control technology required on certain major existing sources (*e.g.*, electric power plants) that may cause (or) contribute to visibility reduction in areas of the US designated as class I under the prevention of significant deterioration provisions of the Clean Air Act. Class I areas include major national parks and wildness areas, mostly in the American West.

Best available technology economically achievable (BAT): In water pollution control, the technological method required for all existing sources of toxic (or) non-conventional pollutants, BAT is set by industrial category (or) sub-category.

Best conventional control technology: The level of water pollution control technology required of existing discharger for the treatment of conventional pollutants by the 1977 Clean Water Act.

Best Demonstrated Achievable Technology (BDAT): A BDAT is a technology that has demonstrated the ability to reduce a particular contaminant to a lower concentration than other currently available technologies. BDATs can change with time as technologies evolve.

Best Management Practice (BMP): Methods that have been determined to be the most effective, practical means of preventing or reducing pollution from non-point sources.

Best practice: A process, technique, or innovative use of technology, equipment or resources or other measurable factors that have a proven record of success.

Bhopal gas tragedy: Refers to an industrial accident at a Union Carbide factory in Bhopal, India in 1984, resulting in the deaths of close to 4,000 humans and thousands of animals and the injury of 300,000 persons.

Bimetal: Beverage containers with steel bodies and aluminum tops; handled differently from pure aluminum in recycling.

Bioaccumulants: Substances that increase in concentration in living organisms as they take in contaminated air, water, or food because the substances are very slowly metabolized or excreted.

Bioaccumulation: The accumulation of a substance, such as a toxic chemical, in the tissues of a living organism. Bioaccumulation occurs in the case of substances excreted or broken down. Sometimes referred to as bioconcentration.

Bioassay: A test to determine the relative strength of a substance by comparing its effect on a test organism with that of a standard preparation.

Bioavailability: Degree of ability to be absorbed and ready to interact in organism metabolism or in other words. The proportion of a contaminant present in the environment in a form that can be potentially assimilated by an organism.

Biocapacity: A measure of the biological productivity of an area. This may depend on natural conditions or human inputs like farming and forestry practices; the area needed to support the consumption of a defined population.

Biocentric ethic: The idea that nature, not humankind, is the measure of all things.

Biochemical Oxygen Demand (BOD): A measure of the amount of oxygen consumed in the biological processes that break down organic matter in water. An indication of the extent to which water is polluted by sewage or other organic waste. It is a measure of the dissolved oxygen consumed by microorganisms as they break down the organic matter, the greater the oxygen required for its decomposition. It is expressed in parts per million of dissolved oxygen consumed. The greater the BOD, the greater the degree of pollution. Also called Biological Oxygen Demand.

Biocide: A biocide is a chemical substance capable of killing living organisms, usually in a selective way. A biocide can be pesticide or germicide or antibiotics or antibacterials or antivirals, etc. Biocides are commonly used in medicine, agriculture, forestry, and in industry where they prevent the fouling of water and oil pipelines. Some substances used as biocides are also employed as anti-fouling agents or disinfectants under other circumstances; chlorine, for example, is used as short-life biocide in industrial water treatment but as a disinfectant in swimming pools. Many biocides are synthetic, put a class of natural biocides, derived from *e.g.*, bacteria and plants, includes Brassica oleracea, and clostridium botulinum bacteria.

Bioconcentration: The accumulation of a chemical in tissues of a fish or other organism to levels greater than in the surrounding medium.

Biodegradable: Material that can be broken down into basic elements or natural substances (such as carbon dioxide and water) by biological processes through microorganisms under natural conditions such as bacteria and fungi; includes most organic waste.

Biodiversity: Refers to the degree of species richness and natural genetic variation. Diversity can be defined as the variety of life in all its forms, levels and combinations; includes ecosystem diversity, species diversity, and genetic diversity. Foı biological diversity, these items are organized at many levels, ranging from complete ecosystems to the biochemical structures that are the molecular basis of heredity. Thus, the term encompasses different ecosystems, species, and genes.

Bioelement: An element required by a living organism.

Bioenergy: Used in different senses; in its most narrow sense it is a synonym for biofuel, fuel derived from biological sources. In its broader sense it encompasses also biomass, the biological material used as a biofuel, as well as the social, economic, scientific and technical fields associated with using biological sources for energy.

Biofuel: The fuel produced by the chemical and/or biological processing of biomass. Biofuel will either be a solid (*e.g.*, charcoal), liquid (*e.g.*, ethanol) or gas (*e.g.*, methane).

Biogas: Methane gas produced by animal and human dung, crop residues, and other organic matter; can be used as a fuel or fertilizer. It is also called biomass gas.

Biogeochemical cycle: A circuit or pathway by which a chemical element or molecule moves through both biotic ("bio - ") and abiotic ("geo – ") parts of an ecosystem. Elements involved in biogeochemical cycles include carbon, nitrogen, sulfur, and phosphorus.

Biogeographical cycle: The chemical interactions among the atmosphere, biosphere, hydrosphere and lithosphere.

Biological amplification: The accumulation of higher levels of pollutants in organisms higher up in the food chain.

Biological cost or metabolic price: In biology, it is defined as a measure of the increased energy metabolism that is required to achieve a function. Drug resistance in microbiology, for instance, has a very high metabolic price, especially for antibiotic resistance.

Biological contaminants: Living organisms or derivates (*e.g.*, viruses, bacteria, fungi, and mammal and bird antigens) that can cause harmful health effects when inhaled, swallowed, or otherwise taken into the body.

Biological control: The use of natural enemies or diseases to control pests. In pest control, the use of animals and organisms that eat or otherwise kill or out-compete pests.

Biological evolution: Changes in the gene pool of a species over time.

Biological exposure indexes (BEI): Numerical values based on procedures to determine the amount of a material absorbed into the human body by measuring it or its metabolic products in tissue, fluid or exhaled air.

Biological Integrity: The ability to support and maintain balanced, integrated, functionality in the natural habitat of a given region. Concept is applied primarily in drinking water management.

Biological magnification (Biomagnification): The phenomenon whereby contaminant concentrations increase in organisms at successively higher trophic levels along a food chain. Refers to the process whereby certain substances such as pesticides or heavy metals move up the food chain, work their way into rivers or lakes, and are eaten by aquatic organisms such as fish, which in turn are eaten by large birds, animals or humans. The substances become concentrated in tissue or internal organs as they move up the chain.

Biological measurement: A measurement taken in a biological medium. For exposure assessment, it is related to the measurement is taken to related it to the established internal dose of a compound.

Biological medium: One of the major component of an organism; *e.g.*, blood, fatty, tissue or lymph nodes, in which chemicals can be stored or transformed.

Biological monitoring: Periodic examination of body substances, such as blood or urine, to determine the extent of hazardous material absorption as opposed to mere exposure.

Biological oxidation: Decomposition of complex organic materials by microorganisms. Occurs in self-purification of water bodies and in activated sludge wastewater treatment.

Biological pest control: A method of controlling pests (including insects, mites, weeds and plant diseases) that relies on predation, parasitism, herbivory, or other natural mechanisms.

Biological pesticides: Certain microorganism, including bacteria, fungi, viruses, and protozoa that are effective in controlling pests. These agents usually do not have toxic effects on animals and people and do not leave toxic or persistent chemical residues in the environment.

Biological productivity (bioproductivity): is the capacity of a given area to produce biomass; different ecosystems (*i.e.*, pasture, forest, etc.) will have different levels of bioproductivity. Biological productivity is determined by dividing the total biological production (how much is grown and living) by the total area available.

Biological stressors: Organisms accidentally or intentionally dropped into habitats in which they do not evolve naturally; *e.g.* gypsy moths, Dutch elm disease, certain types of algae, and bacteria.

Biological treatment: A treatment technology that uses bacteria to consume organic waste.

Biologically effective dose: The amount of a deposited or absorbed compound reaching the cells or target sites where adverse effect occur, or where the chemical interacts with a membrane.

Biologically productive land: is land that is fertile enough to support forests, agriculture and/or animal life. All of the biologically productive land of a country comprises its biological capacity. Arable land is typically the most productive area.

Biologicals: Vaccines, cultures and other preparations made from living organisms and their products, intended for use in diagnosing, immunizing, or treating humans or animals, or in related research.

Biomass: The materials derived from photosynthesis (fossilized materials may or may not be included) such as forest, agricultural crops, wood and wood wastes, animal wastes, livestock operation residues, aquatic plants, and municipal and industrial wastes; the quantity of organic material present in unit area at a particular time mostly expressed as tons of dry matter per unit area; organic matter that can be used as fuel.

Biomass energy: Energy derived from plant matter.

Biome: A climatic and geographically defined area of ecologically similar communities of plants, animals, and soil organisms, often referred to as ecosystems.

Biophysical: The living and non-living components and processes of the ecosphere. Biophysical measurements of nature quantify the ecosphere in physical units such as cubic metres, kilograms or joules.

Biopsy: Removal and examination of tissue from the living body.

Bioregion: (ecoregion) an area comprising a natural ecological community and bounded by natural borders, *e.g.*, river watershed.

Bioremediation: The use of for the decontamination of soil or groundwater.

Bioremediation refers to treatment processes that use living organisms (usually naturally occurring) such as plants, bacteria, yeast, or fungi to break down hazardous substances into less toxic or nontoxic substances. Bioremediation can be used to clean up contaminated soil and water. *In situ* bioremediation treats the contaminated soil or groundwater in the location in which it is found. For *ex situ* bioremediation processes, contaminated soil must be excavated or groundwater pumped before they can be treated.

Biosensor: Analytical device comprising a biological recognition element (*e.g.*, enzyme, receptor, DNA, antibody, or microorganism) in intimate contact with an electrochemical, optical, or acoustic signal transducer that together permit analyses of chemical properties or quantities. Shows potential development in some areas, including environmental monitoring.

Biosolids: Nutrient- rich organic materials derived from wastewater solids (sewage sludge) that have been stabilized through processing.

Biosphere: The natural system of the earth and the atmosphere that supports life. The part of the Earth, including air, land, surface rocks, and water, within which life occurs, and which biotic processes in turn alter or transform. Is the interface between the hydrosphere, geosphere and atmosphere.

Biostabilizer: A machine that converts solid waste into compost by grinding and aeration.

Biota: The animal and plant life a given region (or) the plant and animal life of an ecosystem; often referred to as flora and fanna.

Biotechnology: Biological manipulation of living organisms to produce foods, drugs, and other products for humans. Techniques that use living organisms or parts of organisms to produce a variety of products (from medicines to industrial enzymes) to improve plants or animals or to develop microorganisms to remove toxics from bodies of water, or act as pesticides.

Biotic: Relating to, produced by, or caused by living organisms. (*See* also abiotic).

Biotic community: A naturally occurring assemblage of plants and animals that live in the same environment and are mutually sustaining and interdependent. (*See* biome.)

Biotic potential: The maximum reproductive capacity of a population under optimum environmental conditions.

Biotransformation: Conversion of a substance into other compounds by organisms; includes biodegradation.

Bioventing: Bioventing is an *in situ* remediation tehnology that combines soil vapor extraction methods with bioremediation. It uses vapor extraction wells that induce air flow in the subsurface through the use of a vacuum. Bioventing can be effective in remediating releases of petroleum products, such as gasoline, jet fuels, kerosene, and diesel fuel. See also Bioremediation and soil Vapor Extraction.

Birth chart: A group of people born during a specific time period.

Birth rate: Number of people born as a percentage of the total population in any given period of time ; number of live births in a given year divided by the mid year population or per 1000 people.

Bison: A large grazing animal that once reigned over the grasslands of North America but was slaughtered in great numbers during the 1870s

Black death: The name for plague that swept Europe and Asia during the Middle Ages

Black water: Household wastewater to water that contains, human, or food waste (solid waste) *i.e.*, toilet discharge.

Blocked development: Economic development in less developed countries that is impeded by developed countries.

Blood products: Any product derived from human blood, including but not limited to blood plasma, platelets, red or white corpuscles, and derived licensed products such as interferon.

Bloom: A proliferation of algae and/or higher aquatic plants in a body of water; often related to pollution, especially when pollutants accelerate growth.

Blue water: The water from rivers and aquifers we use for irrigation is termed as blue water. Is the collectable water from rainfall; the water that falls on roofs and hard surfaces usually flowing into rivers and the sea and recharging the ground water. In nature the global average proportion of total rainfall that is blue water is about 40% blue water productivity in the garden can be increased by improving irrigation techniques, soil water storage, moderating the climate, using garden design and water–conserving plantings; also safe use of grey water. Blue water can also be befined as groundwater and stream flow, supporting aquatic ecosystems and that can be tapped for use elsewhere: for domestic aquatic and livestock, irrigation, industrial and urban use.

BOD5: The amount of dissolved oxygen consumed in five days by biological processes in breaking down organic matter.

Body burden: The amount of a chemical stored in the body at a given time, especially a potential toxin in the body as the result of exposure.

Bog: A type of wetland that accumulates appreciable peat deposits, Bogs depend primarily on precipitation for their water source, and are usually acidic and rich in plant residue with a conspicuous mat of living green moss.

Boiler: A vessel designed to transfer heat produced by combustion or electric resistance to water. Boilers may provide hot water or steam.

Boiling liquid expanding vapor explosion (BLEVE): Condition in which liquids are excessively heated, which may result in the violent rupture of a container, and the rapid vaporization of the material . The possibility of a BLEVE increases with the volatility of the material .

Boiling point (BP): Temperature at which a liquid changes to a vapor state at a given pressure. Flammable materials with low boiling points generally present special fire hazards.

Bonding: Safety practice where two objects are interconnected with clamps and bare wire. This equalizes electrical potential between the objects and helps prevent static sparks that could ignite flammable materials.

Boom: 1. A floating device used to contain oil on a body of water. 2. A piece of equipment used to apply pesticides from a tractor of truck.

Borehole: A borehole is a hole cut in to the ground by means of a drilling rig.

Borehole geophysics: Borehole geophysics are nuclear or electric technologies used to identify the physical characteristics of geologic formations that are intersected by a borehole.

Botanical pesticide: A pesticide whose active ingredient is a plant-produced chemical such as nicotine or strychnine. Also called a plant –derived pesticide.

Bottle bill: Proposed or enacted legislation which requires a returnable deposit on beer or soda containers and provides for retail store or other redemption. such legislation is designed to discourage use of throw- away containers.

Bottom ash: The non-airborne combustion residue from burning pulverized coil in a boiler; the material which falls to the bottom of the boiler and is removed mechanically; a concentration of non-combustible materials, which may include toxics, unburnable ash residue remaining in an incinerator or energy recovery facility after the burnables have been incinerated.

Bounding estimate: An estimate of exposure, dose, or risk that is higher than that incurred by the person in the population with the currently highest exposure, dose, or risk. Bounding estimates are useful in developing statements that exposures, doses, or risks are not greater than an estimated value.

Brackish water: Mixed fresh and salt water.

Bradycardia: A slow heartbeat with pulse rate below 60/minute.

Brandt commission: The Independent Commission on International Development held in 1980, which documented the link between environment and development. Members coined the terms North (developed countries) and South (less developed countries) and called for the amortization of old debts among Southern countries.

Breakpoint chlorination: Addition of chlorine to water until the chlorine demand has been satisfied.

Breathing zone: Area of air in which an organism inhales.

Breeder nuclear fission: A nuclear fission process in which new radioactive fuel is produced. It is considered dangerous because the chain reaction is difficult to control.

Brine mud: Waste material, often associated with well drilling or mining, composed of mineral salts or other inorganic compounds.

British thermal unit: Unit of heat energy equal to the amount of heat required to raise the temperature of one pound of water by one degree Fahrenheit at sea level.

Broadcast application: The spreading of pesticides over an entire area.

Brown goods: Items that are typically used for audio, video, telecommunications, or computing purposes, including any item that contains a picture tube or cathode ray tube, or electronic circuitry and circuit boards.

Brownfields: Abandoned, idled, or under used industrial and commercial facilities/sites where expansion or redevelopment is complicated by real or perceived environmental contamination. They can be in urban, suburban, or rural areas.

Brundtland commission: The World Commission on Environment and Development chaired by Norwegian Prime Minister Gro Harlem Brundtland. The Commission's report, Our Common Future (1987), popularized the notion of sustainable development.

Brundtland commission report: A UN report, Our Common Future, published in 1987 and dealing with sustainable development and the policies required to achieve it. The definition of the term in the report is "sustainable development is development that meets the needs of the present without compromising the ability of future generations to meet their own needs."

Bubble: A system under which existing emissions sources can propose alternate means to comply with a set of emissions limitations; under the bubble concept, sources can control more than required at one emission point where control costs are relatively low in return for a comparable relaxation of controls at a second emission point where costs are higher.

Bubble policy: A policy that allows polluters to discharge more pollutants at one source, if an equivalent reduction occurs at other sources.

Buffer: A substance whose chemical makeup is such that it minimizes changes in pH when acids or bases are added to it.

Buffer strips: Strips of grass or other erosion-resisting vegetation between or below cultivated strips or fields.

Building cooling load: The hourly amount of heat that must be removed from a building to maintain indoor comfort (measured in British thermal units).

Building envelop: The exterior surface of a building's construction – the walls, windows, floors, roof, and floor. Also called building shell.

Building related illness: Diagnosable illness whose cause and symptoms can be directly attributed to a specific pollutant source within a building (*e.g.* Legionnaire's disease, hypersensitivity, pneumonitis).

Bulk density: The mass (weight) per unit volume (including voids) of a solid material. Usually expressed as lb/ft^3 or g/cm^3.

Bulk sample: A small portion (usually thumbnail size) of a suspect asbestos-containing building material collected by an asbestos inspector for laboratory analysis to determine asbestos content.

Bulky waste: Large items of waste materials, such as appliances, furniture, large auto parts, tree, stumps.

Buna: Trademark for synthetic rubber and rubber like materials such as Buna – N (Nitrile) or Buna – S (Styrene).

Burial ground (Graveyard): A disposal site for radioactive waste materials that uses earth or water as a shield.

Buy – Back center: Facility where individuals or groups bring recyclables in return for payment.

Buying in bulk: The purchasing practice of buying large amounts of products at one time to avoid excess packaging and transportation costs.

By-product: Material, other than the principal product, generated as a consequence of an industrial process or as a breakdown product in a living system.

C

C3 and C4 plants: C4 plants comprise about 5% of all plants, are most abundant in hot and arid conditions, and include crops like sugarcane and soyabeans. During photosynthesis they form molecules with 4 – carbon atoms and saturate at the given level of CO_2. C3 plants, the other 95%, photosynthesise to form 3 carbon molecules and increase photosynthesis with as CO_2 levels increase.

Cadmium (Cd): is a heavy metal. A silvery white metallic chemical element resembling Zinc.

California condor: A bird once found in the mountains of California but now virtually extinct.

Calorie: A basic measure of energy or heat that has been replaced by the SI unit the joule. A calorie is the amount of heat or energy needed to increase the temperature of 1 gram of water by 1°C which is about 4.184 joules. The Calories in food ratings (spelled with a capital C) and nutrition are 'big C' Calories or kcal.

Calorific value: The energy content of a fuel measured as the heat released on complete combustion.

Cancer: The breakdown of the normal process of cell growth in which cancerous cells invade and destroy other cells and tissues. A large proportion of cancers is thought to be linked to environmental factors, including diet, chemicals, and other substances.

Cap: A layer of clay, or other impermeable material installed over the top of a closed landfill to prevent entry of rainwater and minimize leachate. It is designed to protect the landfill from the infiltration of rain water that can cause landfill contaminants to leach out into the groundwater. A landfill cap also prevents direct contact with the waste.

Capacity assurance plan: A statewide plan which supports a state's ability to manage the hazardous waste generated within its boundaries over a twenty year period.

Capillary action: Movement of water through very small spaces due to molecular forces called capillary forces.

Capillary fringe: The porous material just above the water table which may hold water by capillarity (a property of surface tension that draws water upwards) in the smaller void spaces. Also refers to the zone above the water table within which the porous medium is saturated by water under less than atmospheric pressure.

Capital goods: Accumulated items used to produce other goods and services.

Capitalism: An economy based on private enterprise and the use of markets for allocating economic resources.

Capitalist economy: An economic system in which the means of production are owned privately.

Capping: (in Waste Control context) is the process of adding a 1 to 3 ft thick clay layer or synthetic material over a contaminated area or landfill area as aprotective cover *i.e.*, to prevent aeration of toxic chemicals found in the contaminated region, leaking landfills and disposal sites.

Capture efficiency: The fraction of organic vapors generated by a process that are directed to an abatement or recovery device.

Car emissions: Chemicals produced by the internal combustion engine and considered hazardous. These include carbon monoxide, hydrocarbons, and nitrogen oxides.

Car pooling: Giving people lifts to help reduce emissions and traffic.

Carbon absorber: A pollution control device that uses activated carbon to absorb organic pollutants from a gas stream (or) from a waste water stream. The volatile organic compounds can be recovered later from the carbon.

Carbon adsorption: Carbon adsorption is a treatment system that removes contaminants from groundwater o surface water by forcing it through tanks containing activated carbon.

Carbon budget: A measure of carbon inputs and outputs for a particular activity.

Carbon credit: A market-driven way of reducing the impact of greenhouse gas emissions; it allows an agent to benefit financially from an emission reduction. There are two forms of carbon credit, those that are part of national and international trade and those that are purchased by individuals. Internationally, to achieve Kyoto Protocol objectives, 'caps' (limits) on participating country's emissions are established. To meet these limits countries, in turn, set 'caps' (allowances or credits: 1 convertible and transferable credit =1 metric tonne of CO_2 e-emissions) for operators. Operators that meet the agreed 'caps' can then sell unused credits to operators who exceed 'caps'. Operators can then choose the most cost-effective way of reducing emissions. Individual carbon credits would operate in a similar way cf. carbon offset.

Carbon cycle: The biogeochemical cycle by which carbon is exchanged between the biosphere, geosphere, hydrosphere, and atmosphere of the Earth. The cyclic pathway through which carbon circulates throughout nature.

Carbon dioxide: (CO_2) A heavy, colorless gas produced by combustion of fossil fuels and decomposition of organic substances and as by-product of chemical processes. Plants absorb CO_2 during photosynthesis, and plants and animals produce it as an end product of respiration. It plays an important role in controlling the earth's surgace temperature. Is useful as fire-extinguishing agent to block oxygen and smother fire.

Carbon dioxide equivalent (CO_{2e}): The unit used to measure the impacts of releasing (or avoiding the release of) the seven different greenhouse gases; it is obtained by multiplying the mass of the greenhouse gas by its global warming potential. For example, this would be 21 for methane and 310 for nitrous oxide.

Carbon equivalent: (C - e) Obtained by multiplying the CO_2–e by the factor 12/44.

Carbon footprint: A measure of the carbon emissions that are emitted over the full life cycle of a product or service and usually expressed as grams of CO_2 – e.

Carbon labeling: Use of product labels that display greenhouse emissions associated with goods.

Carbon Monoxide (CO): A colorless, odorless, flammable, and very toxic or poisonous gas produced by the incomplete combustion of carbon compounds (incomplete fossil fuel combustion) and as a by-product of many chemical processes.

Carbon neutral: Generally refers to activities where net carbon inputs and outputs are the same. For example, assuming a constant of vegetation on the planet, in the short term burning wood will add carbon to the atmosphere but this carbon will cycle back into new plant growth.

Carbon pool: A storage reservoir of carbon.

Carbon sink: Any carbon storage system that causes a net removal of greenhouse gases from the atmosphere *e.g.*, oceans and rain forests.

Carbon source: Opposite of carbon sink ; a net source of carbon for the atmosphere.

Carbon stocks: The quantity of carbon held within a carbon pool at a specified time.

Carbon tax: A tax or a surcharge imposed on fossil fuels according to the amount of carbon contained in them or that aims to reduce carbon dioxide emissions.

Carbon Tetrachloride (CCl_4): Compound consisting of one carbon atom and four chlorine atoms, once widely used as a industrial raw material, as a solvent, and in the production of CFCs. Use as a solvent ended when it was discovered to be carcinogenic.

Carboxyhemoglobin: Hemoglobin in which the iron is bound to carbon monoxide instead of oxygen.

Carcinogen: Any substance or agent that can cause or aggravate cancer.

Carcinoma: Malignant tumor or cancer; a new growth made up of epithelial cells tending to grow rapidly, infiltrate other cells, and give rise to metastasis (spreading).

Carnivore: An animal or plant that feeds on and digests animals.

Carrier: 1. The inert liquid or solid material in pesticide product that serves as a delivery vehicle for the active ingredient. Carriers do not have toxic properties of their own. 2. Any material or system that can facilitate the movement of a pollutant into the body or cells.

Carrying capacity: 1. The total population that a particular area can support at a subsistence level. 2. In wildlife management, the maximum number of animals an area can support during a given period. 3. In recreation management, the amount of use a recreation area can sustain without loss of quality.

CAS: Stands for Chemical Abstracts Service, an organization that indexes information published in Chemical Abstracts by the American Chemical Society and that provides index guides by which information about particular substances may be located in the abstracts.

CAS registration number: A number assigned by the Chemical Abstract Service to identify specific chemicals, except when followed by an asterisk (*) which signifies a compound (often naturally occurring) of variable composition. The numbers have no chemical significance. The CAS number is a concise, unique means of material identification.

Case study: A brief fact sheet providing risk, cost, and performance information on alternative methods and other pollution prevention ideas, compliance initiatives, voluntary efforts, etc.

Cash crop: Agricultural produce marketed for cash rather than retained for household use.

Cask: A thick-walled container (usually lead) used to transport radioactive material. Also called a coffin.

Catalyst: Any substance that causes or speeds up a chemical reaction without itself undergoing a permanent chemical change.

Enzymes are proteins produced by the body that act as catalysts in many physiological processes, such as digestion.

Catalytic converter: An air pollution control device that removes pollutants from motor vehicle exhaust, either by oxidation (oxidation of hydrocarbons to CO_2 and water vapour and the oxidation of CO to CO_2) or reduction (nitric oxide and nitrogen dioxide to nitrogen gas).

Catalytic incinerator: An air pollution control device that oxidizes volatile organic compounds (VOCs) by using a catalyst to promote the combustion process. Catalytic incinerators require lower temperatures than conventional thermal incinerators, thus saving fuel and other costs.

Cataract: Loss of transparency of the crystalline lens of the eye or its capsule.

Catchment area: The area that is the source of water for a water supply whether a dam or rain water tank.

Categorical exclusion: A class of actions which either individually or cumulatively would not have a significant effect on the human environment and therefore would not require preparation of an environmental assessment or environmental impact statement under the National Environmental Policy Act (NEPA).

Categorical pretreatment standard: A technology – based effluent limitation for an industrial facility discharging into a municipal sewer system. Analogous in stringency to Best Availability Technology (BAT) for direct dischargers.

Cathodic protection: A technique to prevent corrosion of a metal surface by making it the cathode of an electrochemical cell.

Cavitation: The formation and collapse of gas pockets or bubbles on the blade of an impeller or the gate of a valve; collapse of these pockets or bubbles drives water with such force that it can cause pitting of the gate or valve surface.

Ceiling: Maximum allowable human exposure limit for airborne substances; not to be exceeded even momentarily.

Cells: 1. In solid waste disposal, holes where waste is dumped, compacted, and covered with layers of dirt on a daily basis. 2. The smallest structural part of living matter capable of functioning as an independent unit.

Cementitious: Densely packed and nonfibrous friable material.

Centigrade: A unit of temperature.

Centimeter, (CM) = 1/100 meter. A cm = approximately 0.4 inches.

Centipoise: CGS unit of the measure of viscosity equal to 1/100 poise. Viscosity of water at 20^{0} c is approximately 1 centipoise.

Central collection point: Location where the wastes are gathered together for treatment on – site or for transportation elsewhere for treatment and/or disposal. This term could also apply to community hazardous waste collections, industrial and other waste management systems.

Centrally planned economy: An economy whose investment and production are coordinated by a central government body.

Centrifugal collector: A mechanical system using centrifugal force to remove aerosols from a gas stream or to remove water from sludge.

Centrifugation: A process that uses a centrifuge to separate hazardous constituents.

CERLCA (Comprehensive Environmental Response, Liability and Compensation Act): Passed by Congress in 1980 and usually referred to as superfund, a fund to help pay for the management and cleanup of hazardous waste sites.

CERCLIS: The federal Comprehensive Environmental Response, Compensation, and Liability Information System is a database that include all sites which have been nominated for investigation by the Superfund program.

CFC chloroflorocarbon: CFCs are potent greenhouse gases which are not regulated by the Kyoto Protocol since they are covered by the Montreal Protocol. Associated with damage to the Earth's ozone layer.

Check – Valve tubing pump: Water sampling tool, also referred to as a Water Pump.

Chelating agent: Chemical compound capable of forming multiple chemical bonds to a metal ion. Used to treat metal poisoning.

Chemical cartridge respirator: Respirator using various chemical substances to purify inhaled air of certain contaminative gases or vapors. Typically effective for concentrations no more than

10 times the TLV of the contaminant if it has warning properties (odor or irritation) below the TLV.

Chemical case: For purposes of review and regulation, the grouping of chemically similar pesticide active ingredients (*e.g.* salts and esters of the same chemical) into chemical cases.

Chemical compound: A distinct and pure substance formed by the union of two or more elements in definite proportion by weight.

Chemical dehalogenation: Chemical dehalogenation is a chemical process that removes halogens (usually chlorine) from a chemical contaminant, rendering the contaminant less hazardous. It can be effective in removing halogens from hazardous organic compounds, such as dioxins, PCBs, and certain chlorinated pesticides. The treatment time is short, energy requirements are moderate, and operation and maintenance costs are relatively low. This technology can be brought to the site, eliminating the need to transport hazardous wastes.

Chemical element: A fundamental substance comprising one kind of atom; the simplest form of matter.

Chemical family: Group of single elements or compounds with a common general name.

Chemical formula: Gives the number and kinds of atoms that comprise a molecule of a material.

Chemical hygiene plan (CHP): Per 29 CFR 1910.1450, OSHA standard; "Occupational Exposures to Hazardous Chemicals in Laboratories." Effective 5/1/90. A written plan that includes specific work practices, standard operating procedures, equipment, engineering controls, and policies to ensure that employees are protected from hazardous exposure levels to all potentially hazardous chemicals in use in their work areas. The OSHA standard provides for training, employee access to information, medical consultations, examinations, hazard identification procedures, respirator use, and record keeping practices.

Chemical oxygen demand (COD): A measure of the oxygen required to oxidize all compounds, both organic and inorganic, in water.

Chemical pneumonitis: Inflammation of the lunge caused by accumulation of fluids due to chemical irritation.

Chemical reactivity: Ability of a material to chemically change. Undesirable and dangerous effects such as heat, explosions, or the production of noxious substances can result.

Chemical stressors: Chemicals released to the environment through industrial waste, auto emissions, pesticides, and other human activity that can cause illnesses and even death in plants and animals.

Chemical treatment: Any one of a variety of technologies that use chemicals or a variety of chemical processes to treat waste.

Chemical: Any element, chemical compound, or mixture of elements and/or compounds naturally occurring or created by humans.

Chemiluminescence: Emission of light during a non-combustible chemical reaction.

Chemnet: Mutual aid network of chemical shippers and contractors that assigns a contracted emergency response company to provide technical support if a representative of the firm whose chemicals are involved in an incident is not readily available.

Chemosterilant: A chemical that controls pests by preventing reproduction.

Chernobyl: A nuclear power plant in the former Soviet Union that suffered a serious accident in 1986. Estimates are that 5,000 to 150,000 people who lived in the area will die prematurely.

Chiller: A device that generates a cold liquid that is circulated through an air handling unit's cooling coil to cool the air supplied to the building.

Chipko movement: A local movement that began in India in early 1980s and is opposed to governmental and other deforestation programs.

Chisel Plowing: Preparing croplands by using a special implement that avoids inversion of the soil. Chisel plowing can leave a protective cover or crops residues on the soil surface to help prevent erosion and improve infiltration.

Chloracne: Acne – like eruption caused by excessive contact with certain compounds.

Chlorinated hydrocarbons: Chemicals containing only chlorine, carbon and hydrogen. Synthetic organic molecules in which one or more hydrogen atoms have been replaced by chlorine atoms. These include a class of persistent, broad-spectrum insecticides that linger in the environment and accumulate in the food chin. Among them are DDT, aldrin, dieldrin, heptachlor, chlordane, lindane, endrin, Mirex, hexachloride, and toxaphene. Other examples include TCE, used as an industrial solvent. Any chlorinated organic compounds including chlorinated solvents such as dichloromethane, trichloroethylene, chloroform. Chlorinated hydrocarbons are hazardous compounds because they tend to be non-biodegradable and therefore to bioaccumulate. Many have been shown to be carcinogenic. Also called organochlorides.

Chlorinated solvent: An organic solvent containing chlorine atoms (*e.g.*, methylene chloride and 1,1,1-trichloromethane). Uses of chlorinated solvents are include aerosol spray containers, in highway paint, and dry cleaning fluids.

Chlorination: The application of chlorine to drinking water, sewage, or industrial waste to disinfect or to oxidize undesirable compounds.

Chlorinator: A chlorinator is a device that adds chlorine, in gas or liquid form, to water or sewage to kill bacteria.

Chlorine – contact chamber: That part of a water treatment plant where effluent is disinfected by chlorine.

Chlorofluclocarbons (CFCs): Synthetic organic molecules that contain one or more of both chlorine and fluorine atoms and that are known to cause O^3 destruction. One of the more widely known family of haloalkanes. A family of chemical compounds, some of which are volatile, nontoxic, and easily liquefied chemicals used in refrigeration, air conditioning, packaging, insulation or as solvent and aerosol propellants. CFCs are not destroyed in the lower atmosphere and rise readily into the upper layers, where their chlorine components destroy ozone.

Chlorosis: Discoloration of normally green plants parts caused by disease, lack of nutrients, or various air pollutants.

Cholinesterase: An enzyme found in animals that regulates nerve impulses by the inhibition of acetylcholine. Cholinesterase inhibition is associated with a variety of acute symptoms such as nausea, vomiting, blurred vision, stomach cramps, and rapid heard rate.

Chromium: is a heavy metal.

Chronic effect: An adverse effect on a human or animal in which symptoms recur frequently or develop slowly over a long period of time.

Chronic exposure: Long term contact with a substance. Multiple exposures occurring over an extended period of time or over a significant fraction of an animal's or human's lifetime (Usually seven years to a lifetime.)

Chronic toxicity: The capacity of a substance to cause long-term poisonous heath effect in humans, animals, fish, and other organisms. Adverse effects resulting from repeated doses of or exposures to a material over a relatively prolonged period of time.

Circle of influence: The circular outer edge of a depression produced in the water table by the pumping of water from a well.

Circle of poison: The use of pesticides banned in the developed countries on crops that are produced in less developed countries for export to developed countries.

Circular metabolism: A system in which wastes, especially water and materials, are reused and recycled.

Cistern: Small tank or storage facility used to store water for a home or farm; often used to store rain water.

Civil suit: A lawsuit centered on an individual seeking damages for injury or loss.

Clarification: The process of removing particulate matter from waste water. Normally, the water is allowed to stand, thereby facilitating the settling of the particles. This is often aided by centrifugal action and chemically induced coagulation in wastewater.

Clarifier: In a drinking water facility (0r) waste water treatment plant, the unit that removes suspended particular matter. A tank in which solids settle to the bottom and are subsequently removed as sludge.

Class action suit: A law suit centered on a group seeking damages for injury or loss.

Class I area: Under the clean Air Act, a class I area is one in which visibility is protected more stringently than under the national ambient air quality standards; includes national parks, wilderness areas, monuments, and other areas of special national and cultural significance.

Class I substance: One of several groups of chemicals with an ozone depletion potential of 0.2 or higher, including CFCs, Halons, Carbon Terachloride, and methyl chloroform (listed in the clean Air Act), and HBFCs and Ethy Bromide (added by EPA regulations).

Class II substance: A substance with an ozone depletion potential of less than 0.2. All HCFCs are currently included in this classification.

Class system: A system of social inequality based on the unequal distribution of economic resources.

Class: A person' s ranking in a social hierarch; based on access to wealth and other scarce resources.

Clay soil: Soil material containing more than 40 per cent clay, less than 45 per cent sand, and less than 40 per cent silt.

Clean air act: Originally passed by the U.S. Congress in 1963 to assure air safe enough to protect the public health and amended in 1970 to give the Environmental Protection Agency (EPA) the responsibility of setting air quality standards for each pollutant.

Clean coal technology: Any technology not in widespread use prior to the Clean Air Act Amendments of 1990. This Act will achieve significant reductions in pollutants associated with the burning of coal.

Clean fuels: Blends or substitutes for gasoline fuels, including compressed natural gas, methanol, ethanol, and liquefied petroleum gas.

Clean water act: An act passed by the U.S. Congress in 1972 to protect the nation's water resources including regulation of pollution and sewage treatment.

Cleaner production: The continual effort to prevent pollution, reduce the use of energy, water and material resources and minimize waste – all without reducing production capacity.

Cleaner technologies substitutes assessment: A document that systematically evaluates the relative risk, performance, and cost trade – offs of technological alternatives; serves as a repository for all the technical data (including methodology and results) developed prevention or education project.

Cleanup: Cleanup is the term used for actions taken to deal with a release or threat of release of a hazardous substance that could affect humans and or the environment. The term sometimes is used interchangeably with the terms remedial action, removal action, response action, or corrective action.

Clear well: A reservoir for storing filtered water of sufficient quantity to prevent the need to vary the filtration rate with variations in demand. Also used to provide chlorine contact time for disinfection.

Clear-cutting (clear felling): A logging practice in which a majority of the trees in an area or in a forest sector are cut down.

Climate: The long – term average of weather conditions in an area (OR) The general variations of weather in a region over long periods of time.

Climate change (also referred to as 'global climate change'): A change in weather over time and/or region; usually relating to changes in temperature, wind patterns and rainfall; although may be natural or anthropogenic it generally refers to anthropogenic global warming. The term 'climate change' is sometimes used to refer to all forms of climatic inconsistency, but because the Earth's climate is never static, the term is more properly used to imply a significant change from one climatic condition to another.

Cloning: In biotechnology, obtaining a group of genetically identical cells from a single cell; making identical copies of a gene.

Closed cup: Identifies one of the methods used to measure flash points of flammable liquids.

Closed system: A system in which there is an exchange of energy with the environment, but no exchange of matter.

Closed-Loop recycling: Reclaiming or reusing wastewater for non-potable purposes in an enclosed process.

Closure: The procedure a landfill operator must follow when a landfill reaches its legal capacity for solid ceasing acceptance of solid waste and placing a cap on the landfill site.

Coagulation: In colloid chemistry, a process in which dispersed colloidal particles agglomerate. In waste water treatment it is clumping of particles to settle out impurities, often induced by chemicals such as lime, alum, and iron salts.

Coal cleaning technology: A precombustion process by which coal is physically or chemically treated to remove some of its sulfur so as to reduce sulfur dioxide emissions.

Coal gasification: Conversion of coal to a gaseous product by one of several available technologies.

Coastal erosion: The erosion of the land by the constant battering of waves. The force of the waves creates a hydraulic effect compressing air to form explosive pockets in the rocks and cliffs. Rocks and pebbles may be flung against the cliff face (the process of corrosion) and wear it away. Cliffs of chalk and limestone may be dissolved by solution.

Coastal Zone: Lands and waters adjacent to the coast that exert an influence on the uses of the sea and its ecology, or whose uses and ecology are affected by the sea.

Con-dominant: Two or more species that jointly are the most prevalent or significant species within a plant community.

Coefficient of haze: A measurement of visibility interference in the atmosphere.

Coefficient of water/oil distribution: Also called the partition coefficient, it is the ratio of the solubility of a chemical in water to its solubility in oil. Used to indicate how easily human or other organisms can absorb or store a material. Sometimes abbreviated Ko/w; may also be expressed as its logarithm, log Ko/w.

Co-fire: Burning of two fuels in the same combustion unit; *e.g.*, coal and natural gas, or oil and coal.

Cogeneration: The simultaneous production of electricity and useful heat from the combustion of the same fuel source.

Cognitive process: A mental process involved in human learning and reasoning.

Cokes: These are the solid carbonaceous material derived from destructive distillation low – ash, low-sulfur bituminous coal. Cokes from coal are grey, hard, and porous.

Coke Oven: An industrial process which converts coal into coke, one of the basic materials used in blast furnaces for the conversion of iron ore into iron. New generation non-recovery type coke oven is energy efficient, cost competitive and environment friendly.

Cold temperature CO: A standard for automobile emissions of carbon monoxide (CO) emissions to be met a low temperature (*i.e.* 20 degrees Fahrenheit). Conventional automobile catalytic converters are not efficient in cold weather unit they warm up.

Coliform Index: A rating of the purity of water based on a count of fecal bacteria.

Coliform Organism: Microorganisms found in the intestinal tract of humans and animals. Their presence in water indicates fecal pollution and potentially adverse contamination by pathogens.

Collection efficiency: An expression of the performance of air pollution control equipment (or) the percentage reduction in pollution control concentration entering the device compared with the pollutant concentration in the exhaust air. It is calculated by substracting the pollutant concentration in the exhaust from the pollutant concentration in the incoming air and dividing by the incoming air concentration. This result is multiplied by 100 for expression as percentage efficiency (by weight).

Collector sewers: Pipes used to collect and carry wastewater from individual sources to an interceptor sewer that will carry it to a treatment facility.

Collector: Public or private hauler that collects nonhazardous waste and recyclable materials from residential, commercial, institutional and industrial sources.

Colloids: Very small, finely divided solids (that do not dissolve) that remain dispersed in a liquid for a long time due to their small size ($10^{-9} - 10^{-6}$ m dia) and electrical charge.

Colonialism: The practice of economic and political domination of less developed countries by developed countries.

Colorimetric: Colorimetric refers to chemical reaction – based indicators that are used to produce compound reactions to individual compounds, or classes of compounds. Visible color changes are used to detect and quantify contaminants.

Combined sewer overflows: Discharge of a mixture of storm water and domestic waste when the flow capacity of a sewer system is exceeded during rainstorms.

Combined Sewers: A sewer system that carries both sewage and storm-water runoff. Normally, its entire flow goes to a waste treatment plant, but during a heavy storm, the volume of water may be so great as to cause overflows of untreated mixtures of storm water and sewage into receiving waters. Storm – water runoff may also carry toxic chemicals from industrial areas or streets into the sewer system.

Combined waste water: A mixture of sanitary or municipal waste water storm water created by rainfall.

Combustion chamber: The actual compartment where waste is burned in an incinerator.

Combustion: 1. Burning, or rapid oxidation, accompanied by release of energy in the form of heat and light. 2. Refers to controlled burning of waste, in 'which heat chemically alters organic compounds, converting into stable inorganics such as carbon dioxide and water.

Command post: Facility located at a safe distance upwind from an accident site, where the on-scene coordinator, responders, and technical representatives make response decisions, deploy manpower and equipment, maintain liaison with news media, and handle communications.

Command – and – Control regulations: Specific requirements prescribing how to comply with specific standards defining acceptable levels of pollution.

Commercial and industrial waste – (waste management): Solid waste generated by the business sector as well as that created by government and schools. Does not include that from the construction and demolition industry.

Commercial waste management facility: A treatment, storage, disposal, or transfer facility which accepts waste from a variety of sources, as compared to a private facility which normally manages a limited waste stream generated by its own operations.

Commercial waste: Waste material that originates in wholesale business establishments such as office buildings or stores.

Commingled materials: (waste management) Materials mixed together, such as plastic bottles, glass, and metal containers. Commingled recyclable materials require sorting after collection before they can be recycled.

Commingled recyclables: Mixed recyclables that are collected together.

Comminuter: A machine that shreds or pulverizes solids to make waste treatment easier.

Comminution: Mechanical shredding or pulverizing of waste. Used in both solid waste management and wastewater treatment.

Commodities: Goods and services that can be bought and sold.

Community water system: A public water system which serves at least 15 service connections used by year round residents or regularly serves at least 25 year – round residents.

Community: A group interacting at a specific time and place, as well as sharing a similar cultural background. All the groups of organisms living together in the same area, usually interacting or depending on each other for existence. In ecology, as assemblage of populations of different species within a specified location in space and time. Sometimes, a particular sub-grouping may be specified, such as the fish community in a lake or the soil arthropod community in a forest. Also called biological community.

Compact Fluorescent Lamp (CFL): Small fluorescent lamps used as more efficient alternatives to incandescent lighting. Also called PL, CFL, Twin – Tube, or BIAX lamps.

Compaction: Reduction of the bulk of solid waste by rolling and tamping.

Comparative risk assessment: Process that generally uses the judgment of experts to predict effects and set priorities among a wide range of environmental problems.

Compensation point: The compensation point is the amount of light intensity on the light curve where the rate of photosynthesis exactly matches the rate for respiration.

Compliance coal: Any coal that emits less than 1.2 pounds of sulfur dioxide per million Btu when burned. Also known as low sulfur coal.

Compliance coating: A coating whose volatile organic compound content does not exceed that allowed by regulation.

Compliance cycle: The 9-year calendar year cycle, beginning January 1, 1993, during which public water systems must monitor. Each cycle consists of three 3-year compliance periods.

Compliance monitoring: Collection and evaluation of data, including self-monitoring reports, and verification to show whether pollutant concentrations and loads contained in permitted discharges are in compliance with the limits and conditions specified in the permit.

Compliance order/action: An official communication from an administrative agency that requires a pollution source to confirm to environmental protection regulations by following a compliance schedule. Usually issued after a notice of violation.

Compliance Schedule: A negotiated agreement between a pollution source and a government agency that specifies dates and procedures by which a source will reduce emissions and, thereby, comply with a regulation.

Compost: Decayed organic and animal matter used as fertilizer.

Composting: The biological decomposition of organic materials in the presence of oxygen that yields carbon dioxide, heat, and stabilized organic residues that may be used as a soil additive.

Composting facility: 1. An offsite facility where the organic component of municipal solid waste is decomposed under controlled conditions; 2. An aerobic process in which organic materials are ground or shredded and then decomposed to humus in windrow piles or in mechanical digesters, drums, or similar enclosures.

Compressed gas: Material contained under pressure (dissolved gas, liquefied by compression or refrigeration).

Compressed Natural Gas (CNG): An alternative fuel for motor vehicles; considered one of the cleanest because of low hydrocarbon emissions and its vapors are relatively non-ozone producing. However, vehicles fueled with CNG do emit a significant quantity of nitrogen oxides.

Concentration: The relative amount of a substance mixed with another substance. An example is five ppm of carbon monoxide in air or 1 mg/l of iron in water.

Condensate return system: System that returns the heated water condensing within steam piping to the boiler and thus saves energy.

Condensate: 1. Liquid formed when warm landfill gas cools as it travels through a collection system. 2. Water created by cooling steam or water vapor.

Conditionally exempt generators (CE): Persons or enterprises which produce less than 220 pounds of hazardous waste per month. Exempt from most regulation, they are required merely to determine whether their waste is hazardous, notify appropriate state or local agencies, and ship it by an authorized transporter to a permitted facility for proper disposal.

Conditions to avoid: Conditions encountered during handling or storage that could cause a substance to become unstable.

Conductivity: A measure of the ability of a solution to carry an electrical current.

Cone of depression: A depression in the water table that develops around a pumped well.

Cone of influence: The depression, roughly conical in shape, produced in a water table by the pumping of water from a well.

Cone Penterometer Testing (CPT): A direct push system used to measure lithology based on soil penetration resistance. Sensors in the tip of the cone of the direct push rod measure tip resistance and side – wall friction, transmitting electrical signals to digital processing equipment on the ground surface.

Confidential statement of formula (CSf): A list of the ingredients in a new pesticide or chemical formulation . The list is submitted at the time for application for registration or change in formulation.

Confined aquifer: An aquifer in which ground water is confined under pressure which is significantly greater than atmospheric pressure.

Confined space: Any area that has limited openings for entry and exit that would make escape difficult in an emergency, has a lack of ventilation, contains known and potential hazards, and is not intended nor designated for continuous human occupancy.

Confining layer: A "confining layer" is a geological formation characterized by law permeability that inhibits the flow of water.

Confluent growth: A continuous bacterial growth covering all or part of the filtration area of a membrane filter in which the bacteria colonies are not discrete.

Conjunctivitis: Inflammation of conjunctiva, the delicate membrane that lines eyelid and covers the eyeball.

Consent decree: A legal document, approved by a judge, that formalizes an agreement reached between EPA and potentially responsible parties (PRPs) through which PRPs will conduct all or part of a cleanup action at a Superfund site; cease or correct actions or processes that are polluting the environment; or otherwise comply with EPA initiated regulatory enforcement actions to resolve the contamination at the Superfund site involved. The consent decree describes the actions PRPs will take and may be subject to a public comment period.

Conservation tillage: The practice of reducing or eliminating tillage operations and leaving crop residues on the soil to prevent erosion.

Conservation: Preserving and renewing, when possible, human and natural resources. The use, protection, and improvement of natural resources according to principles that will ensure their highest economic or social benefits.

Conservationalists: Those wanting to preserve natural resources.

Conspicuous consumption: A term used to describe the lavish spending on goods and services that are acquired mainly for the purpose of displaying income or wealth rather than to satisfy basic needs of the consumer.

Constituents of concern: Specific chemicals that are identified for evaluation in the assessment of the pollution status of an area (or) location.

Construction and demolition waste: Waste building materials, dredging materials, tree stumps, and rubble resulting from construction, remodeling, repair, and demolition of homes, commercial buildings and other structure and pavements. May contain lead, asbestos, or other hazardous substances. It excludes construction waste which is included in the municipal waste stream and does not generally include waste from the commercial and industrial waste stream.

Consumer: Someone who uses or buys resources, natural or manufactured. Organism, human being, or industry that maintains itself by transforming a high-quality energy source into a lower one cf. Producer, primary production.

Consumer: An organism that ingests other organisms, either living or dead. All animals and most microorganisms are consumers (herbivores, carnivores, parasites, and detritivores). Fungi and a few plants are consumers, including saprophytes (Indian pipes) and carnivorous plants (pitcher plant, Venus flytrap). See producer, decomposer.

Consumer democracy: Using your economic capacity to promote your values.

Consumption (ecology): The use of resources by a living system, the inflow and degradation of energy that is used for system activity.

Consumptive water use: Water removed from available supplies without return to a water resources system, *e.g.*, water used in manufacturing, agriculture, and food preparation.

Contact pesticide: A chemical that kills pests when it touches them, instead of by ingestion, also, soil that contains the minute skeletons of certain algae that scratch and dehydrate waxy-coated insects.

Contagious diseases: Diseases that are transmitted by physical contact, *e.g.*, tuberculosis and measles. They are the leading causes of death in many less developed countries.

Contaminant: A contaminant is any physical, chemical, biological, or radiological substance or matter present in any media (air, water, or soil) at concentrations that may lower its quality.

Contamination (of recyclable materials): Inadequate sorting of recyclable materials so as to interfere with the clean processing of the recyclables. Contamination of office paper occurs, for example, when food, carbon paper, metal cans or anything that is not paper is mixed with it.

Contamination source inventory: An inventory of contaminant sources within delineated State Water-Protection Areas. Targets likely sources for further investigation.

Contamination: Introduction into water, air, and soil, of microorganisms, chemicals toxic substances, wastes, or wastewater in a concentration that makes the medium unfit for its next intended use. Also applies to surfaces of objects, buildings, and various household and agricultural use products.

Contingency plan: A document setting out an organized, planned, and coordinated course of action to be followed in case of a fire, explosion, or other accident that releases toxic chemicals, hazardous waste, or radioactive materials that threaten human health or the environment.

Contingent valuation: is a survey – based economic technique for the valuation of non-market resources, such as *e.g.*, clean air, life expectancy, wildlife, environmental preservation or the impact contamination.

Continuous discharge: A routine release of a pollutant into the water (or) air during normal operations at a facility. Such discharges are usually made under a permit authorizing the release.

Continuous emission monitoring: The continuous sampling and analysis of gases, particulate matter by monitors placed inside smoke stocks.

Contour Ploughing (contour farming): Soil tilling method that follows the shape of the land to discourage erosion. A soil conservation technique in which cultivation follows the contours of the land. The farming practice of plowing across a slope following its contours. The rows formed have the effect of slowing water run-off during rainstorms so that the soil is not washed away and allows the water to percolate into the soil.

Contour strip farming: A kind of contour farming in which row crops are planted in strips, between alternating strips of close-growing, erosion-resistant forage crops.

Controlled burning: A technique sometimes used in forest management, farming, prairie restoration or greenhouse gas abatement.

Controlled reaction: A chemical reaction under temperature and pressure conditions maintained within safe limits to produce a desired product or process.

Convention: A multilateral agreement between countries, usually a legal agreement on international environmental issues.

Conventional agriculture: Agricultural practices involving the use of chemical fertilizers, pesticides, and machinery.

Conventional pollutants: Statutorily listed pollutants understood well by scientists. These may be in the form of organic waste, sediment, acid, bacteria, viruses, nutrients, oil and grease, or heat.

Conventional systems: Systems that have been traditionally used to collect municipal wastewater in gravity sewers and convey it to a central primary or secondary treatment plant prior to discharge to surface waters.

Conventional tilling: Tillage operations considered standard for a specific location and crop and that tend to bury the crop residues; usually considered as a base for determining the cost effectiveness of control practices.

Conveyance loss: Water loss in pipes, channels, conduits, ditches by leakage or evaporation.

Cooling electricity use: Amount of electricity used to meet the building cooling load.

Cooling towers: Are heat removal devices used to transfer process waste heat to the atmosphere. Common applications include cooling the circulating water used in oil refineries, chemical plants, power plants and building cooling. The towers vary in size from small roof-top units to very large hyperboloid structures that can be up to 200 meters tall and 100 meters in diameter, or rectangular structures that can be over 40 meters tall and 80 meters long. Smaller towers are normally factory – built, while larger ones are constructed on site.

Core countries: Developed countries having the most technologically advanced, capital-intensive, and high-wage economies, *e.g.*, the United States, Japan, and the western European countries.

Core program cooperative agreement: An assistance agreement whereby EPA supports states or tribal governments with funds to help defray the cost of non-item-specific administrative and training activities.

Core: The uranium – containing heart of a nuclear reactor, where energy is released.

Cornucopian: A view that natural resources are unlimited.

Corporate social responsibility: Integration of social and environmental policies into day – to – day corporate business.

Corporation: A large private organization with multiple owners.

Corrosion: The dissolution and wearing away of metal caused by a chemical reaction such as between water and the pipes, chemicals touching a metal surface, or contact between two metals.

Corrosive: A chemical agent that reacts with the surface of a material causing it to deteriorate or wear away. Able to eat into, wear away or dissolve flesh or metal. One of the four hazardous waste characteristics.

Corrosivity: Corrosive wastes include those that are acidic and capable of corroding metal such as tanks, containers, drums, and barrels.

Cost recovery: A legal process by which potentially responsible parties who contributed to contamination at a Superfund site can be required to reimburse the Trust Fund for money spent during any cleanup actions by the federal government.

Cost/Benefit analysis: Evaluating projects, policies, and programs in terms of economic costs and benefits. A quantitative evaluation of the costs which would have incurred by implementing an environmental regulation versus the overall benefits to society of the proposed action.

Cover crop: Plants used to hold the soil during the fallow season. A crop that provides temporary protection for delicate seedlings and/or provides a cover canopy for seasonal soil protection and improvement between normal crop production periods.

Cover material: The soil used to cover garbage in a landfill.

Cowboy economy: An economy that behaves as if natural resources are infinite in supply and nature can absorb all waste.

Cradle –to – Grave system or manifest system: A procedure in which hazardous materials are identified and followed as they are produced, treated, transported and disposed of by a series of permanent, linkable, descriptive documents (*e.g.*, manifests.)

Criteria pollutant: Criteria pollutants are air pollutants which, at certain lands of exposure, do not threaten human health and which meet National Ambient Air Quality Standards. There are two types of standards for most pollutants. National Primary Ambient air anxiety standards are levels of air quality with a margin of safety adequate to protect public health. National Secondary Ambient Air Quality Standards are levels of air quality which are necessary to protect the public welfare from any known or anticipated adverse affects a pollutant.

Criteria pollutants: The six air pollutants constituting the national Ambient Air quality standards set by the USEPA using criteria documents. The criteria pollutants are carbon monoxide, nitrogen dioxide, particulate matter, lead, sulphurdioxide and ozon

Criteria pollutants: The 1970 amendments to the Clean Air Act required EPA to set National Ambient Air Quality Standards for certain pollutants known to be hazardous to human heath. EPA has identified and set standards to protect human health and welfare for six pollutants: ozone, carbon monoxide, total suspended particulates, sulfar dioxide, lead, and nitrogen oxide. The term, "criteria pollutants" derives from the requirement that EPA must describe the characteristics and potential health and welfare effects of these pollutants. It is on the basis of these criteria that standards are set or revised.

Criteria pollutants: Substances that result in the most air pollution: carbon monoxide, sulfur, hydrocarbons, nitrogen oxides, ozone, and lead.

Critical pressure/temperature: Temperature above which a gas connot be liquefied by pressure. The critical pressure is that pressure is that pressure required to liquefy a gas at its critical temperature.

Crop coefficient (KC) – (water management): A variable used to calculate the evapotranspiration of a plant crop based on that of a reference crop.

Crop consumptive use: The amount of water transpired during plant growth plus what evaporated from the soil surface and foliage in the crop area.

Crop Rotation (crop sequencing): The practice of growing a series of dissimilar types of crops in the same space in sequential seasons for various benefits such as to avoid nutrient depletion of the soil the build up of pathogens and pests that often occurs when one species is continuously cropped.

Cross- media pollution: The transfer of chemical contaminants from one environmental medium (air, water, soil) to another For example, water –borne wastes being deposited or adsorbed onto soils, causing contamination of underground area, or wastes in soils contaminating groundwater

Cross Contamination: Cross contamination is the physical movement or transfer of harmful bacteria from one person, object or place to another. 1. The movement of underground contaminants from one level or area to another due to invasive subsurface activities. 2. The movement of chemicals constituents from a sewer system, pipeline, surface drainage (or) other source into a distribution system used to transport drinking water to the customers of a contaminants from one aquifer (or) groundwater level to another as a result of the drilling of a well.

Cross-connection: Any actual or potential connection between a drinking water system and an unapproved water supply or other source of contamination

Crude oil: Naturally occurring mixture of hydrocarbons under normal temperature and pressure

Crumb rubber: Is a term usually applied to recycled rubber from automotive and truck scrap tyres. During the recycling process steel and fluff is removed leaving tyre rubber with a granular consistency. This refers to ground rubber fragments the size of sand or silt used in rubber or plastic products, or processed further into reclaimed rubber or asphalt products.

Cryogenic: Relating to extremely low temperature as for refrigerated gases

Cullet: The term used to describe crushed glass that is suitable for recycling by glass manufactures.

Cultural eutrophication: This is an accelerated process of nutrient and sediment concentration in a body of water caused by excessive human activity. Thus increases the rate at which water bodies "die" by pollution from human activities compared to natural eutrophication

Cultural services: The non-material benefits of ecosystems including refreshment, spiritual enrichment, knowledge, artistic satisfaction

Culture: A commonly held set of beliefs, attitudes, and rules for behavior in a society

Cumulative ecological risk assessment: Consideration of the total ecological risk from multiple stressors to a given eco-zone

Cumulative exposure: The sum of exposure of an organism to a pollutant over a period of time

Cumulative working level months: The sum of lifetime exposure to radon working levels expressed in total working level months

Curb stop: A water service shutoff valve located in a water service pipe near the curb and between the water main and the building

Curbside collection: Method of collecting recyclable materials at homes, community districts or businesses.

Curbside recycling program: A program that collects recyclable materials from our homes. The materials often are left in special containers on the curbside to be picked up by a recycling truck.

Cutie–Pie: An instrument used to measure radiation levels.

Cyanosis: Dark purplish coloration of skin and mucous membrane caused by deficient oxygenation of the blood.

Cyclone: Intense low pressure weather systems; mid – latitude cyclones are atmospheric circulations that rotate clockwise in the Southern Hemisphere and anti-clockwise in the Northern Hemisphere and are generally associated with stronger winds, unsettled conditions, cloudiness and rainfall. Tropical cyclones (which are called hurricanes in the Northern Hemisphere) cause storm surges in coastal areas.

Cyclone Collector: A device that uses centrifugal force to remove large particles from polluted air.

D

Day tank: Another name for deaerating tank. A tank used to store a chemical solution of known concentration for feed to a chemical feeder. Also called 'age tank'.

DDT: Dichlorodiphenyltrichloroethane (DDT) is a first chlorinated hydrocarbon contact insecticide that kills by acting as a nerve poison; is a persistent organic pollutant. The insecticide; is very fat-soluble and is therefore found in fatty foods such as meat and dairy products. DDT and its breakdown products have widespread persistence in the environment, and a high potential to bioaccumulate. It has a reported half – life in the environment of 2-15 years in most soils. Many governmental and inter-governmental organizations regard DDT as a major hazard to the environment.

De minimis: It is a Latin expression meaning 'about minimal things'. Quantities of pollutants (or) other chemical substances that are small enough to be exempted from environmental regulations. In risk assessment it refers to a level of risk that is too small to be concerned with. Some refer to this as a "virtually safe" level.

Dead end: The end of a water main which is not connected to other parts of the distribution system.

Deadmen: Anchors drilled or cemented into the ground to provide additional reactive mass for DP sampling rigs.

Death rate: The ratio of the number of deaths to the total number of individuals in a specific population during a particular time period, usually one year (or) the number of deaths in a given year divided by the midyear population. For humans, the rate is expressed as deaths per thousand persons per year.

Decant: To draw off the upper layer of liquid after the heaviest material (a solid or another liquid) has settled.

Decay products: Degraded radioactive materials, often referred to as "daughters" or "progeny"; radon decay products of most concern from a public health standpoint are polonium-218.

Decay: 1. Decomposition of dead organic matter, usually by bacterial or fungal activity. 2. Short form for radioactive decay.

Decompose: Break down into simpler components to change form. To break down, to rot or decay, especially through microbial action.

Decomposer: An organism that derives its nourishment from dead organic matter (animal and plant bodies), breaking down the complex molecules into simpler organic molecules. Decomposers include earthworms, mushrooms, and other fungi, and bacteria. Also called as microconsumer, detritivore, saprobe, or reducer.

Decomposers: Consumers, mostly microbial, that change dead organic matter into minerals and heat.

Decomposition: The breakdown of a substance or material or matter by bacteria and fungi, changing the chemical makeup and physical appearance of materials.

Decontamination: Removal of harmful substances such as noxious chemicals, harmful bacteria or other organisms, or radioactive material from exposed individuals, rooms and furnishings in buildings, or the exterior environment.

Deep-well injection: Deposition of raw or treated, filtered hazardous waste by pumping it into deep wells, where it is contained in the pores of permeable subsurface rock.

Deflocculating agent: A material added to a suspension to prevent settling.

Defluoridation: The removal of excess fluoride in drinking water.

Defoliant: An herbicide that removes leaves from trees and growing plants.

Deforestation: A forest losing 40 per cent or more of the trees (or) the conversion of forested areas to non-forest land for agriculture, urban use, development, or wasteland.

Degradable: Substance that can be broken down into smaller, less complex molecules.

Degree-Day: It is a measure of heating or cooling. A rough measure used to estimate the amount of heating required in a given area; is defined as the difference between the mean daily temperature and 65 degrees Fahrenheit. Totalized degree days from an appropriate starting date are used to plan the planting of crops and management of pests. Degree – days are also calculated to estimate cooling requirements.

De-inking: Removing most of the ink, filler and other materials from waste paper before using it to manufacture new paper. There are two distinct types of paper recovery systems: 1. recovery based upon de-inking and intended for printing grade or other white papers, accounting for about 5 to 6 percent of the total, and 2. recovery without de-inking, intended for boxboards and coarse papers, accounting for the remainder, can eventually form a solution.

Delist: Use of the petition process to have a facility's toxic designation rescinded.

Demand-side waste management: Prices whereby consumers use purchasing decisions to communicate to product manufacturers that they prefer environmentally sound products packaged with the least amount of waste, made from recycled or recyclable materials, and containing no hazardous substances.

Dematerialization: Decreasing the consumption of materials and resources while maintaining quality of life.

Demolition debris: Waste materials produced during construction or remodeling including items such as used timber, masonary, sheetrock, shingles, insulation, etc.

Denitrification: The biological reduction of nitrate to nitrogen gas by denitrifying bacteria in soil.

Dense Non-Aqueous Phase Liquid (DNAPL): Non-aqueous phase liquids such as chlorinated hydrocarbon solvents or petroleum fractions with a specific gravity graeter than 1.0 that sink through the water column until they reach a confining layer. Because they are at the bottom of aquifers instead of floating on the water table, typical monitoring wells do not indicate their presence.

Density: 1. The size of population of given species in a given area. 2. The mass per unit of volume of a substance, usually stated in grams per cubic centimeter (or pounds per cubic foot). Density gives an indication of how closely packed the molecules of a substance are; for example, liquid water has a density of 1 gram per cubic centimeter, while liquid mercury has a density of 13.59 grams per cubic centimeter.

Depletion curve: In hydraulics, a graphical representation of water depletion from storage-stream channels, surface soil, and groundwater. A depletion curve can be drawn for base flow, direct runoff, or total flow.

Depressurization: A condition that occurs when the air pressure inside a structure is lower that the air pressure outdoors. Depressurization can occur when household appliances such as fireplaces or furnaces, that consume or exhaust house air, are not supplied with enough makeup air. Radon may be drawn into a house more rapidly under depressurized conditions.

Dermal absorption/penetration: Process by which a chemical as an internal dose penetrates the skin and enters the body.

Dermal exposure: Contact between a chemical and the skin.

Desalination: Is removing salts from salty or brackish water to produce potable or recyclable water for human purposes. This is done by three methods; distillation / freezing; reverse osmosis using membranes and electrodialysis; ion exchange. At present, all these methods are energy intensive. Removal of salts from soil by artificial means, usually leaching.

Desiccant: A chemical agent that absorbs moisture; some desiccants are capable of drying out plants or insects, causing death.

Design capacity: The average daily flow that a treatment plant or other facility is designed to accommodate.

Designated pollutant: An air pollutant which is neither a criteria nor hazardous pollutant, as described in the Clean Air Act, but for which new source performance standards exist. The Clean Air Act does require states to control these pollutants, which include acid mist, total reduced sulfur and fluorides.

Designated uses: Those water uses identified in state water quality standards that must be achieved and maintained as required under the Clean Water Act. Uses can include cold water fisheries, public water supply, and irrigation.

Designer bugs: Popular term for microbes developed through biotechnology that can degrade specific toxic chemicals at their sources in toxic waste dumps or in ground water.

Desquamation: It is the detachment of cells from the surface of an epithelium.

Destination facility: The facility to which regulated medical waste is shipped for treatment and destruction, incineration, and/or disposal.

Destratification: Vertical mixing within a lake or reservoir to totally or partially eliminate separate layers of temperature, plant, or animal life.

Destroyed medical waste: Regulated medical waste that has been ruined, torn apart, or mutilated through thermal treatment, melting, shredding, grinding, tearing, or breaking, so that it is no longer generally recognized as medical waste, but has not yet been treated (excludes compacted regulated medical waste).

Destruction and Removal Efficiency (DRE): A percentage that represents the number or molecules of a compound removed or destroyed in an incinerator relative to the number of molecules entering the system (*e.g.*, a DRE of 99.99 percent means that 9.999 molecules are destroyed for every 10,000 that enter; 99.99 per cent is known as "four nines". For some pollutants, the RCRA removal requirement may be as stringent as "six nines").

Destruction facility: A facility that destroys regulated medical waste.

Detectable leak rate: The smallest leak (from a storage tank), expressed in terms of gallons-or liters-per-hour, that a test can reliably discern with a certain probability of detection or false alarm.

Detection criterion: A predetermined rule to ascertain whether a tank is leaking or not. Most volumetric tests use a threshold value as the detection criterion.

Detention time: The theoretical calculated time required for a small amount of water to pass through a tank at a given rate of flow. The actual time that a small amount of water is in a settling basin, flocculating basin, or rapid – mix chamber. In storage reservoirs, the length of time water will be held before being used.

Detergent: Synthetic washing agent that helps to remove dirt and oil. Some contain compounds which kill useful bacteria and encourage algae growth when they are in wastewater that reaches receiving waters.

De-tinning: Removing and recovering the thin tin coating from steel cans before the steel is recycled.

Detritus: Waste material. Non-living particulate organic material (as opposed to dissolved organic material). Detritus is usually organic material, such as dead or partially decayed plants and animals, or excrement. Detritus can also be small particles of minerals from weathered rock, such as sand and silt.

Dewater: Remove or separate a portion of the water in a sludge or slurry to dry the sludge so it can be handled and disposed of. Remove or drain the water from a tank or trench.

dfE: Design for the environment; dfE considers 'cradle to grave' costs and benefits associated with material acquisition, manufacture, use, and disposal.

dfM: Design for manufacturing; designing products in such a way that they are easy to manufacture.

dfS: Design for sustainability; an integrated design approach aiming to achieve both environmental quality and economic efficiency through the redesign of industrial systems.

DGSA: It is an acronym for Dangerous Goods Safety Advisor.

Diatomaceous earth (Diatomite): A chalk – like material (fossilized diatoms) used to filter out solid waste in wastewater treatment plants; also used an active ingredient in some powdered pesticides.

Diazinon: An insecticide. In 1986, EPA banned its use on open areas such as sod farms and golf courses because it posed a danger to migratory birds. The ban did not apply to agricultural, home lawn or commercial establishment uses.

Dieback: A sudden decline in the population of an area after the carrying capacity of environment is reached.

Diet transition: A change in diet that is associated with increased affluence. It typically moves from traditional grains (millet and sorghum) to rice and wheat and on to a combination of meats and grains

Diffusion: The movement of suspended or dissolved particles (or molecules) from a more concentrated to a less concentrated area. The process tends to distribute the particles or molecules more uniformly. Diffusion varies in response to changes in the concentration of the material.

Digester: In wastewater treatment, a closed tank; in solid-waste conversion, a unit in which bacterial action is induced and accelerated in order to break down organic matter and establish the proper carbon to nitrogen ratio.

Digestion: The process in which food is broken down into compounds that cells can absorb, store, or oxidize and use as nourishment. 2. The use of anaerobic bacteria to produce biogas (methane) from refuse such as manure. 3. The use of bacteria (usually aerobic) to break down raw sewage into simpler compounds, used in secondary wastewater treatment.

Diluent: Any liquid or solid material used to dilute or carry an active ingredient.

Dilution ratio: The relationship between the volume of water in a stream and the volume of incoming water. It affects the ability of the stream to assimilate waste.

Dimictic: Lakes and reservoirs that freeze over and normally go through two stratifications and two mixing cycles a year.

Dinocap: A fungicide used primarily by apple growers to control summer diseases. EPA proposed restrictions on its use in 1986 when laboratory tests found it caused birth defects in rabbits.

Dixin: Any of a family of compounds known chemically as dibenzo-p- dioxins.Concern about them arises from their potential toxicity as contaminants in commercial products. Tests on laboratory animals indicate that it is one of the more toxic anthropogenic (man –made) compounds.

Direct discharger: A municipal or industrial facility which introduces pollution through a defined conveyance or system such as outlet pipes; a point source.

Direct filtration: A method of treating water which consists of the addition of coagulant chemicals, flash mixing, coagulation, minimal flocculation, and filtration.

Direct push: Technology used for performing subsurface investigations by driving, pushing, and/or vibrating small – diameter hollow stell rods into the ground. Also known as direct drive, drive point, or push technology.

Direct runoff: Water that flows over the ground surface or through the ground directly into streams, rivers, and lakes.

Direct solar energy: Usable energy obtained directly from sunlight, as through solar cells, solar collectors, and passive solar structures . See indirect solar energy.

Discharge: Flow of surface water in a stream or canal or the outflow of ground water from a flowing artesian well, ditch, or spring. Can also apply to discharge of liquid effluent from a facility or to chemical emissions into the air through designated venting mechanisms.

Disinfectant: A chemical or physical process that kills pathogenic organisms in water, air, or on surfaces. Chlorine is often used to disinfect sewage treatment effluent, water supplies, wells, and swimming pools.

Disinfectant by-product: A compound formed by the reaction of a disinifectant such as chlorine with organic material in the water supply; a chemical by-product of the disinfection process.

Dispersant: A chemical agent used to break up concentrations of organic material such as spilled oil.

Displacement savings: Saving realized by displacing purchases of natural gas or electricity from a local utility by using landfill gas for power and heat.

Display Screen Equipment (DSE) Regulations 1992: In the U.K. specify the steps employers must take to help avoid harm to heavy users of computers. The regulations also cover other types of equipment such as microscopes. The aim of the regulations is to minimize the possibility of users developing back or neck injuries, or suffering from RSI- type disorders through incorrect posture, poor screen positioning, or other inadequacies in the working environment. Employers are under a legal obligation to meet the standards set out in the regulations.

Disposables: A material designed to be thrown away after use. Consumer products other items, and packaging used once or a few times and discarded.

Disposal facilities: Repositories for solid waste, including landfills and combustors intended for permanent or destruction of waste materials. Excludes transfer stations and composting facilities.

Dissolved oxygen (DO): Amount of oxygen gas contained in water or sewage usually given in parts per million at a specified temperature and atmospheric pressure. It is a measure of the ability of water to support aquatic organisms. Water with very low dissolved oxygen content (less than 5 ppm), which is usually caused by too much or improperly treated organic

Dissolved solids: Disintegrated organic and inorganic material in water. Excessive amounts make water unfit to drink or use in industrial processes.

Distillation: The act of purifying liquids through boiling, so that the steam or gaseous vapors condense to a pure liquid. Pollutants and contaminants may remain in a concentrated residue.

Disturbance: Any event or series of events that disrupt ecosystem, community, or population structure and alters the physical environment.

Diversion rate: The percentage of waste materials diverted from traditional disposal such as landfilling or incineration to be recycled, composted, or re-used.

Divertible resource: (water management) The proportion of water runoff and recharge that can be accessed for human use.

DNA hybridization: Use of a segment of DNA, called a DNA probe, to identify its complementary DNA; used to detect specific genes.

Dobson Unit (DU): Units of ozone level measurement. Measurement of ozone levels. If, for example, 100 DU of ozone were brought to the earth's surface they would form a layer one millimeter thick. Ozone levels vary geographically, even in the absence of ozone depletion.

Domestic application: Pesticide application in and around houses, office buildings, motels, and other living or working areas.

Domestic sanitary waste water: Waters used during the course of residential life or those discharged from restorms.

Dominance: The condition of a species contributing a major portion of the biomass or numbers within a community, and thereby influencing the abundance of other species in the community. The ability of one gene to prevail over another different gene in causing a particular trait (phenotype). Condition in which behavior establishes more important social positions for some individual within a population (creating a dominance hierarchy).

Dosage/Dose: The actual quantity of a chemical administered to an organism or to which it is exposed. The amount of a substance available for interaction with metabolic processes after crossing the outer boundary of an organism

Doss rate: In exposure assessment, dose per time unit (*e.g.* mg/day), sometimes also called dosage.

Dose response assessment: The determination of the relationship between dose or exposure and the intensity of the adverse effect. Estimating the potency of a chemical. In exposure assessment, the process of determining the relationship between the dose of a stressor and a specific biological response. Evaluating the quantitative relationship between the dose of a stressor and a specific biological response. Evaluating the quantitative relationship between dose and toxicological responses.

Dose-response: Shifts in toxicological responses of an individual (such as alterations in severity) or populations (such as alterations in incidence) that are related to changes in the dose of any given substance.

Dose-response curve: Graphical representation of the relationship between the dose of a stressor and the biological response thereto. This response is measured by the percentage of the exposed population that shows the defined effect. If that effect is death, such a curve may be used to estimate an LD%) value.

Dosimeter: An instrument to measure dosage; many so-called dosimeters actually measure exposure rather than dosage. Dosimetry is the process or technology of measuring and/or estimating dosage.

Downcycling: (waste management) Recycling in which the quality of an item is diminished with each recycling.

Downgradient: The direction that groundwater flows; similar to "downstream" for surface water.

Downstream: Those processes occurring after a particular activity *e.g.*, the transport of a manufactured product from a factory to the wholesale or retail outlet.

Downstream processors: Industries dependent on crop production (*e.g.*,canneries and food processors).

Drainage: Improving the productivity of agricultural land by removing excess water from the soil by such means as ditches or subsurface drainage tiles.

Drainage well: A well drilled to carry excess water off agricultural fields. Because they act as a funnel from the surface to the groundwater below. Drainage wells can contribute to groundwater pollution.

Drawdown: The drop in the water table or level of water in the ground when water is being pumped from a well: the amount of water used from a tank or reservoir, the drop in the water level of a tank or reservoir.

Dredging: (water management) the repositioning of soil from an aquatic environment, using specialized equipment, in order to initiate infrastructural and/or ecological improvements. Dredging of contaminated muds can expose biota to heavy metals and other toxics.

Drinking eater: (potable water) Water fit for human consumption in accordance with World Health Organisation guidelines.

Drinking water equivalent level: Protective level of exposure related to potentially non-carcinogenic effects of chemicals that are also known to cause cancer.

Drive casing: Heavy duty steel casing driven along with the sampling tool in cased DP systems. Keeps the hole open between sampling runs and is not removed until last sample has been collected.

Drive Point (DP) Profiler: An exposed groundwater DP system used to collect multiple depth- discrete groundwater sampling . Ports in the tip of the probe connect to an internal stainless steel or Teflon tube that surface. Samples are collected via suction or air lift methods. Deionized water is pumped down through the ports to prevent plugging while driving the tool to the next sampling depth.

Driver: (ecology) Any natural or human –induced factor that directly or indirectly causes a change in an ecosystem. A direct driver is one that unequivocally influences ecosystem processes and that can be measured.

Drop off box or center: (waste management) Facility to which individuals and businesses bring garbage or recyclable materials for collection.

Drop-off: Recyclable materials collection method in which individuals bring them to a designated collection site.

Dump: A site used to dispose of solid waste without environmental controls.

Dumpster: A large container to keep waste until it is collected by the trash hauler often used by stores, apartment buildings and restaurants.

Duplicate: A second aliquot or sample that is treated the same as the original sample in order to determine the precision of the analytical method.

Dust fall jar: An open container used to collect large particles from the air for measurement and analysis.

Dystrophic lakes: Acidic, shallow bodies of water that contain much humus and/or other organic matter; contain many plants but few fish.

E

Earth Day: An event established in 1970 by Gaylord Nelson and held every year on April 22 to raise/promote environmental awareness.

Earth Summit: The United Nations Conference on Environment and Development which took place at Rio de Janeiro, Brazil in June, 1992.

Eco: A prefix derived from the Greek word for house and now added to many words indicating a general consideration for the environment *e.g.* ecohousing, ecolabel, ecomaterial, ecotourism.

Eco-asset: A biological asset that provides financial value to priate land owners when they are maintained in or restored to their natural state.

Ecocide: can be used to refer to any large-scale destruction of the environment. An early reference in 1969 described it as "Ecocide–the murder of the environment is everybody's business."

Ecofeminism: Ecofeminism is the social movement that regards the oppression of women and nature as interconnected. It is one of the few movements and analyses that actually connects two movements. It is now better understood as a movement working against the interconnected oppressions of gender, race, class and nature.

Ecolabel: Seal or logo indicating a product has met a certain environmental or social standards. Ecolabel is a labeling system for consumer products (excluding medicines) that are made in a fashion that avoids detrimental effects on the environment. Usually both the precautionary principle and the substitution principle are used when defining the rules for what products can be ecolabelled.

Ecological debt: Since mid 1980s when human demand on biocapacity first exceeded the available biocapacity, mankind has been in condition of overshoot, with its Footprint each year exceeding the planet's biocapacity. By demanding more than the biocapacity can supply, humanity is accruing an ecological debt. This debt is the sum of all the annual deficits.

Ecological deficit/reserve: The difference between the biocapacity and Ecological Footprint of a region or country. An ecological deficit occurs when the Footprint of a population exceeds the biocapacity of the area available to that population. Conversely, an ecological reserve exists when the biocapacity of a region exceeds its population's Footprint.

Ecological economics: Nontradional economics that focus attention not only on allocation and distribution of resources but also on the larger ecosystem or environment. The economy is viewed as a subsystem of the larger and finite ecosystem. Its attention centers on the flow of matter and energy from the environment as raw materials and back to the environment as waste.

Ecological entity: In ecological risk assessment, a general term referring to a species, a group of species, an ecosystem function or characteristic, or specific habitat or biome.

Ecological footprint (Eco-footprint, Footprint): A measure of how much biologically productive land and water an individual, population or activity requires to produce all the resources it consumes and to absorb the waste it generates using prevailing technology and resource management practices. The Ecological Footprint is usually measured in global hectares.

Ecological impact: The effect that a man-caused or natural activity has on living organisms and their non-living (abiotic) environment.

Ecological indicator: A characteristic of an ecosystem that is related to, or derived from, a measure of biotic or abiotic variable, that can provide quantitative information on ecological structure and function. An indicator can contribute to a measure of integrity and sustainability.

Ecological integrity: A living system exhibits integrity if, when subjected to disturbance, it sustains and organizes self – correcting ability to recover toward a biomass end-state that is normal for that system. End-states other than the pristine or naturally whole may be accepted as normal and good.

Ecological limit: The carrying capacity of a given area.

Ecological niche: The habitat of a species or population within its ecosystem. The role of a species in an ecosystem.

Ecological risk assessment: The application of a formal framework, analytical process, or model to estimate the effects of human actions on a natural resource and to interpret the significance of those effects in light of the uncertainties identified in each component of the assessment process. Such analysis includes initial hazard identification, exposure and dose – response assessments, and risk characterization.

Ecological succession: is a fundamental concept in ecology, refers to more – or – less predictable and orderly changes in the composition or structure of an ecological community. Succession may be initiated either by formation of new, unoccupied habitat (*e.g.*, a lava flow or a severe landslide) or by some form of disturbance (*e.g.*, fire, severe windthrow, logging) of an existing community. Succession that begins in areas where no soil is initially present is called primary succession, whereas succession that begins in areas where soil is already present is called secondary succession.

Ecological sustainability: The capacity of ecosystems to maintain their essential processes and function and to retain their biological diversity without impoverishment.

Ecologically sustainable development: Using, conserving and enhancing the human community's resources so that ecological processes, on which all life depends, can be maintained and enriched into the future.

Ecology: The branch of biology that studies the relationships among living organisms and between organisms and their environments. The scientific study of living organisms and their relationships to one another and their environment; the scientific study of the processes regulating the distribution and abundance of organisms; the study of the design of ecosystem structure and function. It is derived from the Greek OIKOS, meaning house, therefore it is literally the study of the house.

Economic depletion: The use of 80 per cent of a nonrenewable resource.

Economic efficiency: Economic efficiency is a term typically used in microeconomic when discussing product. Production of a unit of good is considered to be economically efficient when that unit of good is produced at the lowest possible cost.

Economic externalities: Costs or benefits that are not borne by the producer or supplier of a good or service. In many environmental situations environmental deterioration may be caused by a few while the cost is borne by the community; examples would include over fishing, pollution (*e.g.*, production of greenhouse emissions that are not compensated for in any way by taxes etc.), the environmental cost of land-clearing etc.

Economic growth: It is the increase in the amount of the goods and services produced by an economy over time. It is conventionally measured as the percent rate of increase in real gross domestic product.

Economic indicator: A statistic used for judging the health of an economy, such as GDP per head, the rate of unemployment or the rate of inflation.

Economic poisons: Chemicals used to control pests and to defoliate cash crops such as cotton.

Economic sanctions: A way of punishing errant countries, which is currently more acceptable than bombing or invading them. One or more restrictions are imposed on international trade with the targeted country in order to persuade the target's government to change a policy. Possible sanctions include limiting export or import trade with the target; constraining investment in the government of the target. Sanction can be multilateral, with many countries acting together, perhaps under the auspices of the United Nations, or unilateral, when one country takes action on its own.

Economic system: Organized way in which a state or nation allocates its resources and apportions goods and services in the national community.

Economics: Theories, principles, and models that deal with how the market process (price mechanism) works. It attempts to explain how wealth is created and distributed in communities, how people allocate resources that are scarce and have many alternative uses, and other such matters that arise in dealing with human wants and their satisfaction. However, in the words of the German-born UK economist E. F. Schumaker (1911-77) "..it is inherent in the methodology of economics to ignore man's dependence on the natural world".

Economy: A human system by which resources are produced, distributed, and consumed. Refers to entire network of producers, distributors, and consumers of goods and services in a local, regional, or national community.

Ecoregion: (bioregion) The next smallest ecologically and geographically defined area beneath "realm" or "ecozone". Ecoregion can be defined as a "large unit of land or water containing a geographically distinct assemblage of species, natural communities, and environmental conditions". The biodiversity of flora, fauna and ecosystems that characterize an ecoregion tends to be distinct from that of other ecoregions.

Ecosphere: The "bio-bubble" that contains life on earth, in surface waters, and in the air. In ecology this term refers to the global ecosystem of the planet (essentially the same as biosphere).

Ecosystem: It is a natural dynamic complex of plant, animal and microorganism communities (biotic factors) and their non-living (abiotic) environment all interacting as a functional unit.

Ecosystem altering: Any activity that modifies the biota including species composition, distribution, and abundance and any activity that modifies the physical environment, including soils and the water cycle.

Ecosystem boundary: The spatial delimitation of an ecosystem, typically based on discontinuities in the distribution of organisms, the biophysical environment (soil types, drainage basins, depth in a water body) and spatial interactions (home ranges, migration patterns, fluxes of matter).

Ecosystem dynamics: Those intrinsic ecological functions through which an ecosystem becomes self-regulating, self-sustaining, and capable of recovery from external forces (for example, damaging storm events). These intrinsic processes may cause continual change in biotic composition and structure at specific localities. Collectively, these changes represent internal flux, rather than substantive and permanent alteration of the ecosystem regionally.

Ecosystem function: The collective interspecific interaction of the biota, such as primary and secondary production and production and mutualistic relationships. The interactions between organisms and the physical environment, such as nutrient cycling, soil development, water budgeting, and flammability.

Ecosystem services: These are the processes by which the environment produces resources that we often take for granted such as clean water, timber, and habitat for fisheries, and pollination of native and agricultural plants. Refers to the role played by organisms, without charge, in creating a healthy environment for human beings from production of oxygen to soil formation, maintenance of water quality and much more. These services are now generally divided into four groups, supporting provisioning, regulating and cultural.

Ecosystem structure: The physical and spatial aspects of an ecosystem that are contributed by the biotic composition, examples include species population density, species richness or evenness, and standing crop biomass.

Ecotage: It is portmanteau of the "eco-" prefix and "sabotage". It is used to describe (usually) illegal acts of vandalism and violence, committed in the name of environmental protection. Ecotage is also referred to as ecodefense or monkey wrenching. As a term it goes back to 1972 and predates the more recent neologism eco-terrorism.

Ecoterrorism or green terrorism: It is terrorism/aggressive acts committed in support of ecological, environmental, or animal rights causes. Attacks of these groups on automobile dealerships, housing developments, forestry companies, corporate and university-based medical research laboratories, restaurants, fur farms and other industries are on record.

Ecotone: A habitat created by the juxtaposition of distinctly different habitats; an edge habitat; or an ecological zone or transition area between two adjacent ecological communities.

e-cycling: Recycling electronic waste.

Eddy diffusion: The mixing of clean air and contaminated air by percepts (eddies) at air moving in a random, irregular manner. The eddies that cause this dilution are produced by Mechanical Turbulence (or) Thermal Turbulence. If the eddies are small relative to a plume of contaminated air, they will only dilute the outside edges of the plume, if they are much larger than the plume, the entire plume will be moved without significant dilution.

Edema: Abnormal accumulation of clear, watery fluid in body tissue.

Effective concentration (EC50): Concentration of a material in water, a single dose which is expected to cause a biological effect on 50% of a group of test animals.

Effective rainfall: The volume of rainfall passing into the soil; that part of rainfall available for plant use after runoff, leaching, evaporation and foliage interception.

Efficiency: The amount of product produced per input unit of energy, labor, or material.

Effluent: The etymological meaning of this term "flowing forth out of". A discharge of waste or other noxious material into the environment. Effluent is most commonly used by environmentalists in reference to the Clean Water Act of 1977. It refers to wastewater – treated or untreated – that flows out of a treatment plant, sewer, or industries. Generally refers to wastes discharged into surface waters.

Effluent fee: A fee paid by a polluter to discharge noxious emissions into the air or water.

Effluent standards: The maximal amounts of specific pollutants allowable in waste water discharged by an industrial facility (or) waste water treatment plant. The standards are set for individual pollutants and apply across industrial categories. The term can be contrasted with effluent limitations, which are set for individual pollutants by Industrial Category.

Ejector: A device used to disperse a chemical solution into water being treated.

El Nino – (Spanish: the child) A warm water current which periodically flows southwards along the coast of Ecuador and Peru in South America, replacing the usually cold northwards flowing current; occurs once every five to seven years, usually during the Christmas season (the name refers to the Christ child); the opposite phase of an El Nino is called a La Nina. El Nino results in extensive disruption of weather around the world.

Electrodialysis: A process that uses electrical current applied to permeable membranes to remove minerals from water. Often used to desalinize salty or brackish water.

Electrolyte: Non-metallic substance that conducts electric current in solution by moving ions rather than electrons.

Electromagnetic pollution: Electronic and magnetic fields created by electrical circuits.

Electrostatic precipitation: Process that removes dust or other particles from air and other gases by electrostatic means.

Electrostatic precipitator (ESP): A device that removes particles from a gas stream (smoke) after combustion occurs. The ESP imparts an electrical charge to the particles. An electric discharge is passed through the gas, giving the impurities a negative electric charge. Positively charged plates are used inside the precipitator to attract the charged particles and remove them from the gas flow. Such devices are attached to the Chimmneys of coal burning power stations to remove ash particles. Tapping on the plates causes the particles to fall into a hopper for disposal.

Eleventh commandment: A commandment put forward by the population control movement that "Thou shall not transgress the carrying capacity of the environment."

Eligible costs: The construction costs for wastewater treatment works upon which EPA grants are based.

Embodied energy: It is the energy used during a product's entire life cycle in order to manufacture, transport, use and dispose of the product. Footprint studies often use embodied energy when tracking trade of goods.

Embolism: Obstruction of a blood vessel by a transported clot, a mass of bacteria, etc.

Embryo: Organism in the early stage of development before birth.

Embrotoxin: Material harmful to a developing embryo at a concentration that has no adverse effect on the pregnant female.

Emergency (Chemical): A situation created by an accidental release or spill of hazardous chemicals that poses a threat to the safety of workers, residents, the environment or property.

Emergency removal action: Steps taken to remove contaminated materials that pose imminent threats to local residents (*e.g.*, removal of leaking drums or the excavation of explosive waste.).

Emergency response values: Concentrations of chemicals, published by various groups, defining acceptable levels for short-term exposures in emergencies.

Emergency suspension: Suspension of a pesticide product registration due to an imminent hazarded. The action immediately halts distribution, sale, and sometime actual use of the pesticide involved.

Emergent property: A property that is not evident in the individual components of an object or system.

Emergy: "energy memory" H.T. Odum used 'emergy' to mean both *sequestered* energy and *emergent property of energy use*. Is all the available energy that was used in the work of making a product directly and indirectly, expressed in units of one type required to make energy of another.

Emetic: Agent that induces vomiting.

Emigration: Population migration away from an area. The movement of populations of a species out of a particular area or habitat. Includes the departure of humans from one country in order to relocate in a new country.

Emission cap: A limit designed to prevent projected growth in emissions from existing and future stationary sources from eroding any mandated reductions.

Generally, such provisions require that any emission growth from facilities under the same cap.

Emission factor: The relationship between the amount of pollution produced and the amount of raw material processed. For example, an emission factor for a blast furnace making iron would be the number of pounds of particulates per ton of raw materials.

Emission inventory: A compilation of estimated air, pollutant emissions, by pollutant, from smoke stacks, automobiles and other sources in a given area. The inventory is usually conducted by an environmental regulatory agency.

Emission standard: The maximal legal amount of a particular pollutant that may be released into the air from a pollutant source. The standard applicable to a particular source depends on the type of pollutant, the time when the source began operation, the location of the source of the quality of the air in the surrounding area.

Emission: 1.Waste discharged into the environment by industrial or other human processes. Usually used to refer to gaseous discharges, but can also refer to liquid discharge or radioactivity. 2. The giving off of electrons, electromagnetic energy, or radioactivity. 3. Substances such as gases or particles discharged into the atmosphere as a result of natural processes of human activities, including those from chimneys, elevated point sources, and tailpipes of motor vehicles.

Emissions intensity: Emissions expressed as quantity per monetary unit.

Emissions trading: The creation of surplus emission reductions at certain stacks, vents or similar emissions sources and the use of this surplus to meet or redefine pollution requirements applicable to other emissions sources. This allows one source to increase emissions when another source reduces them, maintaining an overall constant emission level. Facilities that reduce emissions substantially may "bank" their "credits" or sell them to other facilities or industries (See carbon trading).

Employee: A worker who many be exposed to hazardous chemicals under normal operating conditions or in foreseeable emergencies.

Employer: A person engaged in a business where chemicals are either used, distributed, or are produced for use or distribution, including a contractor or subcontractor.

Emulsifier: A chemical that aids in suspending one liquid in another. Usually an organic chemical in an aqueous solution.

Encap solution: A hazardous waste treatment process that permanently encloses the waste with materials such as special adhesives (or) concrete to prevent escape of dangerous into the environment.

Endangered species: Animals birds, fish, plants, or other living organisms threatened with extinction by anthropogenic (man-caused) or other natural changes in their environment. Requirements for declaring a species endangered are contained in the Endangered Species Act.

Endangerment assessment: A study to determine the nature and extent of contamination at a site on the National Priorities List and the risks posed to public health or the environment. EPA or the state conducts the study when a legal action is to be taken to direct potentially responsible parties to clean up a site or pay for it. An endangerment assessment supplements a remedial investigation.

End-of-the-pipe: Occurring or active at the end of a process stream, such as scrubbers in smokestacks and catalytic convertors installed near the end of engine exhaust pipe to reduce emission of pollutants that have already formed in the engine.

Endothermic: A chemical reaction that absorbs heat.

Energetics: The study of how energy flows within an ecosystem: the routes it takes, rates of flow, where it is stored and how it is used.

Energy: The capacity to do work; usable power. Different forms/types of energy are: direct energy, indirect energy, potential and kinetic energy, operational energy, embodied energy, geothermal energy, hydro energy, solar energy, stationary energy etc.

Energy accounting: Measuring value by the energy input required for a good or service.

Energy audit a systematic gathering and analysis of energy use information that can be used to determine energy efficiency improvements.

Energy conservation: Elimination of energy waste.

Energy efficiency: The amount of fuel needed to sustain a particular level of production or consumption. It is typically defined as annual primary energy consumption per dollar of gross domestic product. Using less energy to provide the same level of energy service.

Energy footprint: The sum of areas used to provide non-food and non-feed energy. Refers to the area required to provide or absorb the waste from coal, oil, gas, fuelwood, nuclear energy and hydropower. The Fossil Fuel Footprint is the area required to sequester the emitted CO_2 taking into account CO_2 absorption by the sea etc.

Energy management: A program of well-planned actions aimed at reducing energy use, recurrent energy costs, and detrimental greenhouse gas emissions.

Energy recovery: The amount of energy that can be produced per hectare of ecologically productive land. The units used are gigajoules per hectare and year, or GJ/ha/yr. For fossil fuel (calculated as CO_2 assimilation) the ratio is 100 GJ/ha/yr.

Enforcement Decision Document (EDD): A document that provides an explanation to the public of the cleanup alternative at enforcement sites on the National Priorities List.

Enforcement: EPA, state, or local legal actions to obtain compliance with environmental laws, rules, regulations, or obtain penalties or criminal sanctions for violations. Enforcement procedures may vary, depending on the requirements of different environmental laws and related implementing regulations. Under CERCLA for example, EPA will seek to require potentially responsible parties to clean up a Superfund site, or pay for the cleanup, whereas under the clean Air Act the Agency may invoke sanctions against cities failing to meet ambient air quality standards that could prevent certain types of construction or federal funding. In other situations, if investigations by EPA and state agencies uncover willful violations, criminal trials and penalties are sought.

Engineered controls: Method of managing environmental and health risks by placing a barrier between the contamination and the rest of the site, thus limiting exposure pathways. For *e.g.*, isolation the worker from the hazard or by removing the hazard from work environment. Methods include ventilation, isolation, and enclosure.

Enhanced greenhouse effect: The increase in the natural greenhouse effect resulting from increases in atmospheric concentrations of greenhouse gases due to emissions from activities.

Enhanced Inspection and Maintenance (I and M): An improved automobile inspection and maintenance program – aimed at reducing automobile emissions – that contains, at a minimum, more vehicle types and model years, tighter inspection, and better management practices. It may also include annual computerized or centralized inspections, under-the-hood inspection – for signs of tampering with pollution control equipment – and increased repair waiver cost.

Enrichment: The addition of nutrients (*e.g.*, nitrogen, phosphorus, carbon compounds) from sewage effluent or agricultural runoff to surface water, greatly increases the growth potential for algae and other aquatic plants.

ENSO (El Nino – Southern Oscillation): A suit of events that occur at the time of an El Nino; at one extreme of the cycle, when the central Pacific Ocean is warm and the atmospheric pressure over Australia is relatively high, the ENSO causes drought conditions over eastern Australia.

Entrain: To trap bubbles in water either mechanically through turbulence or chemically through a reaction.

Entropy: The measure of the degree of disorder within a system; derived from thermodynamics.

Environment: A general description of the surroundings of an organism encompassing air, water and land which support its life and receive its waste products.

Environmental accounting: An effort to incorporate into measures of economic output the environmental consequences of economic production, such as soil depletion and air pollution.

Environmental audit: An independent assessment of the current status of party's compliance with applicable environmental requirements or of a party's environmental compliance policies, practices, and controls.

Environmental change: A human – caused decline in the quantity or quality of a renewable resource.

Environmental currency: Monetary values that adequately reflect costs to the environment of human activities, by determining such factors as energy flows.

Environmental degradation: Depletion or destruction of a potentially renewable resource such as soil.

Environmental equity/justice: Equal protection from environmental hazards for individuals, groups, or communities regardless of race, ethnicity, or economic status. This applies to the development, implementation, and enforcement of environmental laws, regulations, and policies, and implies that no population of people should be forced to shoulder a disproportionate share of negative environmental impacts of pollution or environmental hazard due to a lack of political or economic strength levels.

Environmental ethics: Moral relations that hold between humans and the natural world.

Environmental fate data: Data that characterize pesticide's fate in the ecosystems, considering factors that foster its degradation (light, water, microbes), pathways and resultant products.

Environmental fate: The destiny of a chemical or biological pollutant after release into the environment.

Environmental hormones (Environmental estrogen): Chemical pollutants that have the potential to substitute for (or) interfere with, natural hormones. Environmental estrogen is the technical term for any of a group of synthetic substances found in the environment that, when absorbed into a person's system, function in a similar way to estrogen, the generic term for female sex hormones. In both the male and female bodies, environmental estrogens – also known as environmental hormones – disrupt the normal secretion of hormones and disturb the body's metabolism, which can have serious consequences, including damage to the reproductive functions. For *e.g.* a degradation product of dichloro diphenyl trichloro ethane (DDT) limits the reproductive success of birds by promoting a reduction of estrogen levels in the blood of female birds.

Environmental health: It is the branch of public health that is concerned with all aspects of the natural and built environment that may affect human health. Other terms that refer to the discipline of environmental health include environmental public health and environmental health and protection. Environmental health addresses all the physical, chemical, and biological factors external to a person, and all the related factors impacting behaviors. It encompasses the assessment and control of those environmental factors that can potentially affect health. It is targeted towards preventing disease and creating health – supportive environments. This definition excludes behaviour not related to environment, as well as behaviour related to the social and cultural environment, and genetics.

Environmental impact: Effect on the natural environment through any action of specific materials or gross pollution caused by human actions. Includes indirect effects, for example, through pollution, as well as direct effects such as cutting down trees.

Environmental impact assessment: The process of identifying and assessing environmental impacts associated with a project, policy, or program.

Environmental Impact Report (EIR): An EIR is the public document used by governmental agencies and decision-makers to assess the significant environmental effects, compare alternatives, and discuss possible methods to reduce or avoid the environmental impacts of proposed projects.

Environmental impact statement: A report identifying the likely environmental consequences of some project, policy, or program.

Environmental Impairment Liability (EIL): It is a type of insurance policy specifically covering claims made during a policy year for releases of pollutants.

Environmental incident: The release (or) threatened release of a chemical, physical (or) biological agent that may cause harms to human health (or) the environment. Government teams (or) contractors respond to prevent (or) minimize harm.

Environmental indicator: A measurement, statistic or value that provides a proximate gauge or evidence of the effects of environmental management programs or of the state or condition of the environment. Refers to any physical, chemical, biological or socio-economic measure that can be used to assess natural resources and environmental quality.

Environmental insurance: Environmental Insurance is a tool for redevelopment that eliminates or reduces the risk involved in a brownfields property transaction.

Environmental lien: A charge, security, or encumbrance upon a property's title to secure payment of cost of debt maturing out of response actions, cleanup or other remediation of hazardous substances or petroleum products.

Environmental movement (environmentalism): A term that sometimes includes the conservation and green movements; a diverse scientific, social, and political movement. A political movement to reduce resource depletion/destruction and pollution. In general terms, environmentalists advocate the sustainable management of resources and stewardship of the natural environment through changes in public policy and individual behavior. In its recognition of humanity as a participant in ecosystems, the movement is centered around ecology, health, and human rights.

Environmental protection agency: A U.S. regulatory agency established in 1970 to control pollution and conduct research on the environment.

Environmental quality: Characteristics of the environment generally indicative of its desirability for human activity.

Environmental refugees: Migrants from an area devastated by natural or technological hazards.

Environmental science: The study of interactions among physical, chemical, and biological components of the environment.

Environmental sociology: A branch of sociology examining the interaction between humans and the natural environment.

Environmental sustainability: Long – term maintenance of ecosystem components and function for future generations.

Environmental testing: Environmental testing consists of collecting and testing samples from the environment to determine if the samples contain pollutants. Environmental samples typically tested include soil, ground water, and air.

Environmental/Ecological risk: The potential for adverse effects on living organisms associated with pollution of the environment by effluents, emissions, wastes, or accidental chemical release; energy use; or the depletion of natural resources.

Environmentalists: Individuals who attempt to curb resource depletion/destruction and pollution.

Epidemic: Spreading rapidly and affecting a large portion of a population at one time. Also, on outbreak of disease with extensive characteristics.

Epilimnion: Upper waters of a thermally stratified lake subject to wind action.

Equilibrium constant, Keq: The equilibrium constant is a measure of the extent of the reaction between the two chemical species. The value characterizes the point of stability of a chemical reaction at a given temperature. At this point of stability or equilibrium, the system has no tendency to change because the forward rate of the reaction is exactly balanced by the reverse rate of reaction.

Equilibrium: A state of balance in a system, where opposing factors cancel each other out (or are entirely absent). In thermal equilibrium, heat gain balances any heat loss occurring between an object an its surroundings. A body at rest or moving at constant velocity also demonstrates a balance of opposing forces (or an absence of any forces). In relation to radiation, the state at which the radioactivity of consecutive elements within a radioactive series is neither increasing nor decreasing.

Equity: A pattern of fairness.

Equivalent method: Any method of sampling and analyzing for air pollution which has been demonstrated to the EPA Administrator's satisfaction to be, under specific conditions, an acceptable alternative to normally used reference methods.

Ergonomics: Study of human characteristics for the appropriate design of living and work environments.

Erosion: The process of wearing away of a surface by physical means such as running water, wind, glacial ice, and gravity. The term is usually applied to the loss of soil through the agency of wind (or) water.

Escherichia coli (E. coli): A bacterium used as an indicator of faecal contamination and potential disease organisms in water.

Established technology: An established technology is a technology for which cost and performance information is readily available. Only after a technology has been used at many different sites and the results fully documented is that technology considered established. The most frequently used established technology are incineration, solidification and stabilization, and pump-and-treat technologies for groundwater. See also Emerging Technology and Innovative Technology.

Established treatment technologies: Technologies for which cost and performance data are readily available.

Estuary: A semi-enclosed coastal body of water with one more rivers or streams flowing into it, and with a free connection to the open sea. Region of interaction between rivers and near – shore ocean waters, where tidal action and river flow mix fresh and salt water. Such areas include bays, mouths of rivers, salt marshes, and lagoons. These brackish water ecosystems shelter and feed marine life, birds, and wildlife.

Ethanol: An alternative automotive fuel derived from grain and corn; usually blended with gasoline to form gasohol.

Ethical consumerism: buying things that are made ethically *i.e.* without harm to or exploitation of humans, animals or the natural environment. This generally entails favoring products and business that take account of the greater good in their operations.

Ethics: A code of behavior regarding what is appropriate and inappropriate.

Eutrophication: The enrichment of water bodies with nutrients, primarily nitrogen and phosphorus, which stimulates the growth of aquatic organisms. These typically promote excessive growths of algae. As the algae die and decompose, high levels of organic matter and the decomposing organisms deplete the water of available oxygen, causing the death of other organisms, such as fish. Eutrophication is a natural, slow-aging process. The slow aging process during which a lake, estuary, or bay evolves into a bog or marsh and eventually disappears. During the later stages of eutrophication the water body is choked by abundant plant life due to higher levels of nutritive compounds such as nitrogen and phosphorus. Human activities can accelerate the process.

European community: A group of 12 countries; Belgium, Denmark, France, Greece, Ireland, Italy, Luxembourg, Netherlands, United Kingdom, Portugal, Spain, and Germany.

European economic community: An economic union established in 1958 to promote trade in western Europe that currently includes EC countries.

Eutrophic waters: Characterized by nutrient – rich water supporting an abundant growth of algae or other aquatic plants at the surface. Deep eutrophic water has little or no dissolved oxygen.

Eutrophic lakes: Shallow, murky bodies of water with concentrations of plant nutrients causing excessive production of algae. (See: dystrophic lakes.)

Evaporation ponds: Areas where sewage sludge is dumped and dried.

Evaporation rate: Rate at which a particular material will vaporize when compared to the rate of vaporization of a known material. Evaporation rate can be useful in evaluating the health and fire hazards of a material.

Evapotranspiration: The loss of water from the soil both by evaporation and by transpiration from the plants growing in the soil.

Evolution: A process of biological or social change involving adaptation to an ecosystem.

e-waste: Electronic waste, especially mobile phones, televisions and personal computers.

***Ex-Situ*:** The term ex situ or "moved from its original place," means excavated or removed.

Exceedence: The release of a pollutant from an industrial (or) waste treatment facility at levels constituting a violation of permit restrictions.

Exclusionary ordinance: Zoning that excludes classes of persons or businesses from a particular neighborhood or area.

Exempted aquifer: Underground bodies of water defined in the Underground Injection Control program as aquifers that are potential sources of drinking water though not being used as such, and thus exempted from regulations barring underground injection activities.

Exemption: A state (with primacy) may exempt a public water system from a requirement involving a Maximum Contaminant Level (MCL), treatment technique, or both, if the system cannot comply due to compelling economic or other factors, or because the system was in operation before the requirement or MCL was instituted; and the exemption will not will not create a public health risk.

Exotic species: A species that is not indigenous to a region.

Experimental use permit: Obtained by manufactures for testing new pesticides or uses thereof whenever they conduct experimental field studies to support registration on 10 acres or more of land or 1 acre or more of water.

Explosive limits: The amounts of vapor in the air that from explosive mixtures; limits are expressed as lower and upper limits and give the range of vapor concentrations in air that will explode if an ignition source is present.

Explosive: Material that produces a sudden, almost instantaneous release of pressure, gas, and heat when subjected to abrupt shock, pressure, or high temperature.

Exponential growth: A growth pattern in which some entity doubles in size during a given time period.

Export of pollution: Transporting pollutants to another area or country.

Exposure indicator: A characteristic of environment measured to provide evidence of the occurrence or magnitude of a response indicator's exposure to a chemical or biological stress.

Exposure level: The amount (concentration) of a chemical at the absorptive surfaces of an organism.

Exposure limits: Concentration in air of a chemical that is thought to be acceptable.

Exposure pathway: How a pollutant moves from the source of release thought air, water, food (or) soil to natural organisms (or) humans.

Exposure pathway: The path from sources of pollutants via, soil, water, or food to man and other species or settings.

Exposure route: How a pollutant in the environments gains entry into an organism. The possibilities include inhalation, ingestion (or) absorption through the skin.

Exposure-response relationship: The relationship between exposure level and the incidence of adverse effects.

Extensive agriculture: Maximizing the amount of land used for agricultural production.

Externalities: Beneficial or harmful effects associated with production and consumption of a product that are not included in its market price.

Externality (environmental economics): By-products of activities that affects the well-being of people or damage the environment, where those impacts are not reflected in market prices. The costs (or benefits) associated with externalities do not enter standard cost accounting schemes. The environment is often cited as a negatively affected externality of the economy.

Extinction: The disappearance of a species. The cessation of existence of a species of group of taxa, reducing biodiversity.

Exxon valdez: An oil tanker that ran around and spilled 300,000 barrels of oil in Prince William Sound in Alaska on March 24 1989. It is considered one of the most cevastating man-made environmental disasters ever to occur at sea. It ranks well down on the list of the world 's largest oil spills in terms of volume released. This is considered to be the worst oil spill in U.S. history.

F

F-waste: Material from nonspecific sources defined as listed hazardows waste by the USEPA in the code of federal regulations, title 40, part 261.31. See also K-waste, P-waste, Hazardous waste.

Fabric filter: A cloth device that catches dust particles from industrial emissions.

Factory farming: It is the practice of raising farm animals in confinement at high stocking density, where a farm operates as a factory, mechanized hi-tech production of animals for human consumption. Growth hormones and antibiotics used to enhance growth of animals may pose health risks to humans.

Faculatative bacteria: Bacteria that can live under aerobic or anaerobic conditions.

Family planning: Family planning is frequently used to mean that people plan when to have children, using birth control and other techniques to implement that plan. The practice of providing information and contraceptives to help people limit the number of children.

Famine: Malnutrition and starvation resulting from a shortage of food.

Fauna: The animal life of an area.

Fcc or f/cc is an abbreviation for fibers per cubic centimeter of air. (In crystallography, it is an abbreviation for face-centered cubic.)

FDA is an acronym for the US Food and Drug Administration.

Feasibility study: 1. A detailed technical, economic and for legal review of a specific proposed project at a particular location to outline all potential costs, benefits and likely problems. If the project is a hazardows waste site cleanup, the process is called a Remedial investigation/Feasibility study. 2. A small-scale investigation of a problem to ascertain whether a proposed research approach is likely to provide useful data.

Fecal coliform: A type of bacteria whose natural habitat is the colon of warm blooded mammals, such as humans. Specifically, the group includes all of the rod-shaped bacteria that are non-square farming, gram negative and lactose fementing in 24 hours at 44.5°C and grow either with or without oxygen. The presence of this type of bacteria in water, beverages or food is usually taken to mean that the material is contaminated with solid human waste. Bacteria included in this classification represent a sub-group termed Coliform.

Fecal material: Solid waste produced by animals and discharged from the gastrointestinal tract. Also referred as faeces or solid excrement.

Feed lot (feedyard): A confined area for the controlled feeding of animals. Is an open space where animals are patterned before slaughter. Beef cattle usually arrive at the *feed lot* from brith until death. Feed (often grains, alfalfa and molasses) is provided to the animals so they do not have to forage for their food. This feeding regimen promotes the product of higher quality meat more rapidly. There are most standard parameters for the number of animals per acre in a feed lot, but the density cf animals is usually very high. Tends to concentrate large amounts of animal waste that cannot be absorbed by the soil and, hence, may be carried to nearby streams or lakes by rainfall runoff.

Fen is a type of wetland fed by surface and/or groundwater . Fens are characterized by their water chemistry, which is neutral or alkaline. Fens are different from bogs, which are acidic, fed primarily by rainwater (ombrotrophic) and often dominated by *Sphagnum* mosses.

Fenceline concentration: The concentration of an air contaminant measured first outside the perimeter of the property of an industrial facility. The closest possible site of exposure to the public.

Fenceline: The outer perimeter of the property of an industrial facility, the closest possible sets of exposure of the public. In the management of a chemical emergency, as long as hazardous concentrations of a spill or release do not migrate beyond outer property line, the responsibility for managing the incident rests primarily with the personal at the facility. If the impact moves beyond the outer property line, a Hazardous material response team from other industries or local government units may be called in by the first responders. For routine missions, the point of compliance with the National Ambient Air Quality Standards or the estimated levels using an air quality dispersion model is the fenceline.

Ferrous metals: Magnetic metals derived from iron or steel; products made from ferrous metals include appliances, furniture, containers, and packaging like steel drums and barrels . Recycled products include processing tin/stell cans, strapping, and metals from appliances into new products.

Fertilizer: A substance containing chemical elements needed for plant growth, mainly postassium, phosphorus, and nitrogen. They are usually applied either through the soil, for uptake by plant roots, or by foliar feeding, for uptake through leaves.

Fiber: Basic form of matter, usually crystalline, with a high ratio of length to diameter.

Fibrosis: Formation of fibrous tissue, as in reparative or reactive process to particulates, in excess of amounts normally present in lung tissue walls. This reduces the oxygen and carbon dioxide exchange efficiency.

Fill: Man-made deposits of natural soils or rock products and waste materials.

Filter strip: Strip or area of vegetation used for removing sediment, organic matter, and other pollutants from runoff and wastewater.

Filtration: Filtration is a treatment process that removes solid matter from water by passing the water through a porous medium, such as sand or a manufactured filter. Often used to remove particles that contain pathogens.

Final classifier: A gravitation settling tank installed as part of some waste water treatment plants to be used after the biological treatment step. The tank funcation to remove suspended solids. Also called a secondary clarifier.

Fines: Very small airborne particles, usually less than two micrometers in diameter.

Finished water: Water that has completed a purification or treatment process and is ready to be delivered to consumers.

Fire diamond: Symbol designed by the NEPA to give a quick number rating for the particular material's degree of – health (blue), flammability (red), reactivity (yellow), and specific (white) hazard.

Fire point: The lowest temperature at which a liquid vaporizes at a sufficient rate to support continuous combustion.

Firedamp: Methane gas found in under ground coal mines.

Firestorm: An intense fire characterized by rapid burning and strong wind generated by convection currents produced by the fire.

First draw: The water that comes out when a tap is first opened, likely to have the highest level of lead contamination from plumbing materials.

First flush: The first runoff draining from the watershed (fields, streets, parking lots, ditches and so on) when it rains. The concentration of plant nutrients, sediments, sand, oil, trash and other pollutants is much higher than the runoff produced at later times. Controlling the first flush is critical to limiting pollution from runoff.

First law of thermodynamics: A law stating that energy is neither created nor destroyed, but it does change form.

First world: Countries that were the first to industrialize.

Fission: The process of splitting atoms to release energy.

Fissures: Deep, narrow cracks.

Fix a sample: A sample is "fixed" in the field by adding chemicals that prevent water quality indicators of interest in the sample from changing before laboratory measurements are made.

Fixed – Location monitoring: Sampling of an environmental or ambient medium for pollutant concentration a one location continuously or repeatedly.

Flame retardant: A chemical added to cloth to prevent or retard rapid burning.

Flammable: A material that is having capability of igniting easily and burning quickly.

Flammable gas: Gas that at ambient temperature and pressure forms a flammable mixture with air at a concentration of 13% by volume or less; or a gas that at ambient temperature and pressure forms a range of flammable mixtures with air greater than 12% by volume, regardless of the lower limit.

Flammable limits: Refer to the conditions under which a mixture of a flammable material and air may catch fire or explode. If the percentage of flammable materials in the air is between the minimum and maximum limits, the presence of a flame or a source of ignition is likely to lead to rapid combustion or explosion. The material for which flammable limits are quoted may be flammable liquid, such as ether or methanol, or powder or dust such as flour or grain dust. Flammable limits are also sometimes refered to as explosion limits.

Flammable liquid: Under United States Department of Transportation regulations, any liquid that gives off vapors that can be ignited at room temperature with a flash point less than 100° F (38° C).

Flammable material: Any solid, liquid, or gas that burn rapidly when ignited.

Flammable range: The temperature range between a material's lower explosive limit and its upper explosive limit.

Flammable solid: Solid that will ignite readily and continue to burn or is liable to cause fires under ordinary conditions or during transportation through friction or retained heat from manufacturing or processing.

Flare: 1. All stack used for the routine combustion of waste gases or for the burning of materials that must be routed from a chemical reaction vessel or refining process during upset conditions. 2. A control device that burns hazardous materials to prevent their release into the environment; may operate continuously or intermittently, usually on top of a stack.

Flash back: Occurs when a trail of flammable material is ignited by a distant spark or ignition source. The flame then travels along the trail of the material back to its source. Flashback can have several causes: failure to purge a line before use, using incorrect pressures when two gases are being mixed in a touch, or mechanical failure. Flashbacks are potentially extremely dangerous and in many countries flashback arrestors are now compulsory when using cylinders of flammable gases, such as hydrogen or methane.

Flash point: The lowest temperature at which evaporation of a substance produces sufficient vapor to form an ignitable mixture with air. Note that the source of ignition need not be an open flame, but could equally be, for example, the surface of a hot plate, or a steam pipe.

Flexible packing: Packing involving plant materials such as foils, films, paper or flexible sheeting to form a container; for example, wraps, bags, levels and pouches.

Floc: In waste water treatment, the particulates formed by the coagulation of even smaller particles or colloids. A clump of solids formed in sewage biological or chemical action.

Flocculation: Process by which clumps of solids in water or sewage aggregate through biological or chemical action so they can be separated from water or sewage.

Flood plain: A low area along a river or stream or in a tidal area considered to be at risk of flooding.

Floor sweep: It is an exhaust register, placed at the floor level in storage room, that is designed to capture the vapours of flammable contaminant; because the spill of the flammable liquid initially produces high concentrations near the floor.

Flora: The plant life of an area.

Flow rate: The rate, expressed in gallons or liters per hour, at which a fluid escapes from a hole or fissure in a tank. Such measurements are also made of liquid waste, effluent, and surface water movement.

Flowable: Pesticide and other formulations in which the active ingredients are finely ground insoluble solids suspended in a liquid. They are mixed with water for application.

Flow meter: A gauge indicating the velocity of wastewater moving through a treatment plant or of any liquid moving through various industrial processes.

Flue: A conduit for the passage of smoke or exhaust gases.

Flue dust: Very small particles carried along in the gases that exist a furnace during combustion or from a metal-smelting operation.

Flue gas: Hot gases produced during combustion within a furnace. The gas is exhausted through the flue, carrying flue dust and gaseous combustion products such as nitrogen oxides, carbon oxides, water vapor, sulfur oxides, particles and may chemical pollutants.

Flue gas desulfurization: A technology that employs a sorbent, usually lime or limestone, to remove sulfur dioxide from the gases produced by burning fossil fuels. Flue gas desulfurization is current state-of-the art technology for major SO_2 emitters, like power plants.

Fluidized bed incinerator: An incinerator that uses a bed of hot sand or other granular material to transfer heat directly to waste. Used mainly for destroying municipal sludge.

Fluidized: A mass of solid particles that is made to flow like a liquid by injection of water or gas is said to have been fluidized. In water treatment, a bed of filter media is fluidized by backwashing water through the filter.

Flume: A natural or man-made channel that diverts water.

Fluoridation: The addition of fluoride to drinking water in an effect to reduce tooth decay among the human population.

Fluoridation: The addition of a chemical to increase the concentration of fluoride ions in drinking water to reduce the incidence of tooth decay.

Fluorides: Gaseous, solid, or dissolved compounds containing fluorine that result from industrial processes. Excessive amounts in food can lead to fluorosis.

Fluorocarbons (FCs): Any of a number of organic compounds analogous to hydrocarbons in which one or more hydrogen atoms are replaced by fluorine. Once used in the United States as a propellant for domestic aerosols, they are now found mainly in coolants and some industrial processes. FCs containing chlorine are called chlorofluorocarbons (CFCs). They are believed to be modifying the ozone layer in the stratosphere, thereby allowing more harmful solar radiation to reach the Earth's surface.

Flush: 1. To open a cold – water tap to clear out all the water which may have been sitting for a long time in the pipes. 2. In new homes, to flush a system means to send large volumes of water gushing through the unused pipes to remove loose particles of solder and flux. 3. To force large amounts of water through a system to clean out piping or tubing, and storage or process tanks.

Flux: 1. A continuous flow, as in a stream, or succession of changes, as in a state of flux. 2. Rate of flow, as of a liquid or other fluid or radiation, across a given area.

Fly ash: Non-combustible residual particles expelled by flue gas. The fine ash from combustion processes that becomes dispersed in the air. To the casual observer it might appear as smoke and indeed it is often found mixed with smoke. When managed properly, by ash is prevented from escaping by air pollution control devices such as scrubbers or precipitators.

Fogging: Applying insecticide by the release of fine droplets in a product that resembles smoke. The process can be used to reduce mosquito or fly population in a community. A single-family residence can also be treated by the fogging process. Often the home is enclosed in a plastic material and the insecticides fog blown into the enclosure.

Fomite: It is a physical object that serves to transmit an infectious agent from person to person. For example, a comb infested with one or more head lice would be a fomite. Likewise, the dust particles containing infectious cold virus that remain after droplets of infected saliva are coughed into the air are fomites. Should not to be confused with a vector, which is an organism that transmits infection like mosquitoes is a vector for malarial parasite.

Food and Agriculture Organization (FAO): A United Nations Agency based in Rome and given the charge to improve efficiency of production and distribution of agricultural products.

Food chain: (food webs, food networks and/or trophic networks) Describe the feeding relationships between species within an ecosystem. A hierarchy of organisms, each feeding on the lower one.

Food gap: The difference between what people need to subsist and what is produced by farmers. This occurs in some low-income countries that are unable to produce basic foodstuffs. By the year 2020, the "food gap" – the difference between developing world production and demand – will more than double, making some of the world's poorest people even more vulnerable to hunger and possible famine.

Food miles: An integrating concept related to carbon footprints is that of "food miles"- the distance food travels from where it is grown to where it is ultimately purchased or consumed by the end user. The more food miles that attach to given food, the less sustainable and the less environmentally desirable that food is. The term food miles has become part of the vernacular among food system professionals when describing the farm to consumer pathways of food. For example, for every calorie of carrot flown in from South Africa to Europe, 66 calories of fuel is spent contributing significantly to CO_2 emissions. Simply, it means the emissions produced and resources needed to transport food and drink around the globe.

Food processing waste: Solid waste that remains after the conversion of a raw agricultural item into a commercial product. The discards left after processing for example, corn shucks and cobs left after the corn is prepared for canning at a commercial facility.

Food security: Global food security refers to food produced in sufficient quantity to meet the full requirements of all people *i.e.* total global food supply equals the total global demand. For households it is the ability to purchase or produce the food they need for a healthy and active life (disposable income is a crucial issue). Women are typically gatekeepers of household food security. For national food security, the focus is on sufficient food for all people in a nation and it entails a combination of national production, imports and exports. Food security always has components of production, access and utilization.

Food waste: A class of municipal solid waste consisting of the remains of uneaten food or of the process of food preparation at residences or commercial establishments such as grocery stores, cafeterias, restaurants and industrial kitchens, accounts for nearly seven per cent of the municipal solid waste stream.

Food web: The feeding relationships by which energy and nutrients are transferred from one species to another. The interconnected food chains (feeding relationships) in an ecosystem. Plants, herbivores, and carnivores all form parts of the food web.

Footprint: (Ecological Footprint) in a very general environmental sense a "footprint" is a measure of environmental impact. However, this is usually expressed as an area of productive land (the footprint) needed to counteract the impact.

Forage: Broadly, food for animals, especially that obtained by grazing or browsing. Forage is plant material (mainly plant leaves and stems) eaten by grazing livestock. Historically the term forage has meant only plants eaten by the animals directly as pasture, crop residue, or immature cereal crops, but it is also used more loosely to include similar plants cut for fodder and carried to the animals, especially as hay or silage.

Force mains: Pipes in which waste water is pumped under pressure, the system is used in some areas that have small elevation changes with distance and therefore need to augment gravity flow. The primary consideration in the hydraulic design of force mains is to select a pipe size which will provide the required minimum velocities without creating excessive energy losses due to pipe friction. The size is usually governed by the need to maintain minimum velocities at low flows to prevent deposition of solids and to develop sufficient velocity at least once a day to resuspend any solids which may have settled in the line. The minimum diameters to be used are 1–1¼ inch for pressure sewers at grinder pumping installations, 4-inch for force mains serving small pumping stations and pneumatic ejectors, and 6-inch for all other force mains.

Foreseeable emergency: Potential occurrence such as equipment failure, rupture of containers, or failure of control equipment which could result in an uncontrolled release of a hazardous chemical.

Forest: A biome with enough precipitation to support various tree species; land with a canopy cover greater than 30%.

Formaldehyde: A colorless, pungent, and irritating gas, CH_2O, used chiefly as a disinfectant and preservative and in synthesizing other compounds like resins.

Formula: The scientific expression of the chemical composition of a material (*e.g.*, water H_2O, sulfuric acid H_2SO_4 sulfur dioxide is SO_2).

Formulation: The substances comprising all active and inert ingredients in a pesticide.

Fossil fuel: Fuel, such a coal, oil and natural gas, formed from once – living organisms that have become fossilized over geological time. Fossil fuels are a non-renewable resource and will eventually run out. Extraction of coal and oil causes considerable environmental pollution and burning coal contributes to problems of acid rain by releasing SO_2. All fossil fuel release carbondioxide when burnt, regardless of whatever antipollution measures cause global warming by artificially enhancing the greenhouse effect.

Fossil water: Groundwater that has remained in an aquifer for thousands or millions or years; when geological changes seal the aquifer preventing further replenishment, the water becomes trapped inside and is then referred to as fossil water. Fossil water is limited resource and can only be used once.

Four nines: The destruction requirement for selected principal organic hazard constituents in waste incinerators, 99.99%.

Fracture: A break in a rock formation due to structural stresses; *e.g.*, faults, shears, joints, and planes of fracture cleavage.

Free liquids: Liquids capable of migrating from waste and contaminating ground water. Hazardous waste containing free liquids may not be disposed of in land fills.

Free market: A market in which buyers and sellers are free to have contract on whatever terms they like without government interference.

Free moisture: Liquid that drains freely from solid waste by the action of gravity only.

Free radicals: Unstable atoms or molecules or molecules with at least one unparid electron, these materials are highly reactive and thus are short lived.

Freeboard: Vertical distance from the normal water surface to the top of a confining wall. Vertical distance from the sand surface to the underside of a trough in a sand filter.

Freezing point: Temperature at which a material changes its physical state from liquid to solid. This information is important because a frozen hazardows material may burst its container.

Freon: DuPont' s trade name for its odorless, colorless, non-flammable, and noncorrosive chlorfluorocarbon and hydrochlorofluoro-carbon refrigerants, which are used in air conditioning and refrigeration systems.

Fair trade: A guarantee that a fair price is paid to producers of goods or services; in includes a range of other social and environmental standards including safety standards and the right to from unions.

Fresh water: Water that generally contains less than 1,000 milligrams-per-liter of dissolved solids.

Friable asbestos: Any material containing more than one-per cent asbestos, and that can be crumbled or reduced to power by hand pressure. (May include previously non-friable material which becomes broken or damaged by mechanical force.)

Friable: Capable of being crumbled, pulverized, or reduced to power by hand pressure.

Friends of the Earth (FOE): An environmental advocacy group founded in 1970 by David Brower. Environmental pressure groups that aim to protect the environment and to promote national and sustainable use of the earth's resources. They campaign on issues such as acid rain, air, sea, river and land pollution, recycling disposal of toxic wastes, nuclear power and renewable energy. The destruction rain forests, pesticides and agriculture.

Front end recovery: A centralized solid waste treatment process in which mechanical a manual separation of paper, glass, metals and/or organic matter suitable for composting is performed on the collected waste before further processing.frostbite: Damage to tissue from exposure to extreme cold contact with extremely cold liquids or solids.

Frugivore: An animal that lives/feeds on fruit. Frugivores are typically primates that cannot digest cellulose in leaves.

Fuel economy standard: The Corporate Average Fuel Economy Standard (CAFÉ) effective in 1978. IT enhanced the national fuel conservation effort imposing a miles-per- gallon floor motor vehicles.

Fuel efficiency: The proportion of energy released by fuel combustion that is converted into useful energy.

Fuel gas desulphurization: Removal of sulphur pollution from gases emerging from a boiler. Sulphur compounds such as SO_2 all commonly produced by burning fossil fuels, especially coal in power stations and cause acid rain.

Fuel switching: 1. A precombustion process whereby a low-sulfur coil is used in place of a higher sulfur coal in a power plant to reduce sulfur dioxide emissions. 2. Illegally using leaded gasoline in a motor vehicle designed to use only unleaded.

Fuelwood crisis: A shortage of wood for heating and cooking purposes. This is a pressing problem in many less developed countries.

Fugitive emissions: Emissions not caught by a capture system . Gas, liquid, and solids from any contact with skin and prevents them from being inhaled or ingested.

Full-cost pricing: The pricing of commercial goods—such as electric power—that includes not only the private costs of inputs, but also the costs of the externalities required by their production and use.

Fume: Consists of solid particles of very small size, produced by condensation from a liquid or vapour. Typically fume is produced by condensation of metal vapour produced by, for example, hot metal soldering . Because the metals used may be toxic and the particles are small enough to penetrate deep into the lungs, fume is potentially harmful and it may be necessary to take steps to limit exposure to it. Contray to popular use, gasoline 'fumes' are not fumes.

Fumigant: A pesticide vaporized to kill pests. Used in buildings and greenhouses.

Fungicide: Pesticides which are used to control, deter, or destroy fungi.

Fungi stat. A chemical that keeps fungi from growing.

Fungus (fungi): Parasitic or saprophytic plants lacking the green pigment chlorophyll and thus incapable of photosynthesis. Molds, mildews, yeasts, mushrooms, and puffballs, a group of organisms lacking in chlorophyll (*i.e.* are not photosynthetic) and which are usually non-mobile, filamentous, and multicellular. Some grow in soil, others attach themselves to decaying trees and other plants whence they obtain nutrients. Some are pathogens, others stabilize sewage and digest composted waste.

Furrow irrigation: Irrigation method in which water travels through the field by means of small channels between each groups of rows.

Fusion: The process of combining atomic nuclei in order to release energy.

Future liability: Refers to potentially responsible parties' obligations to pay for additional response activities beyond those specified in the Record of Decision or Consent Decree.

G

G7: Seven of the largest industrial countries, including Canada, France, Germany, Italy, Japan, The United Kingdom, and The United States.

G8: The Group of Eight is an international forum for the world' s major industrialised democracies that emerged following the 1973 oil crisis and subsequent global recession. It includes Canada, France, Germany, Italy, Japan, Russia, the UK and the US which represents about 65% of the world economy .

G10: The 10 most influential environmental organizations in the United States . G15: A group of 15 less developed countries that meet to discuss issues of the Third World.

G77: A group of less developed countries in the Third World and Eastern Europe, originally including 77 countries but later expanded to include 129.

Gaia hypothesis is an ecological hypothesis proposing that the biosphere and the physical components of the Earth (atmosphere, cryosphere, hydrosphere and lithosphere) are coupled to form a complex interacting system that maintains the climatic and biogeochemical conditions on Earth in a preferred homeostasis. Originally proposed by James Lovelock as the earth feedback hypothesis, it was named at the suggestion of his neighbor William Golding-the Gaia Hypothesis, after the Greek supreme goddess of Earth. The hypothesis is frequently described as viewing the Earth as a single organism.

Game fish: Species like trout, salmon, or bass, caught for sport. Many of them show more sensitivity to environmental change than "rough" fish.

Gangrene: Death of tissue combined with putrefaction.

Garbage: Waste material, typically from domestic and commercial sources. Waste includes cans, bottles news papers, paper and plastic packages, uneaten food, broken furniture and appliances. Old tyres, loan clipping and other refues.

Garden organics: Organics derived from garden sources *e.g.*, prunings, grass clippings.

Gas: Formless fluid that occupies the space of its enclosure. Can settle to the bottom or top of an enclosure when mixed with other materials. Can be changed to its liquid or solid state only increased pressure and decreased temperature.

Gasohol: Mixture of gasoline and ethanol derived from fermented agricultural products containing at least nine per cent ethanol. Gasohol emissions contain less carbon monoxide than those from gasoline.

Gasification: Conversion of solid material such as coal into a gas for use as a fuel.

Gasoline volatility: The property of gasoline whereby it evaporates into a vapor. Gasoline vapor is a Mixture of volatile organic compounds.

Gastritis: Irritation of lining of stomach which may be evident as stomach pains, vomiting, or diarrhoea, etc.

GATT: General Agreement on Tariffs and Trade. A multilateral trade agreement regulating the world trade system.

Gavage: Feeding by means of a stomach tube.

GDP-Gross national product: The value of goods and services produced in a country plus remittances received from abroad.

Gene: A locatable region of genomic sequence, corresponding to a unit of inheritance, which is associated with regulatory regions, transcribed regions and or other functional sequence regions.

General permit: A permit applicable to a class or category of dischargers.

General ventilation: Removal of contaminated air and its replacement with clean air from general workplace area as opposed to local ventilation, which is specific to air changing in immediate air of a contamination source.

Generally Recognized as Safe (GRAS): Designation by the FDA that a chemical or substance (including certain pesticides) added to food is considered safe by experts, and so is exempted from the usual FFDCA food additive tolerance requirements.

Generator ID Number: A federally reqired identification number assigned to a generator of hazardous waste and used to track the generation, shipment, and offsite disposal of hazardous wastes.

Generator: A facility or mobile source that emits pollutants into the air or releases hazardous waste into water or soil. Any person, by site, whose act or process produces regulated medical waste or whose act first causes such waste to becomes subject to regulation. Where more than one person (*e.g.*, doctors with separate medical practices) are located in the same building, each business entity is the separate generator.

Generic name: Designation or identification to identify a chemical by other than its chemical name.

Genetic diversity: The genetic variability within a group of related organisms. One of the three levels of biodiversity that refers to the total number of gentic characteristics.

Genetic engineering: The manipulation of genetic material for economic purposes. A process of inserting new genetic information into existing cells in order to modify a specific organism for the purpose of changing one of its characteristics.

Genetic: Pertaining to or carried by genes. Hereditary.

Genome: The total genetic composition of an organism. The genome includes both the genes and the non-coding sequences of the DNA. In modern molecular biology the genome of an organism is its hereditary information encoded in DNA (or, for some viruses, RNA).

Genotoxicity describes a deleterious action on a cell genetic affecting its intergrity.

Genotoxic: Describes a poisonous substance which harms an organism by damaging its DNA. Genotoxic chemicals are those which are capable of causing damage to DNA. Such damage can potentially lead to the formation of a malignant tumor, but DNA damage does not lead inevitably to the creation of cancerous cells.

Geographic Information System (GIS): A computer system designed for storing manipulating, analyzing, and displaying data in a geographic context.

Geological log: A detailed description of all underground features (depth, thickness, type of formation) discovered during the drilling of a well.

Geophysical log: A record of the structure and composition of the earth encountered when drilling a well or similar type of text hole or boring.

Geophysical survey: A survey conducted to determine the presence of buried metal.

Geosphere: The nonliving portion of the earth, excluding the atmosphere, hydrosphere, and biosphere. Refers to the solid part of planet Earth, the main divisions being the crust, mantle, and liquid core. The lithosphere is the part of the geosphere that consists of the crust and upper mantle.

Geothermal energy: Energy derived from the natural heat of the earth contained in hot rocks. Steam and hot water (created by the earth's molted core) used to produce electricity and considered to be a renewable energy.

Geothermal/ground source heat pump: These heat pumps are underground coils to transfer heat from the ground to the inside of a building.

Germicide: Any compound that kills disease-causing microorganisms.

Genotype: 1.in genetics-The genetics constitution of an organism or cell, as distinct from its expressed features or phenotype. 2. in zoology- the genetic constitution of an individual or taxon. Not to be confused with type species of a genus for which it is sometimes erroneously used.

Gestation: Pregnancy in mammals: the development of offspring in the uterus from conception to birth.

Ghost acres: The land on which cheap foodstuffs and agricultural products are produced by less developed countries for export to the industrialized countries.

Giardia lamblia: Protozoan in the feces of humans and animals that can cause severe gastrointestinal ailments. It is a common contaminant of surface waters.

Gingivitis: Inflammation of the gums.

GIS: *See* Geographic Information System.

Glass containers: For recycling purposes, containers like bottles and jars for drinks, food, cosmetics and other products. When being recycled, container glass is generally separated into color categories for conversion into new containers, construction materials or fiberglass insulation.

Glazing: A colorless glass coating applies to the surface of a container or product to reduce porosity. Glazing containing lead has been a source of lead poisoning when used on earthenware in which acidic foods or beverages are stored.

Global 2000: A report on global environmental trends that were commissioned by President Carter in 1977 and were published in 1980. The general conclusion of the report was that the world in 2000 would be more crowded, more polluted, and more fragile if global trends continued.

Global change research program: A unit in the U.S. government established in 1989 to coordinate research on short and long term climate change, including the Earth's radiation balance, ozone layer depletion and marine and continental ecosystem health. Over 300 U.S. and foreign universities and research institutions, offices within the U.S. Departments of Agriculture, Commerce, Defense, Energy, Health and Human Services, Interior, State and Transportation and several executive agencies including the USEPA and the National Aeronautics and space administration, participate in the program.

Global climate change: The cumulative effects of various impacts of rising levels of green house gases on earth's climate. These include global warming, weather changes and a rising sea level.

Global commons: Natural systems and resources that do not belong to any one country, *e.g.*, the atmosphere and the oceans.

Global dimming: A reduction in the amount of direct solar radiation reaching the surface of the earth due to light diffusion as a result of air pollution and increasing levels of cloud. A phenomenon of the last 30-50 years.

Global economy: The emerging international economy characterized by free trade in goods and services, unrestricted capital flows and more limited national powers to control domestic economics.

Global environment facility: A program begun in 1991 and jointly operated by the United Nations Development program, the United Nations environment program and the World Bank to provide grants and loans to activities that help combat climate change, ocean pollution, loss of biodiversity and ozone layer depletion.

Global Environment Monitoring System (GEMS): A part of the United Nations Environment program's Earth watch, based in Nairobi, Kenya; purposes of GEMS are to coordinate environmental monitoring and assessment activities world wide and it provide an environmental data exchange service.

Global Environmental Management Initiative (GEMI): A non-profit organization of around 40 multinational companies based in the United States dedicated to fostering environmental, health and safety excellence world wide.

Global Environmental Monitoring System: A global effort to monitor the earth's environment; established in 1972 and operated by the United Nations Environment Programme.

Global hectares: Acres/hectares that have been adjusted according to world average biomass productivity so that they can be compared meaningfully across regions; 1 global hectare is 1 hectare of biologically productive space with world average productivity.

Global positioning system: A U.S. Department of Defense satellite network used as a surveying and navigational aid and available to the public.

Global Warming: The observable increase in global temperatures considered mainly caused by the human induced enhanced greenhouse effect trapping the Sun's heat in the Earth's atmosphere.

Global warming: Projected imminent climate change attributed to the greenhouse effect. By 2030 the average global temperature would have risen by between 0.7 and 2.0 °C. The predicted excessive warming of the atmosphere resulting from the accumulation of atmospheric carbon di oxide. The atmosphere in normally warmed when infrared radiation emitted by the earth is absorbed by carbon di oxide gas and water vapor in the air. As the amount of carbon di oxide increases as a result of the combustion of fossil fuels and deforestation, especially of tropical rain forests, the hypothesis is that more heat energy will be retained by the Earth's atmosphere, resulting in a change in rainfall and wind patterns and melting of polar ice, thus raising the global sea level. The change in weather patterns could have devastating consequences for prime agricultural areas. A significant rise in sea level could flood many coastal cities and damage ecologically important coastal wetlands. Other heat – absorbing gases that are increasing in the atmosphere as a result of human activities methane, nitrous, oxide and the chlorofluorocarbons. Sciencentists generally agree that the Earth's surface has warmed by about 1 degree Fahrenheit in the past 140 years. The Intergovernmental Panel on Climate Change (IPCC). Recently concluded that increased concentrations of greenhouse gases are causing an increase in the Earth's surface temperature and that increased concentrations of sulfate aerosols have lead to relative cooling in some regions, generally over and downwind of heavily industrialized areas.

Global warming potential: The ratio of the warming caused by a substance to the warming caused by a similar mass of carbon dioxide. CFC-12, for example, has a GWP of 8,500, while water has a GWP of zero.

Globalization: The expansion of interactions to a global or worldwide scale; the increasing interdependence, integration and interaction among people and organizations from around the world. A general term, used since the mid 1940s, referring to a mix economic, social, technological, cultural and political interrelationships.

Globe temperature: The temperature reading that indicates the level of radiant heat or heat transferred by infrared radiation. The temperature is read from a thermometer with the bulb set in the centre of a black six inch sphere, which acts to absorb radiant heat in the surroundings.

Glove bag: A polyethylene or polyvinyl chloride bag – like enclosure affixed around an asbestos – containing source (most often thermal system insulation) permitting the material to be removed while minimizing release of airborne fibers to the surrounding atmosphere.

Gooseneck: A portion of a water service connection between the distribution system water main and a meter. Sometimes called a pigtail.

Grab sample: Sample of rock or sediment taken more or less indiscriminately at any place. Refers to a single sample collected at a particular time and place that represents the composition of the water, air, or soil only at that time and place.

Graded effect: It is one whose severity is related continuously to dose. Increases in dose rate or exposure level thus have a steadily increasing (sometimes linear) effect upon the severity of symptoms.

Grain loading: The rate at which particles are emitted from a pollution source. Measurement is made by the number of grains per cubic foot of gas emitted.

Granular Activated Carbon Treatment (GAC): A filtering system often used in small water systems and individual homes to remove organics. Also used by municipal water treatment plants. GAC can be highly effective in lowering elevated levels of radon in water.

Grass cycling: Source reduction activities in which grass clippings are left on the lawn after mowing.

Grassed waterway: Natural or constructed watercourse or outlet that is shaped or graded and established in suitable vegetation for the disposal of runoff water without erosion.

Grassland: A biome with a moderate level of precipitation and vegetation dominated by grasses. A biome comprised mainly of indigenous grasses or grasslike plants; the predominant vegetation type in semi-arid temperate and tropical latitudes. Grassland is used to refer to semi-natural meadows and cultivated farm meadows as well as to natural prairies.

Gravity sewer: A waste water system or sewer in which flow is controlled by gravity.

Grazer: An herbivore that eats non-woody plant matter.

Grazing: The act of animals feeding on fresh grass and herbaceous plants.

Green consumerism: Purchasing environmentally sound products.

Green design: Environmentally sustainable design.

Green lash: A term that refers to dramatic and widespread changes that can result from seemingly small or localized alterations to the environment. Globalization, which invites the invasion of alien species and pathogens into new environments, could be a catalyst for green lash. Not only invasive species, but also gases and minerals, can be introduced into new environments, altering existing cycles of nutrients and materials. The example of green lash scientist cite is the Dust Bowl of the 1930s, when a local drought led farmers to abandon their land, and the resulting lack of crop cover led to an expansion of the drought, as topsoil blew away. As a result, mass migrations of people, and public health problems from inhaled dust, resulted far from the origins of the drought.

Green manure: A type of cover crop grown primarily to add nutrients and organic matter to the soil.

Green politics: Politics centering on issues surrounding environmental problems.

Green power: Electricity generated from clean, renewable energy sources (such as solar, wind, biomass and hydro power) and supplied through the grid.

Green products and services: Products or services that have a lesser or reduced effect on human health and the environment when compared with competing products or services that serve the same purpose. Green products or services may include, but are not limited to, those which contain recycled content, reduce waste, conserve energy or water, use less packaging, and reduce the amount of toxics disposed or consumed.

Green purchasing: Environmentally Preferable Purchasing (EPP), often referred to as "green purchasing" is the affirmative selection and acquisition of products and services that minimize impacts on the environment and that are socially just. Examples of environmentally preferable characteristics include products and services that conserve energy and water, minimize generation of waste and releases of pollutants; products made from recycled materials and that can be reused or recycled; energy from renewable resources such as biobased fuels, solar and wind power; alternate fuel vehicles; and products using alternatives to hazardous or toxic chemicals, radioactive materials and biohazardous agents.

Green revolution: The term "Green Revolution" was first used in 1968 by former USAID director William Gaud. The Green Revolution refers to the transformation of agriculture that began in 1945 at the request of the Mexican government to establish an agricultural research station to develop more varieties of wheat that could be used to feed the rapidly growing population of the country. In 1943 Mexico imported half its wheat; in 1956, the Green Revolution had made Mexico self-sufficient; by 1964, Mexico exported half a million tons of wheat. The associated transformation has continued as the result of programs of agricultural research, extension, and infrastructural development, instigated and largely funded by the Rockefeller Foundation, along with the Ford Foundation and other major agencies. The consensus among some agronomists is that the Green Revolution has allowed food production to keep pace with worldwide population growth. The Green Revolution has had major social and ecological impacts, making it a popular topic of study among sociologists. The effort organized by the United Nations in the 1960s to increase world food production by introducing high – yield varieties of rice, wheat, and maize and new techniques, including irrigation and use of pesticides. The term "Green Revolution" is a general one that is applied to successful agricultural experiments in many Third World countries. It is NOT specific to India. But it was most successful in India.

Green tax: A tax on activities that pollute, deplete, or degrade the environment.

Green washing: A derogatory term used to describe companies that portray themselves as environmentally friendly when their business practices do not back this up. Greenwashing is the unjustified appropriation of environmental virtue by a company, an industry, a government, a politician or even a non-government organization to create a pro-environmental image, sell a product or a policy, or to try and rehabilitate their standing with the public and decision makers after being embroiled in controversy.

Green waste (green organic material or green organics, sometimes referred to as "green wealth"): Plant material discarded as non-pursuable waste – includes tree and shrub cutting and prunings, grass clippings, leaves, natural (untreated) timber waste and weeds (noxious or otherwise).

Green water: Green water is the water held in soil and available to plants. It is the largest fresh water resource but can only be used in situ, by plants. In nature the global average amount of rainfall becoming green water is about 60%. Of the green water about 55% falls on forests, 25% on grasslands and about 20% on crops. We can increase green water productivity by rainwater harvesting, increased infiltration and runoff collection. Green water cannot be piped or drunk (cannot be sold) and is therefore generally ignored by water management authorities but it is crucial to plants in both nature and agriculture and needs careful management as an important part of the global water cycle.

Grey water (*gray water* or sullage) is non-industrial wastewater generated from domestic processes such *Water discharged from sinks, showers, kitchens, laundry, or other non-industrial operations, but not from toilets. This normally comprises 50-80% of residential "waste" water.* Grey water is distinct from black water in the amount and composition of its chemical and biological contaminants (from feces or toxic chemicals). Grey water gets its name from its cloudy appearance and from its status as being neither fresh (white water from groundwater or potable water), nor heavily polluted (black water). According to this definition wastewater containing significant food residues or high concentrations of toxic chemicals from household cleaners etc. may be considered "dark grey" or black water.

Green Water Credits (GWC): It is a mechanism for transfer of cash to rural people in return for water management activities that determine the supply of all fresh water at source- these activities are presently unrecognized and un-rewarded. Direct payment will enable better management of the resource.

Greenhouse effect: A rise in the earth's surface temperature caused by heart radiated by the sun that becomes trapped by greenhouse gases. The insulating effect of atmospheric greenhouse gases (*e.g.*, water vapor, carbon dioxide, methane, etc.) that keeps the Earth's temperature about 60 °F (16 °C) warmer than it would be otherwise. However, rising atmospheric concentrations of carbon di oxide and other greenhouse gases are predicted to cause global warming.

Greenhouse gases: Any gas that contributes to the greenhouse effect; gaseous constituents of the atmosphere, both natural and from human activity, that are transparent to solar radiation but opaque to long wave radiation. Water vapor (H_2O) is the most abundant greenhouse gas. Greenhouse gases are a natural part of the atmosphere and include carbon dioxide (CO_2), methane (CH_4, persisting 9-15 yrs with a greenhouse warming potential (GWP) 22 times that of CO_2), nitrous oxide (N_2O persists 120 yrs and has a GWP of 310), ozone (O_3), hydrofluorocarbons, perfluorocarbons and sulfur hexafluoride.

Greenpeace: is an international non-governmental organization for the protection and conservation of the environment and for promoting peace.

Grinder pump: A mechanical device that shreds solids and raises sewage to a higher elevation through pressure sewers.

Grit: Sand or fine gravel carried in wastewater.

Grit chamber: The initial treatment device at a sewage treatment facility; dense material from the incoming waste water settles to the bottom of the chamber and is removed.

Grit removal: The process of removing sand and fine gravel from a stream of domestic waste in a grill chamber.

Gross alpha/beta particle activity: The total radioactivity due to alpha or beta particle emissions as inferred from measurements on a dry sample.

Gross power-generation potential: The installed power generation capacity that landfill gas can support.

Gross primary production: The total energy produced by photosynthesis and stored in a given biotic community, total carbon assimilation.

Ground cover: Plants grown to keep soil from eroding. Materials, often plants, used to prevent soil erosion or leaching and to protect from freezing.

Ground water: Water found below the surface, usually in porous rocks, or soil, or in under-ground aquifers, that often supplies eels and springs. Ground water often serves as principal source of drinking water, there is growing concern over contamination from leaching agricultural or industrial pollutants or leaking underground storage tanks.

Grounding: Safety practice to conduct electrical charge to ground, preventing igniting sparks of a material.

Ground-Penetrating radar: A geophysical method that uses high frequency electromagnetic waves to obtain subsurface information.

Ground-water discharge: Ground water entering near coastal waters which has been contaminated by landfill leachate, deep well injection of hazardous wastes, septic tanks, etc.

Growth: Increase in size, weight, power *etc.*

Growth mania: A belief that " bigger is always better."

Growth rate: 1. The annual percentage increase in the gross national product. 2. The speed at which a plant, animal, or microorganisms increases in size. 3. The increase (or decrease) of a population over a period of time. In humans, the rate of change is calculated from the number of births minus the number of deaths per thousand in a population, with net migration added; it is expressed as percentage.

Guinea pig maximization test is a widely-used test in the screeing of contact allergens for the possibility that they may act as sensitizers in humans.

Gully erosion: Severe erosion in which trenches are cut to a depth greater than 30 centimeters (afoot). Generally, ditches deep enough to cross with farm equipment are considered gullies.

H

Habitat: An ecological or environmental area that is inhabited by a particular species or in other words the natural abide of an organism including its total environment.

Habitat indicator: A physical attribute of the environment measured to characterize conditions necessary to support an organism, or community in the absence of pollutants ; *e.g.* salinity of estuarine waters or substrate type in streams or lakes.

Habitat loss: The loss of natural habitat, often by human actions.

Half-Life: 1. The time required for a pollutant to lose one-half of its original concentration, for example, the biochemical half- life of DDT in the environment is 15 years. 2. The time required for the elimination of half a total dose from the body.3. The time required for half of the atoms of a radioactive element to undergo self-transmutation or decay (half –life of radium is 1620 years). 4. A property used to characterize radioactive substances. The half-life is the period of time required for half of a quantity of a radioactive substances. The half-life is the period of time required for half of a quantity of a radioactive nuclide to undergo decay. Half- life is independent of conditions such as temperature and chemical form, and specific to each nuclide.

Halogens or halogen elements: These are a series of nonmetal elements from Group 17 IUPAC Style (formerly: VIIA< or Group7) of the periodic table, comprising fluorine, F: chlorine, C1; bromine, I; and astatine, At. The group of halogens is the only group which contains elements in all three familiar states of matter at standard temperature and pressure.

Halogen lamp is an incandescent lamp with higher energy – efficiency than standard ones in which a tungsten filament is sealed into compact transparent envelope filled with an inert gas, plus a small amount of halogen such as iodine or bromine.

The halogen cycle increases the lifetime of the bulb and prevents its darkening by re-depositing tungsten from the inside of the bulb back onto filament.

Halon: Halons are a subset of a more general class of compounds known as halocarbons. Halons contain carbon, bromine, fluorine, and, in some cases, chlorine halons are entirely human-made and are used primarily in fire extinguishers. Halons and other halocarbons (carbon- and halogen –containing compounds), such as chlorofluorocarbons (CFCs), are responsible for the breakdown of stratospheric ozone and the creation of the Antarctic ozone hole.

Hammer mill: A high- speed machine that uses hammers and cutters to crush, grind, chip, or shred solid waste into fine particles.

Hard waste: Household garbage which is not normally accepted into rubbish bins by local councils, *e.g.* old stoves, mattresses.

Hard water: Water that contains sufficient concentrations of metal ions, primarily those of calcium and magnesium to cause the formation of precipitates with soaps and detergents.

Hardness: A measure of the amount of calcium and magnesium salts dissolved in water.

Hardwood: The wood of broad-leaved, flowering trees *e.g.*, oak, mahogany and walnut.

Hauler: Garbage collection company that offers complete refuse removal service; many will also collect recyclables.

Hazard: A technology, activity, or substance that has adverse effects on the environment or human health. Anything that can cause 1. injury, disease, to humans, 2. damage to property, or 3.

degradation of the environment. Cultural hazards include factors that are after a matter of choice. Such as smoking or sunbathing, biological hazards are pathogens and parasites that impact humans. Physical hazards are natural disasters like earth quakes and tornadoes. Chemical hazards refers to the chemicals in use in different technologies and household products. The term " hazard " is often confused with "risk ". A high voltage power supply, a sample of radioactive material, or a toxic chemical may present a hazard, meaning that they present the potential for harm. Concentrated acids, for example, pose a clear hazard because incorrect handling could lead to serious burns.

(The risk is the probability or change that the hazard posed by the chemical will lead to injury. Thus, concentrated sulfuric acid is a hazardous chemical, because it is very corrosive and reactive. However, provided it is handled cautiously and with a full understanding of the hazards that it presents, the risks it poses may be small)

Hazard assessment: The process of examining evidence linking a particular hazard to its harmful effects. Evaluating the effects of a stressor or determining a margin of safety for an organism by comparing the concentration which causes toxic effects with an estimate of exposure to the organism.

Hazard classes: Categories of hazardous material defined by the U.S. Department of Transportation under the hazardous materials transportation law. A few of the major classes are corrosives, poisons, flammable gases, and flammable liquids.

Hazard evaluation: A component of risk evaluation that involves gathering and evaluating data on the types of health injuries that may be produced by a chemical and on the conditions of exposure under which such health effects are produced.

Hazard index: A summation of the hazard quotients for all chemicals to which an individual is exposed. A hazard index value of 1.0 less than 1.0 indicates that no adverse human health effects (non-cancer) are expected to occur.

Hazard quotient: A comparison of an estimated chemical intake (dose) with a reference dose level below which adverse health effects are unlikely. The hazard quotient is expressed as the ratio of the estimated intake to the reference dose. The value is used to evaluate the potential for non-cancer health effects, such as organ damage, from chemical exposures.

Hazard ratio (HR): Broadly equivalent to relative risk (RR); useful when the risk is not constant with respect to time. It uses information collected at different times. The term is typically used in the context of survival over time. If the HR is 0.5 then the approximately 200 chemicals or chemical categories listed as HAP' s in the 1990 amendments to the Clean Air Act.

Hazardous decomposition: Breaking down or separation of a substance into its constituent parts, elements, or into simpler compounds accompanied by the release of heat, gas, or hazardous materials.

Hazardous ingredients: Hazardous substances that make up a mixture.

Hazardous Materials Identification System (HMIS): Developed by the NPCA to provide information on health, flammability, and reactivity hazards that are encountered in the workplace. A number is assigned to a material indicating the degree of hazard, from 0 for the least up to 4 for the most severe. Letters are used to designate personal protective equipment.

Hazardous substance: 1. Any chemical that poses a threat to human health and/or the environment. Typical hazardous substances are toxic, corrosive, ignitable, explosive, or chemically reactive. 2. Any substance designated by EPA to be reported if a designated quantity of the substance is spilled in the waters of the United States or is otherwise released into the environment.

Hazardous waste: Waste possessing chemical, physical, or biological characteristics that represent a threat to either the environment or human health. By-products of society that can pose a substantial or potential hazard to human health or the environment when improperly managed. Any discarded material containing substances known to be toxic. Mutagenic, carcinogenic or teratogenic to humans or other life forms. Possesses at least one of four characteristics (ignitability, corrosivity, reactivity, or toxicity), or appears on special EPA lists.

Hazardous waste landfill: An excavated or engineered site where hazardous waste is deposited and covered.

Hazardous waste minimization: Reducing the amount of toxicity or waste produced by a facility via source reduction or environmentally sound recycling.

Hazardous waste number: Identification number assigned by the EPA, per RCRA law, to identify and track hazardous wastes.

Hazardous waste site: A location for the treatment, storage or disposal of hazardous waste.

Hazard analysis: Procedures used to 1. identify potential sources of release of hazardous materials from fixed facilities or transportation accidents; 2. determine the vulnerability of a geographical area to a release of hazardous materials; and 3. compare hazards to determine which present greater or lesser risks to a community.

Hazards identification: Providing information on which facilities have extremely hazardous substances, what those chemicals are, how much there is at each facility, how the chemicals are stored, and whether they are used at high temperatures.

Headspace: The vapor mixture trapped above a solid or liquid in a sealed vessel.

Health advisory level: A non-regulatory health based reference level of chemical traces (usually in ppm) in drinking water at which there are no adverse health risks when ingested over various periods of time. Such levels are established for one day, 10 days, long-term and life-time exposure periods. They contain a wide margin of safety.

Health hazard: Chemical from which acute or chronic health effects may occur in exposed individuals.

Heat island effect: A "dome" of elevated temperatures over an urban area caused by structural and pavements heat fluxes, and pollutant emissions.

Heat pump: An electric device with both heating and cooling capabilities. It extracts heat from one medium at a lower (the heat source) temperature and transfers it to another at a higher temperature (the heat sink), thereby cooling the first and warming the second. (*See*: geothermal, water source heat pump.)

Heavy metals: Metallic elements with high atomic weights usually greater than 50 and relative density greater than 7 (*e.g.* iron, zinc, mercury, chromium, cadmium and lead); can damage living things at low concentrations and tend to accumulate in the food chain.

Hematopoietic agent: is a chemical which interferes with the blood system by decreasing the oxygen-carrying ability of hemoglobin. This can lead to cyanosis and unconsciousness. Carbon monoxide is done such agent, familiar to smokers.

Hematuria: Presence of blood in the urine.

Hemolysis: Separation of hemoglobin from red blood corpuscles.

Hepatic: Pertaining to the liver.

Hepatotoxin: A substance that causes injury to the liver.

Heptachlor: An insecticide that was banned on some food products in 1975 and in all of them in 1978. It was allowed for use in seed treatment until 1983. More recently it was found in milk and other dairy products in Arkansas and Missouri where dairy cattle were illegally fed treated seed.

Herbicide: A chemical that kills or inhibits growth of a plant; is used to control or destroy unwanted plants, weeds, or grasses.

Herbivore: An organism that obtains energy from plant consumption. Granivores and Frugivores are special types of herbivores.

Herbivory: Predation in which an organism known as an herbivore, consumes principally autotrophs such as plants, algae and photosynthesizing bacteria.

Heterotroph (chemoorganotrophy): An organisms that requires organic substrates to obtain its carbon for growth and development.

Heterotrophic organisms: Species that are dependent on organic matter for food.

Hierarchy: An organization of parts in which control from the top (generally of few parts), proceeds through a series of levels (ranks) to the bottom (generally of many parts) of hierarchy.

High Density Poly Ethylene (HDPE): A member of the polyethylene family of plastics and is used to make products such as milk bottles, pipes and shopping bags. HDPE may be colored or opaque.

High end exposure (dose) estimate: An estimate of exposure, or dose level received anyone in a defined population that is greater than the 90th percentile of all individuals in that population, but less that the exposure at the highest percentile in that population. A high end risk descriptor in an estimate of the risk level for such individuals. Note that risk is based on a combination of exposure and susceptibility to the stressor.

High intensity discharge: A generic term for mercury vapor, metal halide, and high pressure sodium lamps and fixtures.

High level liquid waste: An aqueous solution or suspension containing mixed fission products derive from the recycling of nuclear fuel rods that have been removed from reactors. The waste normally contains more than 100 micro curies per milliliter of fission products.

High Level Waste (HLW): Very radioactive material requiring perpetual isolation; waste that include untreated spend fuel from a nuclear power plant, the residue from the chemical processing of spent fuel, and much of the waste from nuclear weapons production.

High level waste repository: A place where intensively radioactive wastes can be buried and remain unexposed to ground water and earthquakes for tens of thousands of years.

High-Density polyethylene: A material used to make plastic bottles and other products that products toxic fumes when burned.

Highest dose tested: The highest dose of a chemical or substance tested in a study.

High – Level nuclear waste facility: Plant designed to handle disposal of used nuclear fuel, high-level radioactive waste, and plutonium waste.

High – Level Radioactive Waste (HLRW): Waste generated in core fuel of a nuclear reactor, found at nuclear reactors or by nuclear fuel reprocessing; is a serious threat to anyone who comes near the waste without shielding.

High – Line jumpers: Pipes or hoses connected to fire hydrants and laid on top of the ground to provide emergency water service for an isolated portion of a distribution system.

Highly- toxic: A chemical in any of the following categories: A chemical with a median lethal dose (LD_{50}) of 50 milligrams or less per kilogram of body weight when administered orally to albino rats between 200 and 300 grams each. A chemical with a median lethal dose (LD_{50}) of 200 milligrams or less per kilogram of body weight when administered by continuous contact for 24 hours (or less if death occurs within 24 hours) with the bare skin of albino rabbits weighing between 2 and 3 kilograms each. A chemical that has a median lethal concentration (LC_{50}) in air of 200 parts per million by volume or less of gas or 2 milligrams per liter or less of mist, fume, or dust when administered by continuous inhalation for 1 hour (or less if death occurs within 1 hour) to albino rats weighing between 200 and 300 grams each.

High-Risk community: A community located within the vicinity of numerous sites of facilities or other potential sources of environmental exposure/health hazards which may result in high levels of exposure to contaminants or pollutants.

High – to –Low – Dose extrapolation: The process of prediction of low exposure risk to humans and animals from the measured high to -- exposure – high risk data involving laboratory animals.

Holding pond: A pond or reservoir, usually made of earth, built to store polluted runoff.

Holding time: The maximum amount of time a sample may be stored before analysis.

Hollow stem auger drilling: Conventional drilling method that uses augurs to penetrate the soil. As the augers are rotated, soil cutting are conveyed to the ground surface via augur spirals. DP tools can be used inside the hollow augers.

Homeostasis: The property of either an open system or a closed system, especially a living organism, that regulates its internal environment so as to maintain a stable, constant condition.

Homeowner water system: Any water system which supplies piped water to a single residence.

Hood capture efficiency: Ratio of the emissions captured by a hood and directed into a control or disposal device, expressed as a percent of all emissions.

Hot spot: An informal expression designating a specific area as being contaminated with radioactive substance, having a relatively high concentration of air pollutants, or experiencing an abnormal disease or death rate.

Household Hazardous Wastes (HHW): Hazardous products used and disposed of by residential as opposed to industrial consumers. Includes paints, stains, varnishes, solvents, pesticides, and other materials or products containing volatile chemicals that can catch fire, react or explode, or that are corrosive or toxic. An important component of the hazardous – waste problem in the United States.

Household waste (Domestic Waste): Discards produced in a residence, commonly municipal solid waste. Solid waste, composed of garbage and rubbish, which normally originates in a private home or apartment house. Domestic waste may contain a significant amount of toxic or hazardous waste.

HRS (Hazard Ranking System): The HRS is the primary screening tool used by EPA to assess the risks posed to human health or the environment by abandoned or uncontrolled hazardous waste sites. Under the HRS, sites are assigned scores on the basis of the toxicity of hazardous substances that are present and the potential that those substances will spread through the air, surface, water, or groundwater, taking into account such factors as the proximity of the substance to nearby populations. Scores are used in determining which sites should be placed on the National Priorities List.

Human ecology: An area of inquiry concerned with the relationship of humans and their activities with the physical environment.

Human equivalent dose: A dose which, when administered to humans, produces an effect equal to that produced by a dose in animals.

Human exemptionalism paradigm: A set of assumptions underlying the belief that humans transcend the environment in which they live. Specifically, given their unique cultural heritage, humans are different from all other species on earth, social and cultural facts determine human society, the larger environment is irrelevant, and all problems are soluble by human ingenuity and technology.

Human exposure evaluation: Describing the nature and size of the population exposed to a substance and the magnitude and duration of their exposure.

Human health risk: The likelihood that a given exposure or series of exposures may have damaged or will damage the health of individuals.

Human – equivalent (H-e): The approximate human daily energy requirement of 12, 500 Kg. or its approximate energy generating capacity at basal metabolic rate which is equivalent to about 80 watts. A 100 watt light bulb therefore runs at 1.25 H-e.

Humus: Decomposed organic material in which the identity of the biologic source can not be recognized. A dark brown of black, soft, spongy residue of organic matter that remains after the bulk of dead leaves, wood or other organic matter has decomposed. Humus does oxidize, but relatively slowly. It is extremely valuable in enhancing the physical and chemical properties of soil.

HWRIC: Hazardous Waste Research and Information Center.

HWTC: Hazardous Waste Treatment Council.

Hydraulic conductivity: The rate at which water can move through a permeable medium. (*i.e.*, the coefficient of permeability)

Hydraulic gradient: The measure describing the rate at which water can move through a permeable medium.

Hydrocarbons: Chemicals made up of carbon and hydrogen that are found in raw materials such as petroleum, coal, and natural gas, and derived products such as plastics.

Hydroelectric power: Electricity created by movement of water. The electrical power generate using the power of falling water.

Hydrogen sulfide (H_2S): Gas emitted during organic decomposition. Also a by-product of oil refining and burning. Smells like rotten eggs and, in heavy concentration it causes illness or death.

Hydrogeology: The geology of ground water, with particular emphasis on the chemistry and movement of water.

Hydrologic cycle (water cycle): The natural cycle of water – movement or exchange of water between the atmosphere and earth.

Hydrology: The science dealing with the properties, distribution, and circulation of water.

Hydrolysis: The decomposition of organic compound by interaction with water.

Hydronic: A ventilation system using heated or cooled water pumped through a building. Hydronics is the name for the use of water as the heat – transfer medium in heating and cooling systems.

Hydrophilic: Having a strong affinity for water. Materials having large molecules that absorb and retain water, causing them to swell and frequently to gel.

Hydrophobic: Having a strong aversion for water.

Hydropneumatic: A water system, usually small, in which a water pump is automatically controlled by the pressure in a compressed air tank.

Hydrosphere: The water portion of the planet, this would include water found in the sea, streams, lakes and other water bodies, the soil, groundwater, and in the air, approximately 75 per cent of the earth's surface is covered by water.

Hygroscopic: Hygroscopic material is one which attracts moisture from the atmosphere. If not protected from contact with the atmosphere (by being stored under vacuum or under a dry gas); some hygroscopic materials will eventually attract so much water that they will form solutions; such materials are normally termed deliquescent. Sodium hydroxide is one of the most common hygroscopic materials.

Hypergolic: Self-igniting upon contact of its components without a spark or external aid.

Hypertonic: Solution contains a higher concentration of electrolytes than that found in body cells. If such a solution is allowed to enter the blood stream, the osmotic pressure difference between the blood and the cells will cause water to flow out of the cells, which will then shrink. This may cause serious harm, or even be fatal. Consequently, it is essential when blood transfusions are given, or blood replacement products are used, that the electrolyte concentration in the material to be given to a patient matches that of the body.

Hypersensitivity diseases: Diseases characterized by allergic responses to pollutants; diseases most clearly associated with indoor air quality are asthma an pneumonic hypersensitivity.

Hypocalcemia: Calcium deficiency of the blood.

Hypolimnion: The hypolimnion is the dense, bottom layer of water in a thermally – stratified lake. It is the layer that lies below the thermocline. Typically the hypolimnion is the coldest layer of a lake in summer, and the warmest layer during winter.

Hypotonic solution: is one in which the concentration of electrolyte is below that in cells.

Hypoxia: is a condition defined by a low supply of oxygen, especially applied to body cells.

Hypoxic waters: Waters with dissolved oxygen concentrations of less than 2 ppm, the level generally accepted as the minimum required for most marine life to survive and reproduce. In many cases hypoxic waters do not have enough oxygen to support fish and other aquatic animals. Hypoxia can be caused by eutrophication.

I

Ice Ages: (glacial age) denotes a geological period of long – term reduction in the temperature of the Earth's surface and atmosphere, resulting in an expansion of continental ice sheets, polar ice sheets and alpine glaciers. Within a long-term ice age, individual pulses of extra cold climate are termed "glaciations". Glaciologically, ice age implies the presence of extensive ice sheets in the northern and southern hemispheres; by this definition we are still in an ice age (because the Greenland and Antarctic ice sheets still exist). The Little Ice Age (LIA) was a period of cooling occurring after a warmer era known as the Medieval Warm Period or Medieval Climate Optimum Periods when ice sheets moved from the polar cap and covered areas of North America, Europe, and Asia.

IDLH: Immediately Dangerous to Life and Health: The maximum level to which a healthy individual can be exposed to a chemical for 30 minutes and escape without suffering irreversible health effects or impairing symptoms. Used as a "level of concern".

Ignitable waste: A substance or mixture that meets any of several definitions 1. A liquid with a flash point of 60 ºC or less, 2. A non-liquid that can catch fire through friction, absorption or moisture, or spontaneous chemical change.

Ignitable: Capable of burning or causing a fire.

Ignition temperature: Lowest temperature at which a combustible material will catch fire in air and will continue to burn independently of the source of heat when heated.

IM240: A high-tech, transient dynamometer automobile emissions test that takes up to 240 seconds.

IMDG: is an abbreviation for International Maritime Dangerous Goods.

Inhofe cone: A clear, cone-shaped container used to measure the volume of settleable solids in a specific volume of water.

Immigration: Population migration into another area.

Imminent hazard: One that would likely result in unreasonable adverse effects on humans or the environment or risk unreasonable hazard to an endangered species during the time required for a pesticide registration cancellation proceeding.

Imminent threat: A high probability that exposure is occurring.

Immiscibility: The inability of two or more substances or liquids to readily dissolve into one another, such as soil and water.

Immiserization: An economic decline among poor people, making it difficult for them to meet their basic need for food, health, and shelter.

Imperialism: A system in which one country uses the resources of a less powerful country for its own benefit, typically involving economic and political control.

Impermeable: Not easily penetrated. The property of a material or soil that does not allow, or allows only with great difficulty, the movement or passage or water.

Impervious: Material that does not allow another substance to pass through or penetrate it.

Imports: Municipal solid waste and recyclable that have been transported to a state or locality for processing or final disposition (but that did not originate in the state or locality).

Impoundment: A body of water or sludge confined by a dam, dike, floodgate, or other barrier.

Immune response is the reaction of the body to foreign substances. This response might be the production of antibodies, cell-mediated immunity, immunological tolerance, or hypersensitivity (*i.e.*, the development of symptoms of allergy).

Immunotoxic chemical: is one which is potentially harmful to the immune system.

In situ: In its original place; unmoved unexcavated; remaining at the site or in the subsurface.

In vitro: Testing or action outside an organism (*e.g.*, inside a test tube or culture dish.)

In vivo: Testing or action inside an organism.

Incident command post: A facility located a safe distance from an emergency site, where the incident commander, key staff, and technical representatives can make decisions and deploy emergency manpower and equipment.

Incineration: Burning of solid, liquid, gaseous material under controlled conditions to destroy waste at high temperatures; *e.g.*, burning sludge to remove the water and reduce the remaining residues to a safe, non-burnable ash that can be disposed of safely on land, in some waters, or in under locations.

Incineration at sea: Disposal of waste by burning at sea on specially-designed incinerator ships.

Incinerator: A furnace for burning waste under controlled conditions.

Inclinometers: These are options for to monitor subsurface lateral movement of the waste.

Income: A gain or benefit expressed in money or in goods and services for a specified time period.

Incompatible: Materials which could cause dangerous reactions from direct contact with one another.

Incompatible waste: A waste unsuitable for mixing with another waste or material because it may react to form a hazard. If mixed, they may chemically react to form hazardous products or excessive thermal discharge. Also, waste that may damage or corrode a container or containment structure, lodging to a release.

Indemnification: In the pesticide program, legal requirement that EPA pay certain end-users, dealers, and distributors for the cost of stock on hand at the time a pesticide registration is suspended.

Index of sustainable economic welfare: An alternative measure for assessing the strength of an economy and human well-being. Unlike gross national product and similar indicators, it includes measures of environmental degradation.

Indicator: 1. In biology, any biological entity of processes, or community whose characteristics so the presence of specific environmental conditions. 2. In chemistry, a substance that shows a visible change, usually of colour, at a desired point in a chemical reaction. 3. A device that indicates the result of a measurement; *e.g.*, a pressure gauge or a moveable scale.

Indicator organisms: An organism, the presence or absence of which indicates certain conditions. For example, the presence of *Escherichia coli* indicates that water is contaminated with fecal wastes and that pathogens may be present, the absence of *E. coli* indicates that the water is free of pathogens.

Indirect discharge: Introduction of pollutants from a non-domestic source into a publicly owned waste-treatment system. Indirect dischargers can be commercial or industrial facilities shoes wastes enter local sewers.

Indirect hot air: A system designed to reheat an exhaust gas that has been cooled by passage through a pollution control device. Ambient air is heated with steam then added to the flue gas to give the exhaust buoyant density.

Indirect solar energy: Energy derived from sources that are at least one step removed from the energy of sunlight; this includes most traditional energy sources. Wood is a form of indirect solar energy; trees capture the sun's energy through photosynthesis, producing carbon compounds that release energy when burned. These carbon compounds can, over millions of years, become coal or oil. Hydroelectricity, wind, all biomass energy, and clean thermal energy conversion are examples of indirect solar energy. See direct solar energy.

Indirect source: Any facility or building, property, road or parking area that attracts motor vehicle traffic and indirectly, causes pollution.

Indoor air pollution: The presence of excessive levels of air contaminants inside a home or building from sources such as cigarette smoking, fuel combination for heating or cooking, certain wall boards, carpets, insulation, air handling ducts and the geological characteristics of the area. Emissions are more likely to accumulate in structures having limited air exchange with the outside. Many air pollutants typically have higher concentrations indoors than outdoors.

Indoor air: The breathable air inside a habitable structure or conveyance.

Indoor climate: Temperature, humidity, lighting, air flow and noise levels in a habitable structure or conveyance. Indoor climate can affect indoor air pollution.

Industrial agriculture: A form of modern farming that refers to the industrialized production of livestock, poultry, fish, and crops.

Industrial economy: A production system with relative importance of manufacturing lessens and that of services, information, and research grows, based on machines to produce things of value.

Industrial process waste: Residues produced during manufacturing operations.

Industrial revolution: A period in the late 18th and early 19th centuries when major changes in agriculture, manufacturing, and transportation had a profound effect on socioeconomic and cultural conditions.

Industrial sludge: Semi-liquid residue or slurry remaining from treatment of industrial waste and wastewater.

Industrial source reduction: Practice that reduce the amount of any hazardous substance, pollutant, or contaminant entering any waste stream or otherwise released into the environment. Also reduces the threat to public health and the environment associated with such releases. Term includes equipment or technology modifications, substitution of raw materials, and improvements in housekeeping, maintenance, training or inventory control.

Industrial waste: Unwanted material from an industrial operation; may be liquid, sludge, solid, or hazardous waste.

Indust ation: A stage in societal development when resources ed from agriculture to manufacturing.

lient: Anything other than the active ingredient in a

In t; not having active properties. Pesticide components s solvent, carriers, dispersants, and surfactants that are tive against target pests. Not all inert ingredients are uous.

parator: A device that uses centrifugal force to separate e particles.

mortality rate: The number of infants deaths (children between the ages of 0-12 months) per 1,000 live births. The probable lethal concentration of the undiluted product to 50% of the test animals (LC_{50}) is greater than 200 milligrams per liter by volume when inhaled continuously for one hour or less.

Infection: Condition of disease in animals, humans, or plants from invasion by pathogenic organisms such as bacteria, protozoa, viruses, fungi, or animal parasites. Infection can spread by contact with diseased organisms, airborne microbes, and animal carriers such as mosquitoes.

Infectious agent: Any organism, such as a pathogenic virus, parasite, or bacterium, that is capable of invading body tissues, multiplying, and causing disease.

Infectious waste: Hazardous waste capable of causing infections in humans, includes contaminated animal waste; human blood and blood products; isolation waste, pathological waste; and discarded sharps (needles, scalpels or broken medical instruments).

Infiltration: 1. The penetration of water through the ground surface into subsurface soil or the penetration of water from the soil into sewer or other pipes through defective joints, connections, or manhole walls. 2. The technique of applying large volumes of waste water to land to penetrate the surface and percolate through the underlying soil.

Infiltration gallery: A sub-surface ground water collection system, typically shallow in depth, constructed with open-jointed or perforated pipes that discharge collected water into a water tight chamber from which the water is pumped to treatment

facilities and into the distribution system. Usually lo
to streams or ponds.

Infiltration rate: The quantity of water that can enter the specified time interval.

Inflammable: Capable of being easily set on fire and continu burn, especially violently.

Inflammation: Series of reactions produced in tissue by an irrita injury, or infection. Characterized by swelling and redne caused by an influx of blood and fluids.

Inflow: Entry of extraneous rain water into a sewer system from sources other than infiltration, such as basement drains, manholes, storm drains, and street washing.

Influent: Water, wastewater, or other liquid flowing into a reservoir, basin, or treatment plant.

Informal economy: Business outside of the recognized sector of a country's economy.

Ingestion: Taking in of a substance through the mouth.

Inhalable particles: All dust capable of entering the human respiratory tract.

Inhalation toxicity: Ratings corresponding to the following definitions are derived from the test methods and categories of toxicity described in 16 CFR 1500.3. (see non toxic, toxic and highly toxic)

Inhalation: Breathing in of a substance in the form of a gas, vapor, fume, mist, or dust.

Inhibitor: Chemical which is added to another substance to prevent an unwanted chemical change from occurring.

Initial compliance period (Water): The first full three-year compliance period which begins at least 18 months after promulgation.

Initial study: A preliminary analysis of the potential environment impacts of a proposed project. Where a lead agency finds that the project may cause a significant effect on the environment, and ER must prepared.

Injection well: A well into which fluids are injected for purposes such as waste disposal.

Injection zone: A geological formation receiving fluids through a well.

In – Line filtration: Pre-treatment method in which chemicals are missed by the flowing water; commonly used in pressure filtration installations. Eliminates need for flocculation and sedimentation.

Innovative technologies: New or inventive methods to treat effectively hazardous waste and reduce risks to human health and the environment.

Innovative treatment technologies: Technologies whose routine use is inhibited by lack of data on performance and cost.

Inoculum: 1. Bacteria or fungi injected into compost to start biological action. 2. A medium containing organisms, usually bacteria or a virus, that is introduced into cultures or living organisms.

Inorganic: Not coming from animal or vegetable sources, as in inorganic fertilizers. Describing compounds that either contain no carbon or contain only carbon bound to elements other than hydrogen.

Inorganic chemicals: Chemical substances of mineral origin, not of basically carbon structure.

Inorganic materials: Compounds derived from other than vegetable or animal sources; generally do not contain carbon atoms.

Insecticide: A pesticide compound specifically used to kill or prevent the growth of insects.

***In-situ* flushing:** Introduction of large volumes of water, at times supplemented with cleaning compounds, into soil or ground water to flush hazardous contaminants from a site.

***In-situ* oxidation:** Technology that oxidizes contaminants dissolved in ground water, converting them into insoluble compounds.

***In-situ* stripping:** Treatment system that removes or " strips" volatile organic compounds from contaminated ground or surface water by forcing an air stream through the water and causing the compounds to evaporate.

***In-situ* vitrification:** Technology that treats contaminated soil in place at extremely high temperatures, at or more than 3000 degrees fahrenheit.

Insoluble material: It is one which is incapable of dissolving to any significant extent in a specified solvent.

Inspection and Maintenance (I/M): 1. Activities to ensure that vehicles' emission controls work properly . 2. Also applies to wastewater treatment plants and other anti-pollution facilities and processes.

Institutional waste: Waste generated at institutions such as schools, libraries, hospitals, and other organizations.

In -stream use: The use of freshwater where it occurs, usually within a river or stream *e.g.*, hydro-electric power generation, navigation, water quality improvement, fish propagation, recreation.

Integrated exposure assessment: Cumulative or summation (over time) of the magnitude of exposure to a toxic chemical in all media.

Integrated Pest Management (IPM): A pest control strategy that uses an array of complementary methods: natural predators and parasites, pest-resistant varieties, cultural practices, biological controls, various physical techniques, and the strategic use of pesticides.

Integrated product life-cycle management: Management of all phases of goods and services to be environmentally friendly and sustainable.

Integrated waste management: Using a variety of practices to handle municipal solid waste; can include source reduction, recycling, incineration, and landfilling.

Intensive agriculture: A system to maximize output of land through use of chemicals and machinery.

Interceptor sewers: Large sewer lines that, in a combined system, control the flow of sewage to the treatment plant. In a storm, they allow some of the sewage to flow directly into a receiving stream, thus keeping it from overflowing onto the streets. Also used in separate systems to collect the flows from main and trunk sewers and carry them to treatment points.

Interface: The common boundary between two substance such as a water and a solid, water and a gas, or two liquids such as water and oil.

Interfacial tension: The strength of the film separating to immiscible fluids (*e.g.*, oil and water) measured in dynes per centimeter.

Intergenerational equity: A norm that cells for considering the interests of future generations when dealing with natural resources and a healthy environment.

Internal Water Footprint: The water embodied in goods products within a country (although these may be subsequently exploded) cf. external water footprint.

International Biological Programme (IBP): A worldwide study of biological productivity of terrestrial, fresh water and marine communities, conservation, and human adaptability carried out from 1964 to 1974 by the nongovernmental International Council of Scientific Unions, with participation by our 40 countries.

International regime: An international agreement between nations.

International Union for Conservation of Nature and Natural Resources (IUCN): A nongovernmental organization founded in 1948 and concerned with threats to the quality of the natural environment, especially wilderness airs and endangered species.

Interstate carrier water supply: A source of water for drinking and sanitary use on planes, buses, trains, and ships operating in more than one state. These sources are federally regulated.

Interstate waters: Means all surface waters of the state that cross or form a part of the border between states.

Interstate waters: Means all surface waters of the state that are not interstate waters.

Interstitial monitoring: The continuous surveillance of the space between the walls of an underground storage tank.

Intrinsic value: The value of something that is independent of its utility.

Inversion: A layer of warm air that prevents the rise of cooling air and traps pollutants beneath it; can cause an air pollution episode.

Ion exchange treatment: A common water-softening method often found on a large scale at water purification plants that remove some organics and radium by adding calcium oxide or calcium hydroxide to increase the pH to a level where the metals will precipitate out.

Ion: An electrically charged atom or group of atoms.

Ionizing radiation: Radiation that can strip electrons from atoms; *e.g.*, alpha, beta, and gamma radiation.

IOSH: Institute for Occupational Safety and Health.

IPAT: The model holding that environmental degradation (I) is an interactive function of human population size (P), affluence (A), and technology, (T).

IRIS: EPA's Integrated Risk Information System, an electronic data base containing the Agency's latest descriptive and quantitative regulatory information on chemical constituents.

Irradiation: Exposure to radiation of wavelengths shorter than those of visible light (gamma, x-ray, or ultra-violet), for medical purposes, to sterilize milk or other foodstuffs, or to induce polymerization of monomers.

Irradiated food: Food subject to brief radioactivity, usually gamma rays, to kill insects, bacteria, and mold, and to permit storage without refrigeration.

Irreversible effect: Effect characterized by the inability of the body to partially or fully repair injury caused by a toxic agent.

Irrigation efficiency: The amount of water stored in the crop root zone compared to the amount of irrigation water applied.

Irrigation return flow: Surface and subsurface water which leaves the field following application of irrigation water.

Irrigation: Artificial watering of crops. Applying water or wastewater to land areas to supply the water and nutrients needs of plants.

Irritant: It is a chemical which may cause reversible inflammation of the skin, eyes, or respiratory system on contact. A substance that can cause irritation. Effects may be acute from a single high level exposure, or chronic from repeated low-level exposures to such compounds as chlorine,nitrogen dioxide, and nitric acid.

ISO 14001: International Organisation for Strandardisation (ISO) 14001 standard was first published in 1996 specifying the reqirements for an environmental management system in organization (companies and institutions) with the goal of minimizing harmful effects on the environment and the goal of continual improvement of environmental performance.

Isomers: Compounds that have same molecular weight and atomic composition but differ in molecular structure.

Isotope: A variation of an element that has the same atomic number of protons but a different weight because of the number of neutrons. Various isotopes of the same element may have different radioactive behaviors, some are highly unstable.

Isotropy: The condition in which the hydraulic or other properties of an aquifer are the same in all directions.

J

Jar test: A laboratory procedure that simulates a water treatment plant's coagulation/flocculation units with differing chemical doses, mix speeds, and settling times to estimate the minimum or ideal coagulant dose required to achieve certain water quality goals.

Jaundice: Yellowish discoloration of tissue, whites of the eyes, and bodily fluids with bile pigment caused by any of several pathological conditions that interrupt the liver's normal production and discharge of bile.

Joint and several liability: Under CERCLA, this legal concept relates to the liability for Superfund site cleanup and other costs on the part of more than one potentially responsible party (*i.e.*, if there were several owners or users of a site that became contaminated over the years, they could all be considered potentially liable for cleaning up the site.).

Joint United Nations Programme on HIV/AIDS: "UNAIDS, the Joint United Nations Programme on HIV/AIDS, bring together the efforts and resources of ten UN system organizations to the global AIDS response. Cosponsors include UNHCR, UNICEF, WFP, UNDP, UNFPA, UNODC, ILO, UNESCO, WHO and the

World Bank, Based in Geneva, the UNAIDS secretariat works on the ground the more than 75 countries world wide."

Joule: International unit of energy, equivalent to one watt per second.

J-curve: The term J-curve is used in several fields to refer to a variety of unrelated J-shaped diagrams where a curve initially falls, but then rises to higher than the starting point. In economics, the 'J curve' refers to the trend of a country's tread balance following a devaluation. Exponential growth of population is indicated by a J- shaped growth curve.

K

Karst: A geologic formation of irregular limestone deposits with sinks, underground streams, and caverns.

Kerbside collection: Collection of household recyclable materials (separated or co-mingled) that are left at the kerbiside for collection by local council services.

Ketosis: Condition marked by excessive production or accumulation of ketone bodies in the body caused by disturbed carbohydrate metabolism.

Keystone species: A species that has a disproportionate effect on its environment relative to its abundance, affecting many other organisms in an ecosystem and help in determine the types and numbers of various others species in a community.

Kinetic energy: Energy possessed by a moving object or water body.

Kinetic rate coefficient: A number that describes the rate at which a water constituent such as a biochemical oxygen demand or dissolved oxygen rises or falls, or at which an air pollutant reacts.

Kyoto protocol: An international agreement adopted in December 1997 in Kyoto, Japan. The Protocol sets binding emission targets for developed countries that would reduce their emissions on average 5.2 per cent below 1990 levels.

L

Label: Any written, printed, or graphic sign or symbol displayed on or affixed to containers of hazardous chemicals. That contains identity of the material, appropriate hazard warnings, name and address of the chemical manufacturer, importer, or other responsible party.

Laboratory animal studies: Investigations using animals as surrogates for humans.

Lacrimation: Secretion and discharge of tears.

Lagoon: 1. A shallow pond where sunlight, bacterial action, and oxygen work to purify wastewater; also used for storage of wastewater or spent nuclear fuel rods. 2. Shallow body of water, often separated from the sea by coral reefs or sandbars.

Land application: Discharge of wastewater onto the ground for treatment or reuse. (see; irrigation)

Land ban: Phasing out of land disposal of most untreated hazardous wastes, as mandated by the 1984 RCRA amendments.

Land disposal restrictions: Rules developed by the USEPA that require hazardous waste to be treated before disposal in a landfill to destroy or immobilize dangerous substances that have the potential to migrate into soil and groundwater.

Land disposal restrictions: Rules that require hazardous wastes to be treated before disposal on land to destroy or immobilize hazardous constituents that might migrate into soil and ground water.

Land ethic: A philosophy developed by Aldo Leopold around the idea that land is more than a commodity and deserves to be treated with respect.

Land farming (of Waste): A disposal process in which hazardous waste deposited on or in the soil is degraded naturally by microbes.

Land fill: A site where municipal, industrial or chemical wastes are disposed of by burying in the ground or placing than on the ground and covering than with earth, is the oldest form of waste treatment. Sanitary landfills are disposal sites for non-hazardous solid wastes spread in layers, compacted to the smallest practical volume, and covered by material applied at the end of each operating day. Secure chemical landfills are disposal sites for hazardous waste, selected and designed to minimize the chance of release of hazardous substances into the environment.

Landfill gas: The gas emissions from biodegrading waste in landfill, including CO_2, CH_4 and small amounts of nitrogen, oxygen with traces of toluene, benzene and vinyl chloride.

Landfill prohibition: The banning of a certain material or product type from disposal to landfills. Occurs occasionally, for example, where a preferable waste management option is available.

Landscape: The traits, patterns, and structure of a specific geographic area, including its biological composition, its physical environment, and its anthropogenic or social patterns. An area where interacting ecosystems are grouped and repeated in similar form.

Landscape characterization: Documentation of the traits and patterns of the essential elements of the landscape.

Landscape ecology: The study of the distribution patterns of communities and ecosystems, the ecological processes that affect those patterns, and changes in pattern and process over time.

Landscape indicator: A measurement of the landscape, calculated from mapped or remotely sensed data, used to describe spatial patterns of land use and land cover across a geographic area. Landscape indicators may be useful as measures of certain kinds of environmental degradation such as forest fragmentation.

Langelier Index (LI): An index reflecting the equilibrium pH of a water with respect to calcium and alkalinity; used in stabilizing water to control both corrosion and scale deposition.

Large water system: A water system that services more than 50,000 customers.

Laser induced fluorescence: A method for measuring the relative amount of soil and/or groundwater with an in situ sensor.

Latency: Refers to the period of time it takes for a disease to manifest itself within the human body *i.e.*, time that elapses between exposure and the first manifestations of disease or illness.

Latent heat transfer: The removal or addition of heat when a substances changes its state. In the environment, this always refers to the release of heat from water upon condensations and the absorption of heat by water upon evaporation.

Lateral sewers: Pipes that run under city streets and receive the sewage from homes and businesses, as opposed to domestic feeders and main trunk lines.

LC 50/Lethal concentration 50: Median lethal concentration, the chemical concentration required to kill 50 percent of test subjects.

LCLO Lethal Concentration Low: The lowest concentration of a substance in air reported to have caused death in humans or animals. The reported concentrations amy be entered for periods of exposure that are less than 24 hr (acute) or greater than 24 hr (subacute and chronic).

LD 50/Lethal Dose 50: The single dose of a substance that causes the death of 50% of an animal population from exposure to the substance by any route other than inhalation. The lower the LD 50, the more toxic the compound, is usually expressed as milligrams or grams of material per kilogram of animal weight (me/kg or g/kg). The animal species and means of administering the dose (oral, intravenous, etc.) should also be stated.

LDLO: Lethal Dose Low: The lowest dose of a substance introduced by any route, other than inhalation, reported to have caused death in humans or animals.

Leachate: The mixture of water and materials that are leaching. Water that collects contaminants as it trickles through wastes, pesticides or fertilizers. Leaching may occur in farming areas, feedlots, and landfills, and may result in hazardous substances entering surface water, ground water, or soil.

Leachate collection system: A system that gathers leachate and pumps it to the surface for treatment.

Leaching: The process by which soluble constituents are dissolved and filtered through the soil or sludge deposits or other materials by a percolating fluid.

Lead (Pb): A heavy metal that is hazardous to health if breathed or swallowed. Its use in gasoline, paints, and plumbing compounds has been sharply restricted or eliminated by federal laws and regulations.

Lead – acid battery: Any battery that consists of lead and sulfuric acid, has a capacity of six volts or more, and is used as a power source, such as a car battery.

Lesion: Abnormal change, injury or damage to tissue/organ.

Less developed country: A country with a low per capita income and low to moderate industrialization. LDCs include nearly 150 nations and 80 per cent of the world's population.

Leukemia: Progressive, malignant disease of the blood-forming organs.

Level of Concern (LOC): The concentration in air of an extremely hazardous substance above which there may be serious immediate health effects to anyone exposed to it for short periods.

Life Cycle (of a product): All stages of a product's development, from raw materials, manufacturing through to consumption and ultimate disposal.

Life Cycle Assessment: ('LCA', also known as life cycle analysis, ecobalance, cradle-to-grave-analysis, well-to-wheel analysis, and dust-to-dust energy cost) is the assessment of the environmental impact of a given product or service throughout its lifespan.

Life expectancy: The number of years that an average person can expect to live.

Life support systems: According to the World Conservation Union (IUCN), refer to the biophysical processes "that sustain the productivity, adaptability and capacity for renewal of lands, waters, and/or the biosphere as a whole."

Lifetime exposure: Total amount of exposure to a substance that a human would receive in a lifetime (usually assumed to be 70 years).

Lift: In a sanitary landfill, a compacted layer of solid waste and the top layer of cover material.

Light Non-Aqueous Phase Liquid (LNAPL): A non- aqueous phase liquid with a specific gravity less than 1.0. Because the specific gravity of water is 1.0, most LNAPLs float on top of the water table. Most common petroleum hydrocarbon fuels and lrbricating oils are LNAPLs.

Light-Emitting diode: A long –lasting illumination technology used for exit signs which requires very little power

Lilac water: Recycled water that is unsuitable for drinking.

Limestone scrubbing: Use of a limestone and water solution to remove gaseous stack-pipe sulfur before it reaches the atmosphere.

Limit of Detection (LOD): The minimum concentration of a substance being analyzed text that has a 99 per cent probability of being identified .

Limited aquatic life: A designated use, means the surface water is capable of supporting only a limited community of aquatic life. This subcategory includes surface waters that support aquatic species selectively adapted to take advantage of naturally occurring rapid environmental changes, ephemeral or intermittent water, high turbidity, fluctuating temperature, low dissolved oxygen content or unique chemical characteristics.

Limited degradation: An environmental policy permitting some degradation of natural system but terminating at a level well beneath an established health standard.

Limited water soluble substances: Water pollution chemicals that are soluble in water at less than one milligram of substance per litre of water.

Limiting factor: A condition whose absence or excessive concentration, is incompatible with the needs or tolerance of a species or population and which may have a negative influence on their ability to thrive.

Limnology: The study of the physical, chemical, hydrological, and biological aspects of fresh water bodies.

Linear metabolism: Direct conversion of resources into waste that are often sent directly to landfill

Liner: A relatively impermeable barrier designed it keep leachate inside a landfill. Liner materials include plastic and dense clay. An insert or sleeve for sewer pipes to prevent leakage or infiltration.

Lipid pneumonia: Chronic condition caused by the aspiration of oily substances into the lungs.

Lipid solubility: The maximum concentration of a chemical that will dissolve in fatty substances. Lipid soluble substances are insoluble in water. They selectively disperse through the environment via uptake in living tissue.

Liquefaction: Changing a solid into a liquid.

Liquid injection incinerator: Commonly used system that relies on high pressure to prepare liquid wastes for incineration by breaking them up into tiny droplets to allow easier combustion.

List: Shorthand term for EPA list of violating facilities or firms debarred from obtaining government contracts because they violated certain sections of the Clean Air or Clean Water Acts. The list is maintained by the Office of Enforcement and Compliance Monitoring.

Listed waste: Wastes listed as hazardous under RCRA but which have not been subjected to the toxic Characteristics Listing Process because the dangers they present are considered self-evident.

Lithology: Mineralogy, grain size, texture, and other physical properties of granular soil, sediment, or rock.

Litter: 1. In an ecosystem, the natural cover of dead leaves, twigs, and other dead plant material. This natural litter is subject to rapid decomposition and recycling in the ecosystem, whereas human litter, such as bottles, cans and plastics is not. 2. Waste materials carelessly or intentionally discarded in an inappropriate place. Litter is waste out of place.

Little Ice Age: A cold period in Asia, Europe, and North America (A.D.1550 to A.D. 1850) Its end coincided with the beginning of the industrial revolution and an increase of greenhouse emissions.

Littoral zone: That portion of a body of fresh water extending from the shoreline lakeward to the limit of occupancy of rooted plants. A strip of land along the shoreline between the high and low water levels.

Livestock watering; Means the use of thermal effluent released into a supply of stream by human or natural source.

Load allocation: The portion of the pollution load of a stream attributable to human non point source is the waste load allocation.

Loading: 1. The air concentration of a pollution in a gas duct before entry into an air pollution control device. 2. Synonym for the pollution load of a stream.

Local ventilation: Drawing off and replacement of contaminated air directly from its source.

Locally unwanted land use: A land use with adverse environmental consequences, *e.g.*, nuclear power plants, hazardous waste facilities, airports, and refineries.

Love canal: The location of a landfill site near Niagara Falls that was used by Hooker Chemicals and Plastics Company as a dumping site for thousands of tons of chemical wastes during the 1950s

Low density polyethylene (LOPE): Plastic material used for both rigid containers and plastic film applications.

Low emissivity (low-E) windows: New window technology that lowers the amount of energy loss through windows by inhibiting the transmission of radiant heat while still allowing sufficient light to pass through.

Low NOx burners: One of several combustion technologies used to reduce emissions of Nitrogen Oxides (NOx).

Lower detection limit: The smallest signal above background noise an instrument can reliably detect.

Lower Explosive (Flammable) Limit (LEL): Lowest concentration (lowest percentage of the substance in air) that will produce a flash of fire when an ignition source (heat, electric arc, or flame) is present, in other words it is the concentration of a compound in air below which the mixture will not catch on fire.

Lowest acceptable daily dose: The largest quantity of a chemical that will not cause a toxic effect, as determined by animal studies.

Lowest achievable emission rate (LAER): The emission rate required for an air pollutant emitted by a new or modified source in an area that does not meet the national ambient standard for that pollutant (non treatment area). The Clean Air Act requires that the emission rate be set as the most stringent emission limitation in use for that pollutant and type of facility.

Lowest Observed Adverse Effect Level (LOAEL): The lowest level of a stressor that causes statistically and biologically significant differences in test samples as compared to other samples subjected to no stressor.

Low-Level Radioactive Waste (LLRW): Wasteless hazardous than most of those associated with a nuclear reactor; generated by hospitals, research laboratories, and certain industries. The Department of Energy, Nuclear Regulatory Commission, and EPA share responsibilities for managing them. (See: high-level radioactive wastes.)

LULU: Locally Unwanted Land Use.

LUST: Leaking Underground Storage Tank.

M

MACT (Maximum achievable control technology): The best technologies available for reducing the output of especially toxic industrial pollutants.

Mad Hatteis disease: A nervous system disorder resulting from chronic exposure to inorganic mercury compounds. Affected responses, tingling sensation of the skin, mental disturbances of allocations. The term derive from the occupational exposure syndrome caused by over exposure to the mercuric nitrate used to soften fuel in the making of felts hats.

Magma: Molten rock that sometimes forms beneath the surface of the Earth (or any other terrestrial planet) that often collects in a magma chamber and is ejected by volcanoes.

Magnetic separation: Use of magnets to separate ferrous materials from mixed municipal waste stream.

Major modification: This term is used to define modifications of major stationary sources of emissions with respect to Prevention of Significant Deterioration and New Source Review under the clean Air Act.

Major stationary sources: Term used to determine the applicability of Prevention of Significant Deterioration and new source regulations. In a non-attainment area, any stationary pollutant source with potential to emit more than 100 tons per year is considered a major stationary source. In PSD areas the cutoff level may be either 100 or 250 tons, depending upon the source.

Majors: Larger publicly owned treatment works (POTWs) with flows equal to at least one million gallons per day (mgd) or servicing a population equivalent to 10,000 persons; certain other POTWs having significant water quality impacts.

Malnutrition: Insufficient nutrition.

Malthusian growth: A population explosion followed a population clash; also called irrespective growth.

Malthusianism: The belief that population outstrips society's ability to provide for its members. It is based on ideas developed by Thomas Malthus (1766-1834) in an essay on the Principles of Population.

Management plan: Under the Asbestos Hazard Emergency Response Act (AHERA), a document that each Local Education Agency is required to prepare, describing all activities planned and undertaken by a school to comply with AHERA regulations, including building inspections to identify asbestos-containing materials, response actions, and operations and maintenance programs to minimize the risk of exposure.

Managerial controls: Methods of non-point source pollution control based on decisions about managing agricultural wastes or application times or rates for agrochemicals.

Mandatory recycling: Programs which by law require consumers to separate trash so that some or all recyclable materials are recovered for recycling rather than going to landfills.

Manifest: A one-page form used by haulers transporting waste that lists EPA identification numbers, type and quantity of waste, the generator it originated from, the transporter that shipped it, and the storage or disposal facility to which it is being shipped. It includes copies for all participants in the shipping process.

Manifest system: Tracking of hazardous waste from "cradle-to-grave" (generation through disposal) with accompanying documents known as manifests.

Man-made (Anthropogenic) beta particle and photon emitters: All radionuclides emitting beta particles and/or photons listed in Maximum Permissible Body Burdens and Maximum Permissible Concentrations of Radionuclides in Air and water for Occupational Exposure.

Manual separation: Hand sorting of recyclable or compostable materials in water.

Manufacturer's formulation: A list of substances or component parts as described by the maker of a coating, pesticide or other product containing chemicals or other substances.

Manufacturing use product: Any product intended (labeled) for formulation or repackaging into other pesticide products.

Manure: Organic matter used as fertilizer in agriculture.

Margin of exposure (MOE): The ratio of the no-observed adverse – effect – level to the estimated exposure dose.

Margin of safety: Maximum amount of the exposure producing no measurable effect in animals (or studied humans) divided by the actual amount of human exposure in a population.

Marine sanitation device: Any equipment or process installed on board a vessel to receive, retain, treat, or discharge sewage.

Market benefits: Benefits of a climate policy that can be measured in terms of avoided market impacts such as changes in resources productivity (*e.g.*, lower agricultural yields, scarcer water resources) and damages to human-built environment (*e.g.*, coastal flooding due to sea-level rise).

Marsh: A type of wetland that does not accumulate appreciable peat deposits and is dominated by herbaceous vegetarian. Marshes may be either fresh or salt water, tidal or non-tidal.

Marxism: The theory of Karl Marx that all societies go through a series of economics of stages as the production system evolves and increases human control over the environment. The final stage of all societies is socialism.

Mass burn: Indicator of municipal solid waste without prior material recovery or processing but often accompanied by an energy recovery systems such as steam and electricity production.

Material category: In the asbestos program, broad classification of materials into thermal surfacing insulation, surfacing material, and miscellaneous material.

Material flow: The cycling of materials, which is driven by the flow of energy.

Material Recycling Facility (MRF): A processing plant in which regionalized cycling is called out. Recyclable municipal solid waste, usually presorted, is prepared in bulk for the recycling market.

Material Safety Data Sheets (MSDS): Documents containing information on the reactivity and toxicity of chemicals and that must accompany the shipping, storage and handling of our 600 chemicals.

Material type: Classification of suspect material by its specific use or application; *e.g.*, pipe insulation, fireproofing, and floor tile.

Materials Recovery Facility (MRF): A facility processes residentially collected mixed recyclables into new products available for market.

Maternal mortality: Death during child birth.

Maximally (or Most) exposed individual: The person with the highest exposure in a given population.

Maximum acceptable toxic concentration: For a given ecological effects test, the range (or geometric mean) between the No Observable Adverse Effect Level and the Lowest Observable Adverse Effects Level.

Maximum Available Control Technology (MACT): The emission standard for sources of air pollution requiring the maximum reduction of hazardous emissions, taking cost and feasibility into account. Under the Clean Air Act Amendments of 1990, the MACT must not be less than the average emission level achieved by controls on the best performing 12 per cent of existing sources, by category of industrial and utility sources.

Maximum Contaminant Level (MCL): A drinking water standard; the highest allowable concentration of a pollutants at drinking water source.

Maximum Contaminant Level Goal (MCLG): Under the Sale Drinking Water Act., a non-enforceable concentration of a drinking water contaminant, set at the level at which no known or anticipated adverse effects on human health occur and which allows an adequate safety margin. The MCLG is usually the starting point for determining the regulated Maximum Contaminant Level.

Maximum contaminant level: The maximum permissible level of a contaminant level water delivered to any user of a public system. MCLs are enforceable standards.

Maximum exposure range: Estimate of exposure or dose level received by an individual in a defined population that is greater than the 98th percentile dose for all individual in that population, but less than the exposure level received by the person receiving the highest exposure level.

Maximum residue level: Comparable to a U.S. tolerance level, the Maximum Residue Level the enforceable limit on food pesticide in some countries. Levels are set by the Codex Alimentarius Commission, a United Nations agency managed and funded jointly by the World Health Organization and the Food and Agriculture Organization.

Maximum tolerated dose: The maximum dose that an animal species can tolerated for a major portion of its lifetime without significant impairment or toxic effect other than carcinogenicity.

MCL: Maximum Contaminant Levels are established for public drinking water supplies under the Safe Drinking Water Act.

Measure of Effect/Measurement endpoint: A measurable characteristic of ecological entity that can be related to an assessment endpoint; *e.g.*, a laboratory test for eight species meeting certain requirements may serve as a measure of effect for an assessment endpoint, such as survival of fish, aquatic, invertebrate or algal species under acute exposure.

Measure of exposure: A measurable characteristic of a stressor (such as the specific amount of mercury in a body of water) used to help quantify the exposure of an ecological entity or individual organism.

Mechanical aeration: Use of mechanical energy to inject air into water to cause a waste stream to absorb oxygen.

Mechanical separation: Using mechanical means to separate waste into various components.

Mechanical turbulence: Random irregularities of fluid motion in air caused by buildings or other non-thermal processes.

Media: Specific environments – air, water, soil – which are the subject of regulatory concern and activities.

Median lethal concentration LC_{50}: The concentration of a chemical substance that causes death in 50% of the test organisms within a specified period of after exposure.

Medical surveillance: A periodic comprehensive review of a worker's health status; acceptable elements of such surveillance program are listed in the Occupational Safety and Health Administration standards for asbestos.

Medium-size water system: A water system that serves 3,300 to 50,000 customers.

Melt down: The event of a nuclear reactions getting out of control or losing its cooling water so that it melts from its own production of heat. The nested reaction would continue to produce heat and would melt its way out of its containment vessel. It would cause a violent eruption of steam that would spread radioactive materials over a wide area.

Melting point: Temperature at which a solid substance changes to a liquid state. For mixtures, a melting range may be given.

Meniscus: The curved top of a column of liquid in a small tube.

Mercury (Hg): Heavy metal that can accumulate in the environment and is highly toxic if breathed or swallowed.

Mesotrophic: Reservoirs and lakes which contain moderate quantities of nutrients and are moderately productive in terms of aquatic animal and plant life.

Metabolism: Chemical and physical processes whereby the body functions. All of the chemical and physical activities that sustain an organism. Metabolism involves the breakdown of organic compounds to create energy, which animals use to grow, repair tissues, provide heat, and engage in physical activity. Plant metabolism involves the creation as well as utilization of organic compounds. In some organisms, metabolism includes metamorphosis.

Metabolites: Any substances produced by biological processes, such as those from pesticides.

Metalimnion: The middle layer of a thermally stratified lake or reservoir. In this layer there is a rapid decrease in temperature with depth. Also called thermocline.

Metastasis: Transmission of a disease from one part of the body to another.

Methane CH_4: A colorless, nonpoisonous, flammable gas created by anaerobic decomposition or organic compounds. Methane is the primary constituent of natural gas used in the home. CH_4 is one of the green house gases.

Methane hydrate: Small bubbles or individual molecules of Methane (natural gas) trapped in a crystalline matrix of frozen water.

Methanol: An alcohol that can be used as an alternative fuel or as a gasoline additive. It is less volatile than gasoline; when blended with gasoline it lowers the carbon monoxide emission but increases hydrocarbon emission. Used as pure fuel, its emissions are less ozone-forming than those from gasoline. Poisonous to humans and animals if ingested.

Methemoglobinemia: Presence of methemoglobin in the bloodstream caused by the reaction of materials with the hemoglobin in red blood cells that reduces their oxygen-carrying capacity.

Method 18: An EPA test method which uses gas chromatographic techniques to measure the concentration of volatile organic compounds in a gas stream.

Method 24: An EPA reference method to determine density, water content and total volatile content (water and VOC) of coatings.

Method 25: An EPA reference method to determine the VOC concentration is a gas stream.

Methoxychlor: Pesticide that causes adverse health effects in domestic water supplies and is toxic to freshwater and marine aquatic life.

Methyl Orange Alkalinity: A measure of the total alkalinity in a water sample in which the color of methyl orange reflects the change in level.

MH: Natural formations consisting of mounds of ice like material on or just below the sea floor containing large amounts of CH_4 trapped within a lattice of ice like crystals.

Microbe: A microorganism, an organism that can only be seen with the aid of a microscope, especially a bacterium producing disease or fermentation.

Microbial growth: The amplification or multiplication of microorganisms such as bacteria, algae, diatoms, plankton, and fungi.

Microbial pesticide: A microorganism that is used to kill a pest, but is of minimum toxicity to humans.

Microclimate: Localized climate conditions within an urban area or neighborhood. The climate around a tree or shrub or a stand or trees.

Micro environmental method: A method for sequentially assessing exposure for a series of micro environmental that can be approximated by constant concentrations of a stressor. Well-defined surroundings such as the home, office, or kitchen that can be treated as uniform in terms of stressor concentration.

Microfiltration: A process for purifying water in which the water is forced under very high pressure through a membrane that is fine enough to filter out ions and molecules in solution, microfiltration used by small desalination plants to filter salt from sea water. Also called reverse osmosis.

Microgram (μg): One millionth (10^{-6}) of a gram.

Micrometer (μm): One – millionth (10^{-6}) of a meter; often referred to as a micron.

Midnight dumping: The illicit dumping of materials, particularly hazardous wastes, frequently under the cover of darkness. The deliberate disposal of hazardous waste at a site other than a permitted disposal facility, often taking place at night.

Midoceanic ridges: Mountain ranges on the ocean floor where magma wells up through cracks and creates new crust.

Migration: The movement of population from one area to another; The seasonal movement of animal populations from one region to another and back again, often covering tremendous distances;

The daily vertical movements of some freshwater plankton several meters downward during the day and back to the surface at night; Movement of individuals into or out of local populations (includes emigration and immigration).

Millimeter (mm): 1/1000 of a meter.

Million – Gallons per day (MGD): A measure of water flow.

Minimata disease: Insidious neurological disorders resulting from the consumption of fish and shellfish contaminated by organomercurials. The most famous case involved a substantial number of people near Minimata Bay, Japan. Fish and shellfish in the bay accumulate Alkyl Mercury discharged into the bay by in industrial facility. Consumption of contaminated sew food resulted in several hundred poisonings during 1950's through 1970's ranging from mild neurological disorders to paralysis and deaths.

Minimization: A comprehensive program to minimize or eliminate wastes, usually applied to wastes at their point of origin.

Mining of an aquifer: Withdrawal over a period of time of ground water that exceeds the rate of recharge of the aquifer.

Mining waste: Residues resulting from the extraction of raw materials from the earth.

Minors: Publicly owned treatment works with flows less than 1 million gallons per day.

Miscellaneous materials: Interior building materials on structural components, such as floor or ceiling tiles.

Miscible liquids: Two or more liquids that can be mixed and will remain mixed under normal conditions.

Missed detection: The situation that occurs when a test indicates that a tank is "tight" when in fact it is leaking.

Mist: Liquid Aerosol; small droplets suspended in air. Liquid particles 40 to 500 microns in diameter that have formed by condensation of vapour in air. By comparison, fog particles are smaller than 40 micrometers (um).

Mitigation: Measures taken to reduce adverse impacts on the environment.

Mixed glass: Recovered container glass not sorted into categories (*e.g.* color grade).

Mixed liquor: A mixture of activated sludge and water containing organic matter undergoing activated sludge treatment in an aeration tank.

Mixed metals: Recovered metals not sorted into categories such as aluminum, tin or steel cans or ferrous or non-ferrous metals.

Mixed municipal waste: Solid waste that has not been sorted into specific categories (such as plastic, glass yard trimmings etc.)

Mixed paper: Recovered paper not sorted into categories such as old magazines, old newspapers, old corrugated boxes, etc.

Mixed plastic: Recovered plastic unsorted by category.

Mixture: Heterogeneous association of materials that cannot be represented by a chemical formula and that does not undergo chemical change as a result of interaction among the mixed materials.

ml: Milliliter 1/1000 of a liter. A metric unit of capacity, for all practical purposes equal to 1 cubic centimeter. One cubic inch is about 16 ml.

mm Hg: A measure of pressure in millimeters of a mercury column above a reservoir, or difference of level in a U-tube.

Mobile incinerator system: Hazardous waste incinerators that can be transported from one site to another.

Mobile source: any non-stationary source of air pollution such as cars, trucks, motorcycles, buses, airplanes, and locomotives.

Model plant: A hypothetical plant design used for developing economic, environmental, and energy impact analyses as support for regulations or regulatory guidelines.

Modern environmentalism: A fusion of conservation of natural resources and preservation of nature with concerns about pollution, environmental heath and social justice.

Modernization: The process accompanying industrialization.

Modified bin method: Way of calculating the required heating or cooling for a building based on determining how much energy the system would use if outdoor temperatures were within a certain temperature interval and then multiplying the energy use by the time the temperature interval typically occurs.

Moisture content: 1. The amount of water lost from soil upon drying to a constant weight, expressed as the weight per unit of dry soil or as the volume of water per unit bulk volume of the soil. For a fully saturated medium, moisture content indicates the porosity.2. Water equivalent of snow on the ground; an indicator of snowmelt flood potential.

Mole: Quantity of a chemical substance that has a weight in a unit numerically equal to the molecular weight.

Molecular weight: The sum of the atomic weights of the atoms in a molecule.

Molecule: The smallest division of a compound that still retains or exhibits all the properties of the substance.

Molten salt reactor: A thermal treatment unit that rapidly heats waste in a heat-conducting fluid bath of carbonate salt.

Monitoring well: Wells drilled to "monitor" ground water quality and movement. These wells do not supply water for drinking or industrial use. Samples from monitoring wells are analyzed to determine whether contaminants are present.Comparing water levels in monitoring wells shows the direction of groundwater flow.

Monitoring: Periodic or continuous surveillance or testing to determine the level of compliance with statutory requirements and/or pollutant levels in various media or in humans, plants, and animals.

Monkey-wrenching: See ecotage.

Monoclonal antibodies (Also called MABs and MCAs): 1. Man-made (anthropogenic) clones of a molecule, produced in quantity for medical or research purposes.2. Molecules of living organisms that selectively find and attach to other molecules to which their structure conforms exactly. This could also apply to equivalent activity by chemical molecules.

Monoculture forestry: Intensive planting of a single species; an efficient wood production approach, but one that encourages pests and disease infestation and conflicts with wild life habitat or recreation uses.

Monoculture: An agricultural practice in which a single species of plant is cultivated in an area. It requires large amounts of fertilizers and pesticides.

Monomictic: Lakes and reservoirs which are relatively deep, do not freeze over during winter, and undergo a single stratification and mixing cycle during the year (usually in the fall)

Montreal protocol: An agreement made in 1987 by a large group of nations to cut back the production of chloroflorocarbons by 505 by the year 2000 in order to protect the ozone shield. A 1990 amendment called for the complete phase out of these chemicals by 2000 in developed nations and by 2010 in less developed nations. Under the Protocol, various research groups continue to assess the ozone layer. The Multilateral Fund provides resources to developing nations to promote the transition to ozone-safe technologies.

Morbidity: Illness or disease. The rate of incidence of disease in a population.

More developed country: A country with high per capita income and a high level of industrialization, *e.g.*, the United States, Canada, Japan, Australia, New Zealand, and the western European countries.

Mortality: Death rate, Deaths in a population. The incidence of death in a population. Death rate in a population, *e.g.*, number of deaths per thousand people per year.

Most probable number: An estimate of microbial density per unit volume of water sample, based on probability theory.

Mppcf: Millions of particles per cubic foot of air, based on impinger samples counted by light-field techniques (OSHA):

MRF: A building design led to facilitate the separation, sorting and preparing of recyclable items derived from households so that items are acceptable to companies that convert the resources back into commercial items.

MSDS: Material Safety Data Sheet is a document required by the U.S., Occupational Health and Safety Administration to be distributed to employees. The MSDS summarizes information about chemicals to which employees may be exposed.

MSW: Solid waste, including garbage to trash, that originates in households, commercial establishments or construction, demolition sites. Non-hazardous sludge from municipal sewage treatment plants or non-hazardous industrial waste can also be placed in this category of solid waste.

Muck soils: Earth made from decaying plant materials.

Mudballs: Round material that forms in filters and gradually increases in size when not removed by backwashing.

Mulch: A layer of material (wood chips, straw, leaves, etc.) placed around plants to hold moisture, prevent weed growth, and enrich or sterilize the soil.

Multigeneration study: In this study several generations of animals (usually at least three generations) are exposed to a toxic chemical to test its effect. Exposure is typically continuous, rather repeated dose.

Multi-Media approach: Joint approach to several environment media, such as air, water, and land.

Multinational corporation: A large corporation that owns property and produces and sells products in a large number of countries.

Multiple chemical sensitivity: A diagnostic label for people who suffer multi-system illnesses as a result of contact with, or proximity to, a variety of airborne agents and other substances.

Multiple use: Use of land for more than one purpose; *e.g.*, grazing of livestock, watershed and wildlife protection, recreation, and timber production. Also applies to use of bodies of water for recreational purposes, fishing, and water supply.

Multistage remote sensing: A stategy for landscape characterization that involves gathering and analyzing information at several geographic scales, ranging from generalized levels of detail at the national level through high levels of detail at the local scale.

Municipal discharge: Discharge of effluent from waste water treatment plants which receive waste water from households, commercial establishments, and industries in the coastal drainage basin. Combined sewer/separate storm overflows are included in this category.

Municipal sewage: Wastes (mostly liquid) originating from a community ; may be composed of domestic wastewaters and/ or industrial discharges.

Municipal sludge: Semi-liquid residue remaining from the treatment of municipal water and wastewater.

Mutagen: An agent that causes a permanent genetic change in a cell other than that which occurs during normal growth.

Mutagenicity is the capacity of a chemical or physical agent to cause such permanent genetic changes in a living cell.

N

Nanogram (ng): is one billionth of gram; *i.e.*, 10^{-9} g.

Narcotic: is a chemical agent which depresses the central nervous system and may lead to unconsciousness, and –potentially – death through respiratory failure. Many chemicals have narcotic properties, and exposure to them may give an initial feeling of euphoria before the depression.

National Ambient Air Quality Standards (NAAQS): The allowable levels of ambient air pollutants set by EPA regulation. Regulations established by the Clean Air Act prescribing levels of pollution that may not be exceeded during a specified time in a defined area. The ambient air concentration standards set for particulate matter, SO_2, NO_2, O_3, CO and Pb to protect human health (primary standards) or welfare (secondary standards).

National Audubon society: An organization founded in 1905 to protect bird populations but now concerned with broader environmental issues.

National environmental performance partnership agreements: System that allows states to assume greater responsibility for environmental programs based on their relative ability to execute them.

National estuary program: A program established under the Clean Water Act Amendments of 1987 to develop and implement conservation and management plans for protecting estuaries and restoring and maintaining their chemical, physical, and biological integrity, as well as controlling point and non-point pollution sources.

National Pollution List (NPL): A list of hazardous waste disposal sites most in need of clean up, the list is updated annually by the USEPA, based primarily on how a site scores using the Hazard Ranking System, also called the super fund list.

National Wildlife Federation: An organization founded in 1936 as an environmental advocacy and educational group.

Natural: The existing air, water, land and energy resources from which all resources derive. Main function include resource production (such as fish, timber or cereals), waste assimilation (such as CO_2 absorption, sewage decomposition), and life support services (UV protection, biodiversity, water cleansing, climate stability). The environmental services that must be maintained so that human development can be sustainable.

Natural capital: Natural resources and ecological processes that are equivalent to financial capital.

Natural causes: Means those casual agents that would affect water quality and the effect is not caused by human activity but is due to naturally occurring conditions.

Natural chemical control: The use of one or more natural chemicals, such as hormones or pheromones, to control a pest.

Natural control methods: Any of many techniques of controlling a pest population without resulting to the use of synthetic organic or inorganic chemicals.

Natural enemies: All the predators and parasites that may feed on a given organism. Organisms used to control a specific pest through predation or parasitism.

Natural gas: Mixture of flammable gases found in the Earth's crust (often in association with petroleum). Natural gas occurs in natural underground deposits. Natural gas is valued as an efficient and clean – burning fuel. It contains essentially CH_4, but also small proportions of other gases. It also covers CH_4 recovered in coal mines.

Natural hazard: A natural event that damages the environment and humans, *e.g.*, floods, volcanoes, and earthquakes.

Natural pollutants: Pollutants created through natural processes, *e.g.*, ozone, dust, and gas of volcanic origin.

Natural Resources: Naturally occurring substances that are considered valuable in their relatively unmodified (natural) form.

Natural Resources Defense Council: An organization founded in 1970 to protect natural resources.

Natural sciences: Sciences that examine the physical environment, *e.g.*, astronomy, biology, chemistry, geology, physics, oceanography, and meteorology. Grow until they reach the carrying capacity of the environment.

Natural selection: The process by which favorable heritable traits become more common in successive generations of a population of reproducing organisms, and unfavorable heritable traits become less common.

Natural services: Functions performed free of charge by natural ecosystems, such as control of runoff and erosion, absorption of nutrients and assimilation of air pollutants.

Nausea: Tendency to vomit, a feeling of sickness at the stomach.

Navigable waters: Traditionally, waters sufficiently deep and wide for navigation by all, or specified vessels; are protected by certain provisions o the Clean Water Act.

NCP: National contingency Plan is the regulation which outlines the activities and procedures to followed in responding to releases of hazardous substances.

Necrosis: Death of plant or animal cells or tissues locally. In plants, necrosis can discolor stems or leaves or kill a plant entirely.

Neighborhood environment improvement plan: Plans developed by a local community including residents, special interest groups, local government, local industry and government agencies.

Nematocide: A chemical agent which is destructive to nematodes.

Neoplasm: New or abnormal tissue growth that is uncontrollable and progressive.

Nephelometric: Method of measuring turbidity in a water sample by passing light through the sample and measuring the amount of the light that is deflected.

Nephrotoxic: Poisonous to the kidney.

Net economic welfare: The gross national product adjusted by subtracting the "bads" such as pollution and by adding the value of beneficial, nonmarket activities such as leisure.

Net Primary Production (NPP): The total amount of energy produced and stored through photosynthesis in a specific community or ecological system (gross primary production), minus the quantity consumed during respiration by the photosynthetic organisms. NPP is expressed as oven dry weight (grams) of tissues/area or energy content of tissues (Kcal)/area. Also called apparent photosynthesis, or net assimilation, and often used interchangeably with net primary productivity (although that is technically the productivity per year).

Netting: A concept in which all emissions sources in the same area that owned or controlled by a single company are treated as one large source, thereby allowing flexibility in controlling individual sources in order to meet a single emissions standard. (See: bubble)

Neuritis: Inflammation of the nerves.

Neurotoxins: A substance that can damage or destroy nerve tissue. Botulism toxin represents a common example.

Neutralization: Decreasing the acidity or alkalinity of a substance by adding alkaline or acidic materials, respectively.

Neutralize: To render chemically harmless; to return the pH to the neutral level 7.

New environmental paradigm: A new set of assumptions about the nature of the relationship between humans and the larger environment. This paradigm holds that humans, despite their exceptional characteristics, represent one among many species on earth; human activities are determined not only by social and cultural factors but by the environment; and humans are dependent on a finite environment.

New forestry: Now part of the forest services management practice; a forestry management strategy that places priority on protecting the ecological health and diversity of forest rather than maximizing the harvest of logs.

New international economic order: A list of demands made by the Group of 77 nations in the 1970s regarding changes in the structure of North – South economic relations.

New social movements: Recent social movements whose origins, political tactics, and goals differ substantially from those of the traditional class and economic-based, social movements, *e.g.*, the environmental movement, the women's movement, and the gay rights movement.

New Source: Any stationary source built or modified after publication of final or proposed regulations that prescribe a given standard of performance.

New Source Performance Standards (NSPS): Uniform national EPA air emission and water effluent standards which limit the amount of pollution allowed from new sources or from modified existing sources.

New Source Review (NSR): A Clean Air Act requirement that State Implementation Plan must include a permit review that applies to the construction and operation of new and modified stationary sources in non attainment areas to ensure attainment of national ambient air quality standards.

Newly industrializing countries: Countries that have been industrialized only recently, *e.g.*, South Korea, Taiwan, and Singapore.

NGO: Non Governmental Organization – An nonprofit organization that is not affiliated with any government but is concerned with problems of global and local environment and development.

NIABY- Not in anybody's backyard: The idea that hazardous activities and substances should not be located in anybody's backyard.

NIMBY – "Not in my Backyard": The idea that hazardous activities and substances should not be located in my backyard. It is often the response of neighbors to the potential sitting of a waste disposal facility in their community. In government circles it's called the "NIMBY problem". Whether the proposal is for AIDS clinics, halfway houses for prison parolees or dumps for toxic and nuclear waste, it is usually met by the opposition of citizens groups who shout NIMBY – "not in my backyard!"

NIMTOF – Not in my term of office: The idea that hazardous activities and substances should not be located in the area during a politician's term of office.

NIOSH: National Institute of Occupational Safety and Health.

Nitrate: A compound containing nitrogen that can exist in the atmosphere or as a dissolved gas in water and which can have harmful effects on humans and animals. Nitrates in water can cause severe illness in infants and domestic animals. A plant nutrient and inorganic fertilizer, nitrate is found in septic systems, animal feed lots, agricultural fertilizers, manure, industrial waste waters, sanitary landfills, and garbage dumps.

Nitric acid (HNO_3): Secondary air pollutant. One of the acids in acid rain. Formed by reactions between nitrogen oxides and the winter vapor in the atmosphere.

Nitric Oxide (NO): A gas formed by combustion under high temperature and high pressure in an internal combustion engine. It is converted by sunlight and photochemical processes in ambient air to nitrogen oxide. NO is a precursor of ground – level ozone pollution, or smog.

Nitrification: The process whereby ammonia in wastewater is oxidized to nitrite and then to nitrate by bacterial or chemical reactions.

Nitrilotriacetic Acid (NTA): A compound now replacing phosphates in detergents.

Nitrite: 1. An intermediate in the process of nitrification. 2. Nitrous oxide salts used in food preservation.

Nitrogen Dioxide (NO_2): The result of nitric oxide combining with oxygen in the atmosphere; major component of photochemical smog.

Nitrogen oxides (NO): A group of nitrogen – oxygen compounds formed when some of the nitrogen gas in all combines with oxygen during high temperature combustion. Nitrogen oxides are a major category of air pollutants and along with hydrocarbons, are a primary factor in the production of Ozone and other photochemical oxidants that are the most harmful components of photochemical smog. Nitrogen oxides also contribute to acid precipitation. The major nitrogen oxides are nitric oxide (NO), nitrogen dioxide (NO_2) and nitrogen tetraoxide (N_2O_2).

Nitrogenous wastes: Animal or vegetable residues that contain significant amounts of nitrogen.

Nitrophenols: Synthetic organopesticides containing carbon, hydrogen, nitrogen, and oxygen.

Nitrous oxide: N_2O; is a colourless gas; introduced to the atmosphere by microbial metabolism, from biomass burning, fossil fuel burning, and the use of chemical fertilizers; used as anesthetic (laughing gas); is of concern because it is a greenhouse gas in the troposphere and it contributes to O_3 destruction in the stratosphere.

NOAEL (No Observable Adverse Effect Level): An exposure level at which there are no statistically or biologically significant increases in the frequency or severity of adverse effects between the exposed population and its appropriate control; some effects may be produced at this level, but they are not considered as adverse, or as precurors to adverse effects. In an experiment with several NOAELs, the regulatory focus is primarily on the highest one, leading to the common usage of the term NOAEL as the highest exposure without adverse effects.

NOPE (Not on our planet): The idea that hazardous activities and substances should not be located on earth.

No Till: Planting crops without prior seedbed preparation, into an existing cover crop, sod, or crop residues, and eliminating subsequent tillage operations.

No till agriculture: The farming practice in which weeds are killed with chemicals and seeds are planted and grown without resorting to plowing of cultivation. The practice is highly effective in reducing soil erosion and recently in carbon trading.

Noble Metal: Chemically inactive metal such as gold; does not corrode easily.

Noise Pollution (environmental noise): Displeasing human or machine created sound that disrupts the activity or happiness of human or animal life.

Non biodegradable: Describing a substance that is not consumed or broken down by biological organisms/by natural processes and so remains in its original form for long periods of time. Non-biodegradable substances include plastics, aluminum and many chemicals used in industry and agriculture. Particularly dangerous are human made non-biodegradable chemicals that are also toxic and tend to accumulate in organisms (*i.e.*, non biodegradable synthetic organic compounds).

Non-Community water system: A public water system that is not a community water system; *e.g.*, the water supply at camp site or national park.

Non-Compliance coal: Any coal that emits greater than 3.0 pounds of sulfur dioxide per million BTU when burned. Also known as high-sulfur coal.

Non-Contact cooling water: Water used for cooling which does not come into direct contact with any raw material, product, byproduct, or waste.

Non-Conventional: Any pollutant not statutorily listed or which is poorly understood by the scientific community.

Non-Degradation: An environmental policy which disallows any lowering of naturally occurring quality regardless of pre established health standards.

Non discharging treatment plant: A treatment plant that does not discharge treated wastewater into any stream or river. Most are pond systems that dispose of the total flow they receive by means of evaporation or percolation to groundwater, or facilities that dispose of their effluent by recycling or reuse (*e.g.*, spray irrigation or groundwater discharge).

Non-Ferrous metals: Nonmagnetic metals such as aluminum, lead, and copper, Products made all or in part from such metals include containers, packaging appliances, furniture, electronic equipment and aluminum foil.

Non-flammable: Incapable of being easily ignited or burning with extreme rapidity when lighted.

Non- ionizing electromagnetic radiation: Radiation that does not change the structure of atoms but does heat tissue and may cause harmful biological effects. Ex: Microwaves, and low-frequency electromagnetic fields from high-voltage transmission lines.

Non-Methane Hydrocarbon (NMHC): The sum of all hydrocarbon air pollutants. Except methane; significant precursors to ozone formation.

Non-Methane Organic Gases (NMOG): The sum of all organic air pollutants. Excluding methane; they account for aldehydes, ketones, alcohols, and other pollutants that are not hydrocarbons but are precursors of ozone.

Non-persistent: Said of chemicals that break down readily to harmless compounds as, for example natural organic compounds break down to CO_2 and H_2O.

Non-point source pollution: Means any source of pollutants not regulated as a point source that degrades the quality or adversely affects the biological, chemical or physical integrity of surface waters of the state. Scattered, diffuse sources of pollutants, such as runoff from farm fields, construction sites etc. Sources of pollution such as the general runoff of sediments, fertilizer, pesticides and other materials from farms and urban areas, opposed to specific points of discharge such as factories. Also called diffuse sources.

Non-potable: Water that is unsafe or unpalatable to drink because it contains pollutants, contaminants, minerals, or infective agents.

Nonrenewable resource: A resource that cannot be replenished. Renewable resources are those that are replenished through biogeochemical and physical cycles. By contrast, non-renewable resources do not replenish themselves, or, as in the case of fossil fuels, do so only at a very slow rate.

Non-Road emissions: Pollutants emitted by combustion engines on farm and construction equipment, gasoline-powered lawn and garden equipment, and powerboats and outboard motors:

Non-sparking tools: These are used when working with flammable liquids, as potential hazard arises because of the possibility that sparks produced by metal tools striking each other might ignite the liquid or its vapour. These are generally made from beryllium-copper or aluminium-bronze and significantly reduce the possibility of igniting dusts, gases, or flammable vapors. Although these tools may emit some sparks when striking metal, the sparks have a low heat content and are not likely to ignite most flammable liquids.

Non-transmissable disease: A disease that is not caused by organisms and cannot be transmitted through human contact, *e.g.*, cancer and heart disease.

No-observed Effect Level (NOEL): Exposure level at which there are no statistically or biological significant differences in the frequency or severity of any effect in the exposed or control populations .

No sphere: That part of the biosphere that is affected by human activities.

North atlantic free trade agreement: A treaty freeing trade restrictions between Canada, the U.S., and Mexico.

North or Northern: Industrialized countries located mainly in the Northern Hemisphere and referred to as the First World, including Canada, the U.S., western European countries, the former Soviet Union, Japan, Australia, and New Zealand. The term was coined by members of the Brandt Commission in 1980.

Notice of deficiency: An EPA request to a facility owner or operator requesting additional information before a preliminary decision on a permit application can be made.

Notice of intent to cancel: Notification sent to registrants when EPA decides to cancel registration of a product containing a pesticide.

Notice of intent to deny: Notification by EPA its preliminary intent to deny a permit application.

Notice of intent to suspend: Notification sent to a pesticide registrant when EPA decides to suspend product sale and distribution because of failure to submit requested date in a timely and/or acceptable manner, or because of imminent hazard.

NOx: A general formula for oxides of nitrogen (NO, NO_2). (See nitrogen oxides)

Noxious: Physically harmful to living organisms.

NSDWRs: National Secondary Drinking Water Regulations.

Nuclear energy: The energy produced through nuclear fission or nuclear fusion.

Nuclear reactors and support facilities: Uranium mills, commercial power reactors, fuel reprocessing plants, and uranium enrichment facilities.

Nuclear Winter: Prediction by some scientists that smoke, dust, debris and other materials rising from massive fires of a nuclear war could block sunlight for weeks or months, cooling the earth's surface and producing climate changes that could, for example, negatively affect world agricultural and weather patterns.

Nuisance material: is one which can cause transient irritation or discomfort, but which has no long-term or systemic effects.

Nuisance particulates: Dusts that do not produce significant organic disease or toxic effect from "reasonable" concentrations and exposures.

Nutrient: Any substance assimilated by living things that promotes growth.

Nutrient pollution: Contamination of water resources by excessive inputs of nutrients. The term is generally applied to nitrogen and phosphorus in wastewater, but is also applied to other essential and trace elements. In surface waters, excess algal production is a major concern.

O

Occlusion: is the trapping of hazardous material next to the skin. This keeps the material in contact with the skin for long periods of time, increasing the chance that dermal damage will occur.

Occupational Hygiene: is an applied science which is concerned with ensuring that standards of health in the workplace are maintained. It deals with the chemical, environmental and physical factors which may affect the health of workers. The practice of occupations hygiene is increasingly dominated by regulatory standards designed to protect the workplace environment.

Occupational Exposure Limit (OEL): A (generally legally - enforceable) limit on the amount or concentration of a chemical to which workers may be exposed.

Ocean discharge waiver: A variance from Clean Water Act requirements for discharges into marine waters.

Ocean dumping: Dumping any substance including hazardous wastes and other wastes in oceans.

Octane: A hydrocarbon containing eight carbon atoms (C_8H_{18}). Octane is a liquid and is found in petroleum; it exists as 18 different isomers. The term octane number is used informally for rating gasoline.

Odor threshold: Lowest concentration of a substance's vapor, in air or water which can be detected by smell. Also called as threshold odor.

Odor: Description of the smell of the substance.

Office paper: High grade papers such as copier paper, computer printout, and stationary almost entirely made of uncoated chemical pulp, although some ground wood is used. Such waste is also generated in homes, schools, and elsewhere.

Offsets: A concept whereby emissions from proposed new or modified stationary sources are balanced by reductions from existing sources to stabilize total emissions. (See: bubble, emissions trading, netting)

Off-Site facility: A hazardous waste treatment, storage or disposal area that is located away from the generating site.

Off stream use: Water withdrawn from surface or groundwater sources for use at another place.

Ogallala aquifer: An underground water source that stretches from Texas to South Dakota and is used for irrigation in the Great Plains. Depletion is thought to have profound implications for agriculture because of the large quantities of grain produced in the area.

Oil and gas waste: Gas and oil drilling muds, oil production brines, and other waste associated with exploration for, development and production of crude oil or natural gas.

Oil desulphurization: Widely used precombustion method for reducing sulfur dioxide emissions from oil-burning power plants. The oil is treated with hydrogen, which removes some of the sulfur by forming hydrogen sulfide gas.

Oil fingerprinting: A method that identifies sources of oil and allows spills to be traced to their source.

Oil spill: An accidental or intentional discharge of oil which reaches bodies of water. Can be controlled by chemical dispersion, combustion, mechanical containment, and/or adsorption. Spills from tanks and pipelines can also occur away from water bodies, contaminating the soil, getting into sewer system and threatening underground water sources.

Old growth forests: Forests dominated by matures trees 250 years or older in age exhibiting unique biological features and with little or no evidence of any disturbance such logging, ground clearing and building.

Olfactory: Relating to the sense of smell.

Oligotrophic lakes: Deep clear lakes with few nutrients, little organic matter and a high dissolved – oxygen level.

Omnivore: An animal that feeds in more than one trophic level, for example, potentially eating animal, plant, and fungal material as its primary food source. Humans are good examples.

Onboard controls: Devices placed on vehicles to capture gasoline vapor during refueling and route it to the engines when the vehicle is starting so that it can be efficiently burned.

One in one million cancer risk: One person out of a population of one million would likely develop cancer as a result of exposure to site contaminants.

One - hit model: A mathematical model based on the biological theory that a single "hit" of some minimum critical amount of a carcinogen at a cellular target such as DNA can start an irreversible series events leading to a tumor.

On – site facility: A hazardous waste treatment, storage or disposal area that is located on the generating site.

Opacity: The amount of light obscured by particulate pollution in the air; clear window glass has zero opacity, a brick wall is 100 per cent opaque. Opacity is an indicator of changes in performance of particulate control systems.

Opaque: Impervious to light rays.

Open burning: Uncontrolled fires in an open dump.

Open dump: An uncovered site used for disposal of waste without environmental controls.

Open-Pit mining: (opencast mining, open – cut mining) A method of extracting rock or minerals from the earth by their removal from an open pit or borrow.

Optimal corrosion control treatment: An erosion control treatment that minimizes the lead and copper concentrations at user's taps while also ensuring that the treatment does not cause the water system to violate any national primary drinking water regulations.

Oral: Used in or taken into the body through the mouth.

Oral toxicity: Ability of a pesticide to cause injury when ingested.

Organic: 1. Referring to or derived from living organisms like plant or animal matter originating from domestic or industrial sources, *e.g.*, grass clippings, tree prunings, food waste. 2. According to chemist, any compound which contains carbon.

Organic agriculture: A holistic production management system that avoids the use of synthetic fertilizers, pesticides and GM organisms, minimizes pollution of air, soil, and water, and optimizes the health and productivity of interdependent communities of plants, animals and people.

Organic agriculture: The practice of growing crops without chemical fertilizers and pesticides, but otherwise similar to alternative agriculture.

Organic beef: Beef from cattle raised without antibiotics, growth hormones, or synthetic chemicals.

Organic chemicals: Naturally occurring (animal, or plant-produced) or synthetic substances containing mainly carbon, hydrogen, nitrogen, and oxygen.

Organic compounds: Complex molecules organized around skeletons of carbon atoms arranged in rings of chains; include biomolecules; molecules synthesized by living organisms. All living matter is organic. The original definition of the team organic related to the source of chemical compounds: organic compounds where those carbon – containing compounds obtained from plant or animal sources. Whereas inorganic compounds were obtained from mineral sources we now know that compounds containing bonds between carbon atoms can be made in the laboratory and industrially.

Organic farming: Farming without the use of artificial fertilizers or pesticides.

Organic fertilizer: Organic matter added to the soil to increase production, *e.g.*, manure, plants plowed into the soil, and compost.

Organic gardening: Gardening that follows, in general principal, the philosophy or organic agriculture.

Organic materials: Compounds composed of carbon, hydrogen, and other elements with chain or ring structures.

Organic matter: Material derived from decaying organic molecules of natural organisms (the remains of plants and of animals). Organic matter is essential for healthy soil. It may be partially recognizable, as in rough compost of leaves on their way to becoming leaf mold; when fully broken down, organic matter in soil is called humus.

Organisms: Any unicellular or multicellular living body whose different components work together as a whole to carry out life processes. Animals, plants, fungi, and microbes are all organisms.

Organization for Economic Cooperation and Development (OECD) Guidelines: Testing guidelines prepared by the OECD of The United Nations. It also assists in preparation of protocol for studies of toxicology, environmental fate, etc.

Organization for economic cooperation and development: An organization founded in 1961 to further economic development and consisting of 13 European countries plus the U.S., Canada, Japan, Australia, and New Zealand.

Organization of Petroleum Exporting Countries: An organization created in 1960 and consisting of 13 countries that control 60 per cent of the world's oil reserves.

Organoleptic: Means the capability to produce a detectable sensory stimulus such as odor or taste.

Organophosphates: is the general name for esters of phosphoric acid. Many of the most important biochemicals are organophosphates, including DNA and RNA as well as many cofactors that are essential for life. Organophosphates are also the basis of many insecticides, herbicides, and nerve gases. Organophosphates are widely used as solvents, plasticizers. Organophosphates are widely employed both in natural and synthetic applications because of the ease with which organic groups can be linked together.

Organophyllic: A substance that easily combines with organic compounds.

Organotins: Chemical compound used in anti-foulant paints to protect the hulls of boats and ships, buoys, and pilings from marine organisms such as barnacles.

Original generation point: Where regulated medical or other material first becomes waste.

OSHA: is the (USA) Occupational Safety and Health Administration, which adopts and enforces health and safety standards.

Osmosis: The passage of a liquid from a weak solution to a more concentrated solution across a semipermeable membrane that allows passage of the solvent (water) but not dissolved solids.

Other ferrous metals: Recyclable metals from strapping, furniture, and,metal found in tyres and consumer electronics but does not include metals found in construction material or cars, locomotives, and ships.

Other glass: Recyclable glass from furniture, appliances, and consumer electronics. Does not include glass from transportation products (cars trucks or shipping containers) and construction or demolition debris.

Other paper: For Recyclable paper from books, third-class mail, commercial printing, paper towels, plates and cups; and other nonpackaging paper such as posters, photographic papers, cards and games, milk cartons, folding boxes, bags, wrapping paper, and paperboard. Does not include wrapping paper or shipping cartons.

Other plastics: Recyclable plastic from appliances, eating utensils, plates containers, toys, and various kinds of equipment. Does not include heavy – duty plastics such as yielding materials.

Other woods: Recyclable wood from furniture, consumer electronics cabinets, and other non packaging wood products. Does not include lumber and tree stumps recovered from construction and demolition activities, and industrial process waste such as shavings and sawdust.

Outdoor air supply: Air brought into a building from outside.

Outfall: The place where effluent is discharged into receiving waters.

Over burden: Soil, gravel or other surface material overlying bedrock; Rock and soil cleared away before mining.

Over consumption: A situation in which some people consume resources at levels beyond their needs, often at the expense of those who cannot meet their basic needs.

Over draft: The pumping of water from a groundwater basin or aquifer in excess of the supply flowing into the basin; result in a depletion or "mining" of the groundwater in the basin.

Overexposure: is the exposure to a hazardous chemical or environment beyond the allowable or recommended or regulatory limits.

Over fire air: Air forced into the top of an incinerator or boiler to fan the flames.

Overflow rate: One of the guidelines for design of the settling tanks and clarifiers in a treatment plant; used by plant operators to determine if tanks and clarifiers are over or under – used.

Overgrazing: Grazing by animals on vegetation at a rate greater than the ability of vegetation to regenerate itself. Causes a gradual change in the vegetation to less desirable species, or causing increased erosion so that portions of the pasture or the rangeland become bare. Although overgrazing may provide short term economic gain, in the long term it produces land that supports fewer animals or even destroys the land for grazing, greatly reducing the potential for economic gain.

Overland Flow: A land application technique that cleanses waste water by allowing it to flow over a sloped surface. As the water flows over the surface, contaminants are absorbed and the water is collected at the bottom of the slope for reuse.

Over nutrition: Excessive food consumption, especially red meat, fats, sugars, and processed foods. This is a problem in developed countries that contributes to high rates of non transmissible diseases, such as coronary heart disease and cancer.

Overpopulation: More organisms in a population than the existing resources can support.

Overshoot: Population growing beyond the carrying capacity of the environment; ecological deficit occurs when human consumption and waste production exceed the capacity of the Earth to create new resources and absorb waste. During overshoot, natural capital is being liquidated to support current use so the Earth's ability to support future life declines.

Oversized regulated medical waste: Medical waste that is too large for plastic bags or standard containers.

Overturn: One complete cycle of top to bottom mixing of previously stratified water masses. This phenomenon may occur in spring or fall, or after storms, and results in uniformity of chemical and physical properties of water at all depths.

Oxidation: Reaction in which a substance combines with oxygen provided by an oxidizer or oxidizing agent. The oxidation of one compound is always coupled with the reduction of another compound, and so the process is sometimes called oxidation-reduction. The chemical addition of oxygen to break down pollutants or organizac waste; *e.g.*, destruction of chemicals such as cyanides, phenols, and organic sulfur compounds in sewage lagoon:

Oxidation- Reduction potential: The electric potential required to transfer electrons from one compound or element (the oxidant) to another compound (the reductant); used as a qualitative measure of oxidation in water treatment systems.

Oxide pox: Dermatitis caused by contact with oxides under poor personal hygienic conditions.

Oxidizer: Substance that yields oxygen readily to simulate the combustion of organic matter.

Oxidizing agent: Any substance that is capable of removing electrons from another substance. An oxidizing agent carries out the process of oxidation and is itself reduced (*i.e.*, it gains electrons.) A variety of substances can act as oxidizing agents. Oxygen on its own is a reasonably strong oxidizing agent, but other materials, such as fluorine, metal nitrates, potassium permanganate, hydrogen peroxide, sodium hypochlorite (bleach), or sodium dichromate are very effective. Strong oxidizing agents are often very reactive chemicals, and, in contact with combustible material such as paper, sawdust, fabrics or powdered metals, may from unstable mixtures which constitute a risk of fire or explosion.

Oxygen sag: Oxygen decline down stream from a pollution source that introduces materials with high biological oxygen demands.

Oxygen defiant: Atmosphere is one in which the level of oxygen is below that of normal air, around 19.5%.

Oxygen sag curve: A graph of the measured concentrations of dissolved oxygen in water samples collected 1. Upstream from a significant point source of readily degradable organic material 2. From the area of the discharge 3. from some distance down stream from the discharge, plotted by sample location. The amount of dissolved oxygen is typically high upstream, diminishes at and just down stream from the discharge location (causing a sag in the line graph) and returns to the upstream levels at some distance down stream from the source of pollution.

Oxygenated fuels: Gasoline which has been blended with alcohols or ethers that contain oxygen in order to reduce carbon monoxide and other emissions.

Oxygenated solvent: An organic solvent containing as part of the molecular structure. Alcohols and ketones are oxygenated compounds often used as paint solvents.

Ozonation: Application of ozone to water for disinfection or for taste and odor control. The ozonator is the device that does this.

Ozone (O_3): A highly reactive molecule containing three oxygen, atoms, a dangerous pollutant in ambient air. In the stratosphere, however, ozone forms an ultraviolet absorbing shield that protects us from mutagenic radiation. In the lower atmosphere (troposphere), however, it forms from combustion gases and is a major air pollutant contributing to photochemical smog. Ozone has commercial uses as a bleaching agent and water purifier . In sufficiently high concentrations at ground level, the gas acts as an irritant to the eyes and respiratory tract.

Ozone hole: A thin break in the stratospheric ozone layer. It was first discovered in 1985 over the Antarctic (south pole) and was of the size of whole America. Designation of amount of such depletion as an "ozone hole" is made when the detected amount of depletion exceeds fifty percent. Seasonal ozone holes have been observed over both the Antarctic and Arctic regions, part of Canada, and the extreme northeastern United States.

Ozone layer: A layer ozone in the earth's upper atmosphere, at about 12 to 30 miles in altitude, where the intensity of short wavelength ultraviolet light from the sun (between 130 and 200 nanometers) is sufficiently high to convert normal diatomic oxygen (O_2) to **Ozone (O_3):** The ozone thus formed shields much of the earth by absorbing ultraviolet radiation at wavelengths that are mutagenic (about 250-270 nm).

Ozone layer depletion: The destruction of ozone molecules in the ozone layer of the stratosphere by chemical ractions with materials released by human activities. The main ozone consuming chemicals are the chlorofluro carbons (CFCS's) and the Halons, life times of 60 to 100 years. If the CFC's or halons migrate to the stratospheric ozone layer, the ultraviolet radiation is strong enough to break the molecules apart, releasing chlorine atoms (CFC's) or bromine atoms (Halons) which react with and destroy ozone.

P

Pacific rim: The 34 countries and 23 islands is and around the Pacific Ocean with an area of 70 million square miles.

Packaging: It is the science art and technology of enclosing or protecting products for distribution, storage, sale, and use. This refers to designing evaluating and producing suitable packages for different situations; components necessary to ensure minimum compliance with a program 's storage and shipment packaging requirements.

Packed bed scrubber: The principal of this type of scrubber is to remove contaminants from the gas stream by passing the stream through a packed structure which provides a large wetted surface area to induce intimate contact between the gas and the scrubbing liquor. The contaminant is absorbed into or reacted with the scrubbing liquor. Example: emission pass through alkaline water to neutralize hydrogen chloride gas.

Packed tower: A pollution control device that forces dirty air through a tower packed with crushed rock or wood chips while liquid is sprayed over the packing material . The pollutants in the air stream either dissolve or chemically react with the liquid.

Packer: is an expandable plug or balloon, used to create a temporary seal in a borehole, probe hole, well, or drive casing. It is made of rubber or non-reactive materials.

Palatable water: Water at a desirable temperature, that is free from objectionable tastes, odors, colors, and turbidity.

Palpitation: Irregular, rapid heartbeat.

PAN: Peroxynitrates. A group of compound present in photochemical smog. One of the photo-chemical oxidants produced by the reaction between nitrogen oxides, hydroactions and UV radiation in the atmosphere, causing irritation of eyes along with other photochemical oxidants.

Pandemic: A widespread epidemic throughout an area, nation or the world.

Paper: is thin material mainly used for writing upon, printing upon or packaging. It is produced by pressing together moist fibers, typically cellulose pulp derived from wood, rags or grasses, and drying them into flexible sheets. In the recycling business, refers to products and materials, include newspapers, magazines, office papers, corrugated containers, bags and some paperboard packaging that can be recycled into new paper products.

Paper Processor/Plastic processor: Intermediate facility where recovered paper or plastic products and material are sorted, decontaminated, and prepared for final recycling.

Para biosphere: Upper levels of atmosphere at the height of 60-80 km which are populated by casual organisms, bearing dormant bacterial and fungal spores only.

Parameter: A variable measurable property whose value is a determinant of the characteristics of a system; *e.g.*, temperature, pressure, and density are parameters of the atmosphere. Also refers to numerical quantity measuring some aspect of population like, mean is measure of central tendency.

Paraquat: Paraquat is non-selective, which means it kills a wide range of annual grasses and broad – leaved weeds and the tops of established perennial weeds. A standard herbicide used to kill various types of crops, including marijuana. Causes lung damage if smoke from the crop is inhaled.

Parasite: An organism that derives its nutrition by feeding on another living organism *i.e.*, derives nourishment at the expense of its host, usually without killing it.

Parathesia: Sensation of pricking, tinkling, or creeping on the skin that has no objective cause with no long term physical effect.

Parshall flume: Device used to measure the flow of water in a ditch, canal, lateral or open channel. It is named after its principal developer, the late Mr. Ralph L. Parshall

Participation rate: Portion of population participating in a recycling program.

Particle count: Results of a microscopic examination of treated water with a special "particle counter" that classifies suspended particles by number and size.

Particulates: Fine liquid or solid particle such as dust, smoke, mist, fumes, or smog, found in air or emissions. Very small solids suspended in water; they can vary in size, shape, density and electrical charge and can be gathered together by coagulation and flocculation.

Particulate loading: The mass of particulates per unit volume of air or water.

Partition coefficient: The partition coefficient is defined as the ratio of the equilibrium concentrations of a dissolved substance in a two-phase system consisting of two largely immiscible solvents. Measure of the sorption phenomenon, whereby a pesticide is divided between the soil and water phase.

Parts Per Billion (ppb): Units commonly used to express contamination ratios, as in establishing the maximum permissible amount of a contaminant in water, land, or air.

Parts Per Million (ppm): Units for measuring low concentration of a contaminants/gas in a medium. Parts of the gas or vapor in a million parts of air. Also used to indicate the concentration of a particular substance in a liquid or solid. Is equivalent to one microgram per gram or one milligram per litre.

Passive Smoking/Secondhand smoke: Inhalation of other's tobacco smoke.

Passive solar heating: Heating a building through the direct, indirect or isolated absorbtion of the sun's energy and convertion of it into usable heat (waste, air, thermal mass), cause air-movement for ventilating, or store heat for future use, with little use of other energy sources.

Passive treatment walls: Technology in which a chemical reaction takes place when contaminated ground water comes in contact with barriers such as limestone or a wall containing iron filings or with a catalyst contained in the porous media of the wall.

Pathogen: A casual agent of a disease such as microorganisms (*e.g.*, bacteria, viruses, or parasites) or an organism that produces disease in a host organism (humans, animals, and plants), disease being an alteration of one or more metabolic functions in response to the presence of the organism.

Pathway: The physical course a chemical or pollutant takes from its source to the exposed organism.

Pay – As – You – Throw/Unit – Based Pricing: Systems under which residents pay for municipal waste management and disposal services by weight or volume collected, not a fixed fee. Where this system is implemented, recyclable waste is usually collected free of charge or variable rate pricing.

Peak electricity demand: It is the greatest amount of electricity that need to be supplied to all its customers at any given time, in any one month. Since electricity cannot be effectively stored, electrical networks must instantaneously balance generation and load, *i.e.*, supply must always equal demand. Therefore, there is a need to build for the peak because sufficient generation capacity must meet maximum instantaneous demand whenever it happens. Meeting varying demands requires a mix of generation capacity including base-load and peak-load generation.

Percent saturation: The amount of a substance that is dissolved in a solution compared to the amount that could be dissolved in it at the same temperature.

Percent volatile: Percent volatile by volume is the percentage of a liquid or solid (by volume) that will evaporate at an ambient temperature of 70 degrees F (unless some other temperature each specified). Example: butane, gasoline, and paint thinner (mineral spirits) are 100 percent volatile; there individual evaporation rates vary, but intime, each will evaporate completely.

Perched water: Zone of unpressurized water held above the water table by impermeable rock or sediment.

Percolating water: Water that passes through rocks or soil under the source of gravity.

Percolation: The downward movement of the water through the soil due to force of gravity is termed as Percolation. The percolation water goes deep into the soil until it meets the free water table. On the one hand, due to rapid percolation, there is practically no danger of soil suffering from bed drainage, but on the other hand, there is a possibility of the dissolved plant nutrients like calcium and magnesium being carried deep into lower layers and depositing beyond the reach of the roots of common field crops.

Percolation tanks: These are the structures for recharging ground water. These are generally constructed across streams and bigger gullies in order to impound a part of the run-off water. This water, in due course, finds its way into subsoil and recharges the ground water.

Performance bond: Cash or securities deposited before a landfill operating permit is issued, which are held to endure that all requirements for operating and subsequently closing the landfill are faithfully performed. The money is returned to the owner after proper closure of the landfill is completed. If contamination or other problems appear at any time during operation, or upon closure, and are not addressed, the owner must forfeit all or part of the bond which is then used to cover clean – up costs.

Performance standards: Operating standards established by EPA for various permitted pollution systems. For example Regulatory requirements limiting the concentrations of designated organic compounds, particulate matter, and hydrogen chloride in emission from incinerators.

Periphery: The lowest position in the three strata hierarchy (periphery, semiperiphery, and core) characterizing the world economic system. The vast majority of less developed countries fall into this category.

Periphyton: Microscopic underwater plants and animals that are firmly attached to solid surface such as rocks, logs and pilings.

Permanent retrievable storage: Placing waste storage containers in a secure location where they can be inspected periodically and retrieved, if necessary for repacking or for transfer if a better means of disposal or reuse of gas through it.

Permissible dose: The dose of a chemical that may be received by an individual without the expectation of a significantly harmful result.

Permissible Exposure Limit (PEL): Legally enforced exposure limit for a substance established by Occupational Safety Health Administration (OSHA). The PEL indicates the permissible concentration of air contaminants to which nearly all workers may be repeatedly exposed and hours a day, 40 hours a week, over a working lifetime (40 years), without adverse effects.

Permit: An authorization, license, or equivalent control document issued by EPA or an approved state agency to implement the requirements of an environmental regulation ; *e.g.* a permit to operate a wastewater treatment plant or to operate a facility that may generate harmful emissions.

Perpetual resource: There are three types of resources in the world: renewable, non-renewable, and perpetual. Perpetual resources are not affected by human use, these are inexhaustible on a human time scale, *e.g.*, solar energy and wind.

Persistence: Refers to the length of time a compound stays in the environment, once introduced. A compound may persist for less than a second or indefinitely.

Persistent Organic Pollutant (POPs): Any member of a class of organic pollutants, that are resistant to biodegradation and that are often toxic; for example, DDT, PCBs and dionin are persistent organic pollutants.

Persistent pesticides: Pesticides that are very resistant to degradation by any means, chemically or biologically, remain in the environment for long periods of time, for more than one growing season or for more than one year after application. These can move over long distances and can build up in the environment leading to greater potential for adverse effects to occur.

Personal air samples: Air samples taken with a pump that is directly attached to the worker with the collecting filter and cassette placed in the worker's breathing zone (required under OSHA asbestos standards and EPA worker protection rule).

Personal hygiene: Precautionary measures taken to maintain good health when exposed to harmful materials.

Personal measurement: A measurement collected from an individual's immediate environment.

Personal protective equipment: Devices or clothing worn to help or isolate a worker from direct exposure to hazardous materials.

Pest: An insect, rodent, nematode, fungus, weed or other form of terrestrial or aquatic plant or animal life that is injurious to health or that reduces the availability, quality or value of a useful resource in the environment. Usually an agricultural pest is detrimental to agricultural production.

Pesticide: Substances or mixture there of intended for preventing, destroying, repelling, or mitigating any pest. In addition, under FIFRA, a substance used as a plant regulator, defoliant, or desiccant is defined as a pesticide and regulated accordingly. Pesticides are classified by the type of pest against which they are active: insecticides (ants, termites, etc.); herbicides (broadleaved weeds, grasses, algae); fungicides (mildew, molds, plant disease, etc.); acaricides (mites, ticks); rodenticides (rats, gophers, ground squirrels); avicides (birds); pisicides (fish); molluscicides (snails, slugs); and nematicides (nematodes or nonsegmented soil worms).

Pesticide regulation notice: Formal notice to pesticide registrants about important changes in regulatory policy, procedures, regulations.

Pesticide resistance: A situation in which pests are not affected by a particular pesticide.

Pesticide tolerance: The amount of pesticide residue allowed by law to remain in or on a harvested crop. EPA sets these levels well below the point where the compounds might be harmful to consumers.

Pesticide treadmill: A situation in which more pesticide doses or new pesticides are applied to maintain previous levels of production and to overcome developing resistance and secondary outbreaks caused by the pesticides applications because their effectiveness has been decreased by development of resistance in pests.

Pest control operator: Person or company that applies pesticides as a business (*e.g.*, exterminator); usually describes household services, not agricultural applications.

Pest resurgence: Rebound of pest population due to acquired resistance to chemicals and nonspecific destruction of natural predators and competitions by broad scale pesticides.

PETE (Polyethylene Terepthalate): Thermoplastic material used in plastic soft drink and rigid containers.

Petrochemicals: Substances created during the refinement of oil and used in the production of plastics, paints, and other products.

Petroleum: Naturally occurring complex liquid hydrocarbons which, after distillation, yield combustible fuels, petrochemicals, and lubricants. Crude oil or any fraction thereof that is liquid under normal conditions of temperature and pressure. In modern technical usage the term petroleum includes gaseous and solid as well as liquid hydrocarbons.

Petroleum derivatives: Chemicals formed when gasoline breaks down in contact with ground water.

pH: A measure of the relative concentration of hydrogen ions in a solution; this value indicates the acidity or alkalinity of the solution in a scale of 0 to 14, 7 indicates a neutral solution; pH values greater than 7 are basic, and those below 7 are acidic. It is calculated as the negative logarithm to the base ten of the hydrogen ion concentration in moles per liter. A pH value of Vinegar has a pH of 3; ocean water has a pH value of approximately 8. Natural waters usually have a pH between 6.5 and 8.5.

Pharmacokinetics: The study of the way that drugs move through the body after they are swallowed or injected.

Phenolphthalein alkalinity: The alkalinity in a water sample measured by the amount of standard acid needed to lower the pH to a level of 8.3 as indicated by the change of color of the phenolphthalein from pink to clear.

Phenols: These are a class of chemical compounds consisting of a hydroxyl functional group (-OH) attached to an aromatic hydrocarbon group. The simplest of the class is phenol (C_6H_5OH). Although similar to alcohols, phenols have unique properties including relatively high acidities due to the aromatic ring tightly coupling with the oxygen and a relatively loose bond between the oxygen and hydrogen. Organic compounds that are byproducts of petroleum refining, tanning, and textile, dye and resin manufacturing. Low concentrations cause taste and odor problems in water; higher concentrations can kill aquatic life and humans.

Phase I A site characterization report: A preliminary report which summarizes the site date collected in the initial portion of the remedial investigation.

Phase II environmental site assessment: A Phase II Assessment involved more detailed evaluation of sites that may be potentially contaminated. Soil, ground water, and building material may be sampled and analyzed. Surveys may be conducted to identify underground tanks and structures.

Phase III environmental site assessment: A Phase III Assessment includes proposing costs effective strategies for dealing with site contamination identified in the Phase II Assessment.

Phlegm: Thick mucous from the respiratory passage.

Phosphogypsum Piles (Stacks): Principal byproduct generated in production of phosphoric acid from phosphate rock. These piles may generate radioactive radon gas.

Phosphorus: A nonmetallic element with atomic weight 30.97 and atomic number 15. It exists in a number of different forms; one form of pure phosphorus is poisonous and burns skin on contact. Phosphorus is an essential nutrient for plants and animals; it is part of DNA molecules. Increased phosphorous levels in soils result from discharge of phosphorous – containing materials into surface waters cause eutrophication. It is used in matches to make them flare up quickly with the heat of friction.

Photochemical oxidants: Air pollutants formed by the action of sunlight on oxides of nitrogen and hydrocarbons.

Photochemical smog: Air pollution caused by chemical reactions among various pollutants emitted from different sources in the atmosphere. Photochemical smog is produced when sunlight acts on nitrogen oxides, (NOx), ozone, and hydrocarbons in a series of complex reactions. Photochemical smog is a respiratory irritant and can kill or alter plant tissues.

Photodegradable plastics: Plastics that break down when exposed to sun light or to a specific wavelength of light.

Photophobia: Intolerance to light.

Photooxidation: Short form for photochemical oxidation, chemical reactions that change (oxidize) compounds using energy from sunlight and some oxidant, such as oxygen or ozone, produce photochemical smog.

Photosynthesis: The series of chemical reactions by which plant cells transform light energy into chemical energy mediated by chlorophyll through producing simple sugars (or other energy compounds) and oxygen, from carbon dioxide and water. It is measured with a variety of units, including milligrams (mg) of carbon dioxide (CO_2) per gram of leaf tissue per hour, mg (CO_2)/ mg chlorophyll/hour, mg (CO_2)/square decimeter of leaf surface/hour, and mg (CO_2)/square meter aquatic or marine surface/day or hour.

Photovoltaic cell: A cell that converts sunlight into electricity.

Physical quality of life: A measure of economic welfare that is more sophisticated than the gross national product. It is based on three variables: percent of population literate, infant mortality rate, and average life expectancy.

Physical hazard: A factor within the environment that can threaten your physical safety or harm the body without necessarily touching it. Vibration, noise, working at heights, electrical hazards, fire, explosion, chemical reactivity are some examples of physical hazards. Peroxides, sulfuric acid, diethyl ether and phosphorus pentachloride are examples of materials which causes both physical hazards and toxic health hazards.

Physical state: Condition of a material (solid, liquid, or gas) at room temperature.

Physical and chemical treatment: Processes generally used in large-scale wastewater treatment facilities. Physical processes may include air – stripping or filtration. Chemical treatment includes coagulation, chlorination, or ozonation. The term can also refer to treatment of toxic materials in surface and ground waters, oil spills, and some methods of dealing with hazardous materials on or in the ground.

Phytoremediation: Low-cost remediation option for sites with widely dispersed contamination at low concentration using plants.

Phytotoxic: Harmful to plants.

Phytotreatment: The cultivation of specialized plants that absorb specific contaminants form the soil through their roots or foliage. This reduces the concentration of contaminants in the soil, but incorporates them into biomasses that may be released back into the environment when the plant dies or is harvested.

Phytoplankton: The portion of the plankton community comprised of tiny plants capable of photosynthesis, such as unicellular algae, diatoms.

Picocuries Per Liter (pCi/L): A unit of measure for levels of radon gas.

Piezometer: A non pumping well, generally of small diameter, for measuring the elevation of a water table.

Pilot tests: Testing a cleanup technology under actual site conditions to identify potential problems prior to full-scale implementation.

PITBY: Put in Their Backyard.

Plankton: Tiny plants and animals that live in water.

Plasmid: A circular piece of DNA that exists apart from the chromosome and replicates independently of it. Bacterial plasmids carry information that renders the bacterial resistant to antibiotics. Plasmids are often used in genetic engineering to carry desired genes into organisms.

Plastics: Non-metallic chemoreactive compounds molded into rigid or pliable construction materials, fabrics, etc.

Plate tower scrubber: An air pollution control device that neutralizes hydrogen chloride gas by bubbling alkaline water through holes in a series of metal plates.

Plateau: Broadly, any comparatively flat area of great extent and elevation; specific an extensive land region considerably elevated (more than 150 to 300m in altitude) above the adjacent country or above sea level; it is commonly limited on at least one side by a scarp.

Plug flow: Type of flow that occurs in tanks, basins, or reactors when a slug of water moves through without ever dispersing or mixing with the rest of the water flowing through.

Plugging: Act or process of stopping the flow of water, oil, or gas into or out of a formation through a borehole or well penetrating that formation.

Plume: A visible or measurable discharge of a contaminant from a given point of origin. Can be visible or thermal in water, or visible in the air as, for example, a column of smoke from a chimney or a substance moving with groundwater. Plumes can be described by the volume or air or water they occupy and the direction they move.

PM-10/PM-2.5: PM 10 is measure of particles in the atmosphere with a diameter of less than ten or equal to a nominal 10 micrometers. PM-2.5 is a measure of smaller particles in the air. PM-10 has been the pollutant particulate level standard against which EPA has been measuring Clean Air Act compliance. On the basis of newer scientific findings, the Agency is considering regulations that will make PM-2.5 the new "standard".

Pneumoconiosis: Health conditions characterized by permanent deposition of substantial amounts of particulate matter in the lungs and by the tissue reaction to its presence; can range from relatively harmless forms of sclerosis to the destructive fibrotic effect of silicosis.

Poaching: Hunting wild life illegally.

Point source pollution: Pollution that can be linked to any single identifiable source, such as a pipe, ditch, ship, ore pit, factory smokestack, discharges from factories, power plants, sewage treatment plants.

Point of Contact measurement of exposure: Estimating exposure by measuring concentrations over time (while the exposure is taking place) at or near the place where it is occurring.

Point of Disinfectant application: The point where disinfectant is applied and water downstream of that point is not subject to recontamination by surface water runoff.

Point of Use treatment device: Treatment device applied to a single tap to reduce contaminants in the drinking water at the one faucet.

Poison: Any substance that is injurious to health and may lead to death when relatively small amounts are taken either internally or externally.

Poison, Class A: DOT term for an extremely dangerous poison such as a poisonous gas or liquid of such a nature that a very small amount of the gas or vapor of the liquid mixed with air is dangerous to life.

Poison, Class B: Term for liquid, solid, paste, or semisolid substances other than class A poisons or irritating materials known or presumed by animal tests to be so toxic to man to a health hazard during transportation.

Poison control center: Provides medical information on a 24 – hour basis for accidents involving ingestion of potentially poisonous materials.

Pollen: Pollen is a fine to coarse powder consisting of pollen grains, which produce the male gametes (sperm cells) of seed plants. Nowadays it is considered as air pollutant and pollen data is also being considered in air quality monitoring at some places.

Pollutant: Generally, any substance, biological, chemical or physical introduced into the environment that adversely affects the usefulness of a resource or the health of humans, animals, or ecosystems.

Pollutant pathways: Avenues for distribution of pollutants. In most buildings, for example, HVAC systems are the primary pathways although all building components can interact to affect how air movement distributes pollutants.

Pollutant Standard Index (PSI): Indicator of one or more pollutants that may be used to inform the public about the potential for adverse health effects from air pollution in major cities.

Pollution: Contamination or introduction of undesirable substances or energies into the environment that lower the quality of the environment, subsequently effects humans and other organisms. It is a problem when pollutants are emitted at rates greater than the rate at which they can be recycled, absorbed, or otherwise rendered harmless. It usually refers to the results of human activity, but volcano eruptions and contamination of a water body by dead animals or by animal excrement are also pollution.

Pollution charges: Fees assessed per unit of pollution based on the "pollutes pays" principle.

Pollution prevention: Techniques or process designed to reduce the volume and toxicity of hazardous materials used in manufacturing, thereby cutting down on (preventing) end – of – pipe pollution.

Polonium: A radioactive element that occurs in pitchblende and other uranium containing ores.

Polychlorinated biphenyl (PCB): Any of a group of a chlorinated isomers of biphenyl $(C_6 H_5)_2$. Valuable for their fire-resisting qualities. They constitute an environmental hazard because of their persistent toxicity.

Polymerization: Chemical reaction in which one or more small molecules combine to form larger molecules. A hazardous polymerization is such a reaction that takes place at a rate that releases large amounts of energy that can cause fires or explosions or burst containers. Materials that can polymerize usually contain inhibitors that can delay the reaction.

Polyelectrolytes: Synthetic chemicals that help solids to clump during sewage treatment.

Polymer: A natural or synthetic chemical structure where two or more like molecules are joined to form a more complex molecular structure (*e.g.*, polyethylene in plastic), polyvinyl Chloride (PVC): Although, environmentally indestructible plastic that releases hydrochloric acid when burned.

Population: Organisms of the same species that inhabit a specific area; the total count of individuals within such a group.

Population at risk: A population subgroup that is more likely to be exposed to a chemical, or is more sensitive to the chemical, than is the general population.

Population crash: A sudden and dramatic decline in the numbers of individuals of a specific species within an area of interest. The reasons are varied and include loss of food source, introduction of a predator species, disease, environmental toxins and extreme weather events or natural disasters.

A sudden population decline caused by predation, waste accumulation and environmental factors that cause these fluctuations.

Population dynamics: The study of fluctuations in population levels and the major biotic and environmental factors that cause these fluctuations.

Population explosion: A rapid increase in the number of individuals of a specific species in a defined location. Commonly the number of individuals exceeds the capacity of the environment to provide resources, leading to a population crash. The result is marked showing in numbers from high to low, back to high and so on.

Growth of a population at exponential rates to a size that exceeds environmental carrying capacity ; usually followed by a population crash.

Acceleration of the rate of population growth, especially after 1800 in industrialized countries and in the 20th century in less developed countries.

Population density: The number of persons per unit of land area.

Porosity: Degree to which soil, gravel, sediment, or rock is permeated with pores or cavities through which water or air can move.

Portal-of-Entry effect: A local effect produced in the tissue or organ of first contact between a toxicant and the biological system.

Post consumer recycling: The reuse of materials from residential or commercial waste.

Post-Chlorination: Addition of chlorine to plant effluent for disinfectant purposes after the effluent has been treated.

Post-Closure: The time period following the shutdown of a waste management or manufacturer. *e.g.* facility; for monitoring purposes, often considered to be 30 years. Post- Consumer materials/waste: Materials or finished products that have served their intended use and have been diverted or recovered from waste destined for disposal, having completed their lives as consumer items. Post consumer materials are part of the broader category of recovered materials.

Post-Consumer Recycling: Use of materials generated from residential and consumer waste for new or similar purposes; *e.g.* converting wastepaper from offices into corrugated boxes or newsprint.

Postindustrial society: A modern economy dominated by services and information, rather than industry.

Potable water: Water that is safe for drinking and cooking.

Potential dose: The amount of a compound contained in material swallowed, breathed, or applied to the skin.

Potentially Responsible Party (PRP): Any individual or company including owners, operators. transporters or generators potentially responsible for, or contributing to a spill or other contamination at a Superfund site. Whenever possible, through administrative and legal actions. EPA requires PRPs to clean up hazardous sites they have contaminated.

Potentiation: The ability of one chemical to increase the effect of another chemical.

Potentiometric surface: The surface to which water in an aquifer can rise by hydrostatic pressure.

Protozoa: One-celled animals that are larger and more complex than bacteria. May cause disease.

Potential energy: Energy stored in a substances or a body because of its position (or the position of its component parts) or state, rather than its current motion. Water pumped up into a storage tank has potential energy because when it is allowed to fall, gravity enables the water to accomplish work. Gravitational energy, electrical energy, nuclear energy, and chemical energy are all forms of potential energy. *See* kinetic energy.

Pour point: Temperature at which a liquid begins to flow . Also, the pour point can be defined as the minimum temperature a lubricant below which it ceases to flow. In case of marine fuel below the pour point temperature wax will begin to separate out and block the filters.

Poverty: A low standard of living in terms of other people (relative poverty) or in terms of basic needs (absolute poverty).

Pox: A general term for the several oxides of phosphorus.

Prairie: A form of grassland vegetation typical of semiarid temperate latitudes in continental interiors, as in the United States. Prairies are typically large, flat areas covered with a mixture of grass species and a few herbs.

Precautionary principle: When information about potential risks is incomplete, basing decisions about the best ways to manage or reduce risks on a preference for avoiding.

Prechlorination: The addition of chlorine the head works of a treatment plant prior to other treatment processes. Done mainly for disinfection and control of tastes, and aquatic growths, and to aid in coagulation and setting.

Precipitate: A substance separated from a solution or suspension by chemical or physical change.

Precipitation: The conversion of dissolved solids into suspended solids which may be concentrated subsequently by flocculation and sedimentation.

Water falling to the earth including snow, sleet and rain.

Precipitator: Pollution control device that collects particles from an air stream.

Pre-consumer material/waste: Material generated in manufacturing and converting processes such as manufacturing scrap and trimmings and cuttings. Includes print overruns, over issue publications, and obsolete inventories.

Precursor: In photochemistry, a compound antecedent to a pollutant. For example, volatile organic compounds (VOCs) and nitric oxides of nitrogen react in sunlight to form ozone or other photochemical oxidants. As such, VOCs and oxides of nitrogen are precursors.

Predation: An interaction between two organisms of different species in which one (usually larger) individual or species hunts, kills, and eats another individual or species.

Predator: An animal that hunts and kills other animals for its food.

Pre-harvest interval: The time between the last pesticide pplication and harvest of the treated crops.

Preliminary assessment: The process of collecting and reviewing available information about a known or suspected waste site or release.

Prescriptive: Water rights which are acquired by diverting water and putting it to use in accordance with specified procedures; *e.g.*, filing a request with a state agency to use unused water in a stream, river, or lake.

Pressed wood products: Materials used in building and furniture construction that are made from wood veneers, particles, or fibers bonded together with an adhesive under heat and pressure.

Pressure sewers: A system of pipes in which water, wastewater, or other liquid is pumped to a higher elevation.

Pressure static: In flowing air, the total pressure minus velocity pressure, pushing equally in all directions.

Pretreatment: Processes used to reduce, eliminate, or alter the nature of wastewater pollutants form non-domestic sources before they are discharged into publicly owned treatment works.

Prevalent level samples: Air samples taken under normal conditions (also known as ambient background samples).

Prevalent levels: Levels of airborne contaminant occurring under normal conditions.

Prevention of Significant Deterioration (PSD): EPA program in which state and/or federal permits are required in order to restrict emission from new or modified sources in places where air quality already meets or exceeds primary and secondary ambient air quality standards.

Primacy: Having the primary responsibility for administering and enforcing regulations.

Primary drinking water regulation: Applies to public water systems and specifies a contaminant level, which, in the judgment of the EPA Administrator, will not adversely affect human health.

Primary effect: An effect where the stressor acts directly on the ecological component of interest, not on other parts of the ecosystem. (See: secondary effect.)

Primary effect: in context to Environment this is defined as an effect where the stressor acts directly on the ecological component of interest, not on other parts of the ecosystem. (See: secondary effect)

Primary pollutants: A general term for a pollutant in the environment that is in the same chemical form as it was when released.

Primary standards: The Clean Air Act established two types of national air quality standards. The Primary standards-set limits to protect public health, including the health of "sensitive" populations such as asthmatics, children, and the elderly.

Primary succession: The sequential development of communities on an area that was not previously occupied by organisms, such as the surface of a recent volcano or deglaciated land or abandoned strip mine.

Primary skin irritant: This is a non corrosive substance which produces nonallergic, inflammatory reaction of the skin if present in sufficient quantities and time.

Primary Treatment: First stage of wastewater treatment in which solids are removed by screening and settling. Primary treatment removes about 30 percent of carbonaceous biochemical oxygen demand from domestic sewage.

Principal Organic Hazardous Constituents (POHCs): Hazardous compounds monitored during an incinerator's trial burn, selected for high concentration in the waste feed and difficulty of combustion.

Prions: Microscopic particles made of protein that can cause disease. Human diseases caused by prions include Creutzfeldt–Jakob syndrome and Gerstmann-Straussler syndrome.

Prior appropriation: A doctrine of water law that allocates the rights to use water on a first-come, first-served basis.

Private good: Item of consumption that, if used by one person or firm, may not be available for others, such as food and clothing. Private goods are provided generally by private firms.

Probability of detection: The likelihood, expressed as a percentage, that a test method will correctly identify a leaking tank.

Process variable: A physical or chemical quantity which is usually measured and controlled in the operation of a water treatment plant or industrial plant.

Process verification: Verifying that process raw materials, water usage, waste treatment processes, production rate and other facts relative to quantity and quality of pollutants contained in discharges are substantially described in the permit application and the issued permit.

Process wastewater: Any water that comes into contact with any raw material, product, byproduct, or waste.

Process weight: Total weight of all materials, including fuel, used in a manufacturing process; used to calculate the allowable particulate emission rate.

Produce: To manufacture, process, formulate, or repackage.

Producer: Another term for autotrophs. Plants that perform photosynthesis and provide food to consumer. Also called primary producer.

Product level: The level of a product in a storage tank.

Product of Incomplete Combustion (PICs): Organic compounds formed by combustion. Usually generated in small amounts and sometimes toxic. PICs are heat-altered versions of the original material fed into the incinerator (*e.g.* charcoal is a P.I.C. from burning wood).

Product water: Water that has passed through a water treatment plant and is ready to be delivered to consumers.

Productivity: The amount of real output produced by input units of labor and capital.

Project XL: An EPA initiative to give states and the regulated community the flexibility to develop comprehensive strategies as alternatives to multiple current regulatory requirements in order to exceed compliance and increase overall environmental benefits.

Propellant: Liquid in a self-pressurized pesticide product that expels the active ingredient from its container.

Proportionate Mortality Ratio (PMR): The number of deaths from a specific cause in a specific period of time per 100 deaths from all causes in the same time period.

Proposed plan: A plan for a site cleanup that is available to the public for comment.

Proteins: Complex nitrogenous organic compounds of high molecular weight made of amino acids; essential for growth and repair of animal tissue. Many but not all, proteins are enzymes.

Protocol: A series of formal steps for conducting a test.

Protoplast: A membrane-bound cell from which the outer wall has been partially or completely removed. The term often is applied to plant cells.

Proteinuria: Presence of protein in the urine.

Protoxin: A nontoxic molecule, which yields a toxin molecule after some modification.

Proximate causes: Those human activities that directly cause some environmental problem. For instance, fossil fuel consumption causes increased levels of carbon dioxide, which, in turn, contribute to global warming.

Public comment period: The time allowed for the public to express its views and concerns regarding an action by EPA.

Public good: A commodity or service that is available to everyone in an area, that cannot be withheld from nonpayers, and whose consumption by one person does not diminish that by others. These include law-and-order enforcement, national defense, national parks, etc. No market exists for such goods, and they must be provided to everyone by the government.

Public health approach: Regulatory and voluntary focus on effective and feasible risk management actions at the national and community level to reduce human exposures and risks, with priority given to reducing exposures with the biggest impacts in terms of the number affected and severity of effect.

Public health context: The incidence, prevalence, and severity of diseases in communities or populations and the factors that account for them, including infections, exposure to pollutants, and other exposures or activities.

Public Hearing: It is a public meeting, the main purpose of most public hearings is to obtain public testimony or comment. It is a "open record hearing" where as public meeting is a "closed record appeal". A formal meeting wherein EPA officials hear the public's views and concerns about an EPA action or proposal. EPA is required to consider such comments when evaluating its action. Public hearing must be held upon request during the public comment period.

Public Notice: 1. Notification by EPA informing the public of Agency actions such as the issuance of a draft permit or scheduling of a hearing. EPA is required to ensure proper public notice, including publication in newspapers and broadcast over radio and television stations. 2. In the safe drinking water program, water suppliers are required to publish and broadcast notices when pollution problems are discovered.

Public water system: A system that provides piped water for human consumption to at least 15 service connections or regularly serves 25 individuals.

Publicly Owned Treatment Works (POTWs): A waste – treatment works owned by a state, unit of local government, or Indian tribe, usually designed to treat domestic wastewaters.

Pulmonary edema: The build-up of fluid in the spaces outside the blood vessels of the lunge is called pulmonary edema, usually because the heart failure.

Pump and Treat: Refers to Pumping the contaminated groundwater and Treating it before it is discharged, hence the name. This is the most common form of groundwater remediation. Pump – and – treat systems often take a very long time (*e.g.*, 50-100 years) to meet cleanup goals. Pumping system, and treatment are dependent on the site characteristics and contaminant type. It is often associated with treatment technologies such as air stripping and liquid – phase granular activated charcoal.

Pumping station: Mechanical device installed in sewer or water system or other liquid – carrying pipelines to move the liquids to a higher level.

Pumping test: A test conducted to determine aquifer or well characteristics.

Purging: Removing stagnant air or water from sampling zone or equipment prior to sample collection.

Push – pull hypothesis: This refers to the concept that migration is due to socioeconomic imbalances between regions, certain conditions (such as poverty) push people out (emigration), whereas other conditions in the area pull or attract people (immigration).

Put in their backyard: The idea that hazardous activities and substances should be located in other's areas.

Putrefaction: Biological decomposition of organic matter; especially protein, associated with anaerobic conditions, resulting in production of foul-smelling matter.

Putrescible: Solid waste that contains organic matter capable of being decomposed by microorganisms and of such a character and proportion as to cause obnoxious odors and to be capable of attracting or providing food for birds or animals.

Pyrophoric: Materials that ignite spontaneously in air at a temperature below 130 °F. Occasionally friction will ignite them.

Pyrolysis: is the chemical decomposition or breakdown of organic materials by heating in the absence of oxygen or any other reagents, expect possibly steam.

Q

QA/QC: Quality Assurance/Quality Control is a program for assuring the reliability of monitoring and measurement date and the routine application of procedure for obtaining prescribed standards of performance in the monitoring and measurement process.

Qualitative use assessment: Report summarizing the major uses of a pesticide including percentage of crop treated, and amount of pesticide used on a site.

Quality Assurance/Quality Control: A system of procedure, checks, audits, and corrective actions to ensure that all EPA research design and performance, environmental monitoring and sampling, and other technical and reporting activities are of the highest achievable quality.

Quality of life: The extent to which basic human needs (including health, education, shelter, and food) are being met.

Quadrat: A boundary structure that includes at least a minimum specified area used for sampling in phytosociological study. It may be round, square or rectangular.

Quench tank: A water-filled tank used to cool incinerator residues or hot materials during industrial processes.

R

Radiation: Transmission of energy through space or any medium. Also known as Radiant energy.

Radiation standards: Regulations that set maximum exposure limits for protection of the public from radioactive materials.

Radical ecology: A variety of philosophic positions that deny the possibility of slowing environmental problems through economic and political reforms, *e.g.*, deep ecology, ecofeminism. and social ecology.

Radical: A group of atoms or a single atom with an unpaired electron and no charge. In general, radicals are quite reactive and as a consequence have short lifetimes. Radical species result from hemolytic bond cleavage reactions such as the decomposition of CFCs when exposed to ultraviolet light.

Radioactive decay: The process of causing radioactivity; the spontaneous or induced disintegration of a substance in which ionizing radiation such as alpha particles, beta particles, or gamma rays is given off. Radioactive substances have characteristic rates of radioactive decay called half lives.

Radioactive substances: Substances that emit ionizing radiation.

Radioactive waste: Any waste that emits energy as rays, waves, streams or energetic particles. Radioactive materials are often mixed with hazardous waste, from nuclear reactors, research institutions, or hospitals.

Radioisotopes: Isotopes that manifest particular electromagnetic radiations. Chemical variants of radioactive elements with potentially oncogenic, teratogenic, and mutagenic effects on the human body.

Radionuclide: Radioactive particle, man-made (anthropogenic) or natural, with a distinct atomic weight number. Can have a long life as soil or water pollutant.

Radius of influence: 1. The radial distance from the center of a wellbore to the point where there is no lowering of the water table or potentiometric surface (the edge to the cone of depression); 2. The radial distance from an extraction well that has adequate air flow for effective removal of contaminants when a vacuum is applied to the extraction well.

Radius of vulnerability zone: The maximum distance from the point of release of a hazardous substance in which the airborne concentration could reach the level of concern under specified weather conditions.

Radon daughters/radon progeny: Short – lived radioactive decay products of radon that decay into longer-lived lead isotopes that can attach themselves to airborne dust and other particles and, if inhaled, damage the linings of the lungs.

Radon decay products: A term used to refer collectively to the immediate products of the radon decay chain. These include Po-218, Pb-214, Bi-214, and Po-214, which have an average combined half – life of about 30 minutes.

Radon: A colorless naturally occurring radioactive gas formed by radioactive decay of radium atoms in soil or rocks.

Rain forest: A dense forest located in areas receiving 80 inches or more of rain each year.

Rainbow report: Comprehensive document giving the status of all pesticides now or ever in registration or special reviews. Known as the "rainbow report" because chapters are printed on different colors of paper.

Rangeland: An area that provides vegetation for grazing animals.

Rasp: A machine that grinds waste into a manageable material and helps prevent odor.

Raw agricultural commodity: An unprocessed human food or animal feed crop (*e.g.*, raw carrots, apples, corn, or eggs).

Raw sewage: Untreated wastewater and its contents.

Raw water: Intake water prior to any treatment or use.

Reaction: Chemical transformation or change; interaction of two or more substances to form new substance.

Reactive material: Chemical substance or mixture that will vigorously polymerize, decompose, condense, or become self-reactive due to shock, pressure, or temperature. Includes explosive materials, organic peroxides, pressure-generating materials, and water-reactive materials.

Reactivity: Tendency of a substance to undergo chemical reaction with the release of energy. Refers to those hazardous wastes that are normally unstable and readily undergo violent chemical change but do not explode.

Re-aeration: Introduction of air into the lower layers of a reservoir. As the air bubbles form and rise through the water, the oxygen dissolves into the water and replenishes the dissolved oxygen. The rising bubbles also cause the lower waters to rise to the surface where they take on oxygen from the atmosphere.

Reagent: Substance used in a chemical reaction to produce another substance or to detect its composition.

Real gross national product: The gross national product adjusted for inflation.

Real-Time monitoring: Monitoring and measuring environmental developments with technology and communications systems that provide time-relevant information to the public in an easily understood format, people can use in day-to-day decision – making about their health and the environment.

Reasonable worst case: An estimate of the individual dose, exposure, or risk level received by an individual in a defined population that is greater than the 90th percentile but less than that received by anyone in the 98th percentile in the same population.

Reasonably Available Control Technology (RACT): Control technology that is reasonably available, and both technologically and economically feasible. Usually applied to existing sources in non-attainment areas; in most cases is less stringent than new source performance standards.

Recarbonization: Process in which carbon dioxide is bubbled into water being treated to lower the pH.

Receiving waters: A river, lake, ocean, stream or other watercourse into which wastewater or treated effluent is discharged.

Receptor: Ecological entity exposed to a stressor.

Recharge area: A land area in which water reaches the zone of saturation from surface infiltration, *e.g.*, where rainwater soaks through the earth to reach an aquifer.

Recharge rate: The quantity of water per unit of time that replenishes or refills an aquifer.

Recharge: The process by which water is added to a zone of saturation, usually by percolation from the soil surface; *e.g.*, the recharge of an aquifer.

Reclamation: The process of restoring natural areas damaged by human activities.

Reclamation: (In recycling) Restoration of materials found in the waste stream to a beneficial use which may be for purposes other than the original use.

Recombinant bacteria: A microorganism whose genetic makeup has been altered by deliberate introduction of new genetic elements. The offspring of these altered bacteria also contain these new genetic elements; *i.e.*, they "breed true".

Recombinant DNA: The new DNA that is formed by combining pieces of DNA from different organisms or cells.

Recommended exposure limit: The highest allowable airborne concentration that is not expected to injure a worker. Expressed as a ceiling limit or as a time weighted average, usually for 10-hour work shift.

Recommended Maximum Contaminant Level (RMCL): The maximum level of a contaminant in drinking water at which known or anticipated adverse effect on human health would occur, and that includes an adequate margin of safety. Recommended levels are nonenforceable health goals.

Reconstructed source: Facility in which components are replaced to such an extent that the fixed capital cost of the new components exceeds 50 per cent of the capital cost of constructing a comparable brand-new facility. New-source performance standards may be applied to sources reconstructed after the proposal of the standard if it is technologically and economically feasible to meet the standards.

Recovery rate: Percentage of usable recycled materials that have been removed from the total amount of municipal solid waste generated in a specific area or by a specific business.

Recyclable: Still retaining useful properties after serving a primary function and, thus, capable of being used again.

Recycle/Reuse: Minimizing waste generation by recovering and reprocessing usable products that might otherwise become waste (*i.e.*, recycling of aluminum cans, paper, and bottles, etc.).

Recycling: Reprocessing of discarded materials into new, useful products, not the same as reuse of materials for their original purpose, but these are often used interchangeably.

Recycling and reuse business assistance centers: Located in state solid-waste or economic-development agencies, these centers provided recycling businesses with customized and targeted assistance.

Recycling economic development advocates: Individuals hired by state or tribal economic development offices to focus financial, marketing, and permitting resources on creating recycling businesses.

Recycling in your backyard: Engaging in recycling.

Recycling mill: Facility where recovered materials are remanufactured into new products.

Recycling technical assistance partnership national network: A national information-sharing resource designed to help businesses and manufactures increase their use of recovered materials.

Recycling: The reuse of scarce raw materials, especially papers, glass, and metals.

Red bag waste: The term "Red Bag Waste" (RBW) is used to identify infectious waste generated by hospitals, surgery centers, dialysis centers, and other health care facilities.

Red border: An EPA document undergoing review before being submitted for final management decision making.

Red tides: Red tides is a colloquial term used to refer to a national phenomenon known as a "harmful algal bloom". A proliferation of a marine plankton toxic and often fatal to fish, perhaps stimulated by the addition of nutrients. A tide can be red, green, or brown, depending on the coloration of the plankton. A population explosion of certain species of dinoflagellates, a kind of protozoan found in plankton, color the water red or reddish-brown and secrete a toxin that kills fish. Red tide usually occurs in warm coastal waters.

Redemption program: Program in which consumers monetarily compensated for the collection of recyclable materials, generally prepaid deposits or taxes on beverage containers. In some states or localities legislation has enacted redemption programs to help prevent roadside litter.

Redshift: Redshift is defined as an increase in the wavelength of electromagnetic radiation received by a detector compared with the wavelength emitted by the source. A redshift is a shift in the frequency of a photon toward lower energy, or longer wavelength. The redshift is defined as the change in the wavelength of the light divided by the rest wavelength of the light.

Reducing agent: Substance that combines with oxygen or loses electrons to the reaction during a reduction reaction.

Reduction: The addition of hydrogen, removal of oxygen, or addition of electrons to an element or compound. The opposite of oxidation; a chemical reaction in which a compound undergoes a decrease in oxidation state by gaining electrons. Reduction of one compound is always accompanied by oxidation of another compound. Many reduction reaction involve removing oxygen atoms from a compound or liberating metals from their compounds.

Reentry interval: The period of time immediately following the application of a pesticide during which unprotected workers should not enter a field.

Re-entry: (In indoor air program) Refers to air exhausted from a building that is immediately brought back into the system through the air intake and other openings.

Reference Dose (RfD): Used in assessing health risk of contaminants. The RfD is a numerical estimate of a daily exposure to the human population that is likely to be without an appreciable risk of deleterious effects during a lifetime.

Reforestation: The restoration (replanting) of a forest that had been reduced by fire or cutting.

Reformer: A change for the better; an improvement; Catalytic reformer - a unit in an oil refinery that reforms lighter hydrocarbons into higher octane molecules and hydrogen. Hydrogen reformer – A device that strips hydrogen from fuels such as natural gas, methanol, ammonia, gasoline or vegetable oil.

Reformulated gasoline: Reformulated gasoline (RFG) is a cleaner gasoline. Gasoline is made up of different hydrocarbon compounds, including aromatics, olefins, and benzene, all of which contribute to ozone and toxic air pollution. The levels at which these components are present in RFG will be lower, resulting in fewer harmful emissions. RFG will not evaporate as easily as conventional gasoline, contain "chemical oxygen" (oxygenates).

Refueling emissions: Emissions released during vehicle re-fueling.

Refuse: All organic solid waste produced by a community. The combustible fraction of solid waste, including garbage, rubbish and trash.

Refuse reclamation: Conversion of solid waste into useful products; *e.g.*, composting organic wastes to make soil conditioners or separating aluminum and other metals for recycling.

Refuse-derived fuel: An energy resource produced by separation of combustible materials (refuse) from municipal solid waste. This material is then combined with coal and burnt as a refuse derived fuel, which is usually 5% to 25% refuse.

Regime: A set of principles, norms, rules, and procedures governing negotiations and other interactions between international participants.

Regional Response Team (RRT): Representatives of federal, local, and state agencies who may assist in coordination of activities at the request of the On-Scene Coordinator before and during a significant pollution incident such as an oil spill, major chemical release, or Superfund response.

Registrant: Any manufacturer or formulator who obtains registration for a pesticide active ingredient or product.

Registration standards: Published documents which include summary reviews of the data available on a pesticide's active ingredient, data gaps, and the agency's existing regulatory position on the pesticide.

Registry of toxic effects of chemical substances: Published by NIOSH. Presents basic toxicity data on thousands of materials. Objective is to identify "all known toxic substances" and to reference original studies.

Regolith: Regolith is a layer of loose, heterogeneous material covering solid rock.

Regulated Asbestos – Containing Material (RACM): Friable asbestos material or nonfriable ACM that will be or has been subjected to sanding, grinding, cutting, or abrading or has crumbled, or been pulverized or reduced to powder in the course of demolition or renovation operations.

Regulated medical waste: Under the Medical Waste Tracking Act of 1988, any solid waste generated in the diagnosis, or immunization of human beings or animals, in research pertaining thereto, or in the production or testing of biologicals. Included are cultures and stocks of infectious agents; human blood and blood products; human pathological body wastes from surgery and autopsy; contaminated animal carcasses from medical research; waste from patient with communicable diseases; and all used sharp implements, such as needles and scalpels, and certain unused sharps.

Relative ecological sustainability: Ability of an ecosystem to maintain relative ecological integrity indefinitely.

Relative permeability: A dimensionless term devised to adapt the Darcy equation to multiphase flow conditions. Relative permeability is the ratio of effective permeability of a particular fluid at a particular saturation to absolute permeability of that fluid at total saturation. If a single fluid is present in a rock, its relative permeability is 1.0. Calculation of relative permeability allows comparison of the different abilities of fluids to flow in the presence of each other, since the presence of more than one fluid generally inhibits flow.

Release: Any spilling, leaking, pumping, pouring, emitting, emptying, discharging, injecting, escaping, leaching, dumping, or disposing into the environment.

Remedial Action (RA): The actual construction or implementation phase of a Superfund site cleanup that follows remedial design.

Remedial Alternative: Option evaluated by EPA to address the source and/or migration of contaminants at a Superfund site to meet health based cleanup goals.

Remedial design: The development of technical drawings and plans for the construction or implementation of the cleanup alternative selected in the Record of Decision. A phase of remedial action that follows the remedial investigation/feasibility study and includes development of engineering drawings and specifications for a site cleanup.

Remedial investigation: An in-depth study designed to gather data needed to determine the nature and extent of contamination a Superfund site; establish site cleanup criteria; identify preliminary alternatives for remedial action; and support technical and cost analyses of alternatives. The remedial investigation is usually done with the feasibility study. Together they are usually referred to as the "RI/FS".

Remedial response: Long-term action that stops or substantially reduces a release or threat of a release of hazardous substances that is serious but not an immediate threat to public health.

Remediation: Process of cleaning up contaminants from a polluted area, water, soil and rock.

Remote sensing: The collection and interpretation of information about an object without physical contact with the object; *e.g.*, satellite imaging, aerial photography, and open path measurements.

Renal: Pertaining to the kidney.

Renewable Energy Production Incentive (REPI): Incentive established by the Energy Policy Act available to renewable energy power projects owned by a state or local government or nonprofit electric cooperative.

Renewable resource: A resource considered to be inexhaustible. Renewable resources are those that are replenished through biogeochemical and physical cycles. Renewable resources are seldom perfectly renewable. If their levels are heavily decreased, they may not be able to completely replenish themselves. For example, if a species is nearly eradicated, it may not be able to avoid extinction. Renewable resources can also be lost through pollution. Though water renews itself, if it is polluted, it is no longer useful for human use. Urban sprawl, cultivation, irrigation, grazing, deforestation, fishing, hunting, and habitat destruction can all be causes of the destruction of an otherwise renewable resource. There are both organic and inorganic types of renewable resources. Examples of renewable organic substances include plant and animal species. Inorganic examples include water and certain gases.

Repeat compliance period: Any subsequent compliance period after the initial one.

Replacement fertility rate: The rate at which the number of people born equals the number of people dying, resulting in a constant population size.

Re-powering: Rebuilding and replacing major components of a power plant instead of building a new one.

Representative sample: A portion of material or water that is as nearly identical in content and consistency as possible to that in the larger body of material or water being sampled.

Reproductive health hazard: Any agent that has a harmful effect on the adult male or female reproductive system or the developing fetus or child.

Re-registration: The reevaluation and re-licensing of existing pesticides originally registered prior to current scientific and regulatory standards. EPA reregisters pesticides through its Registration Standards Program. The goal of the re-registration program is to mitigate risks associated with the use of older pesticides while preserving their benefits. Pesticides that meet today's scientific and regulatory standards may be declared "eligible" for re-registration.

Reserve capacity: Extra treatment capacity built into solid waste and wastewater treatment plants and interceptor sewers to accommodate flow increase due to future population growth.

Reservoir: Any natural or artificial holding area used to store, regulate, or control water.

Residence time: The average length of time a specified amount or quantity of a substance stay in a designed compartment of a system. For example, 1. the residence time of water in a lake is calculated by dividing the volume of the lake by the mean annual water flux into the lake. 2. time a nutrient takes to pass through a specific compartment of a model such as a watershed or ecosystem.

Residential use: Pesticide application in and around houses, office buildings, apartment buildings, motels, and other living or working areas.

Residential waste: Wastes are generated by people in their everyday environment. Familiar examples of such wastes are household trash, such as yard trimmings and food scraps, used batteries, and scrap tyres.

Residual risk: The extent of health risk from air pollutants remaining after application of the Maximum Achievable Control Technology.

Residual saturation: Saturation level below which fluid drainage will not occur.

Residual: Amount of a pollutant remaining in the environment after a natural or technological process has taken place; *e.g.*, the sludge remaining after initial wastewater treatment, or particulates remaining in air after it passes through a scrubbing or other process.

Residue: The dry solids remaining after the evaporation of a sample of water or sludge.

Resistance: The ability of an organism to live in the presence of environmental stress, chemicals, pathogens, or pests. May be inborn or acquired. Natural resistance of crops can be increased through breeding to reduce use of pesticides and irrigation.

Resource depletion: Using a resource at a non replacement rate.

Resource mobilization theory: A theory arguing that social movements arise and act when resources are available.

Resource recovery: The process of obtaining matter or energy from materials formerly discarded.

Resources: Things obtained from the biosphere by humans to meet their basic needs and wants.

Respiration: The process in which a living organism or cell utilizes oxygen to convert food compounds into carbon dioxide and water, releasing energy.

Respiration protection: Devices that will protect the wearer's respiratory system from overexposure by inhalation to airborne contaminants. Respiratory protection is used when a worker must work in an area where he/she might be exposed to concentration in excess of the allowable exposure limit.

Respiratory system: Breathing system, including the lungs and air passages, as well as the associated system of nerves and circulatory supply.

Response action: Generic term for actions taken in response to actual or potential health-threatening environmental events such as spills, sudden releases, and asbestos abatement/management problems. A superfund action taken to recognize, evaluate, and control an incident, including both short-term removal actions and long-term remedial actions.

Responsible party: Someone who can provide additional information on the hazardous chemical and appropriate emergency procedures, if necessary.

Responsiveness summary: A summary of oral and/or written public comments received by EPA during a comment period on key EPA documents, and EPA's response to those comments.

Restricted entry interval (REI): The period of time after a field is treated with a pesticide during which restrictions on entry are in effect to protect persons from potential exposure to hazardous levels of pesticide residues.

Restricted use: A pesticide may be classified for restricted use if it requires special handling because of its toxicity, and if so, it may be applied only by trained, certified applicators or those under their direct supervision.

Restriction enzymes: Any of a group of enzymes that catalyze the cleavage of DNA at specific sites to produce discrete fragments, used especially in genetic engineering. Also called restriction endonuclease.

Retrofit: An instance of modernizing or expanding with new or modified parts, devices, systems, or equipment. Ex: Addition of a pollution control device on an existing facility without making major changes to the generating plant. Also called backfit.

Retrovirus: Any several viruses that contain RNA and which are able to manufacture DNA copies of their RNA and to insert these into the chromosomes of their human or animal hosts. AIDS is cause by a retrovirus, and these microorganisms often produce cancerous tumors.

Reuse: Using a product or component of municipal solid waste in its original form more than once; *e.g.*, refilling a glass bottle that has been returned or using a coffee can to hold nuts and bolts.

Reverse osmosis: A treatment process used in water systems by adding pressure to force water through a semi-permeable membrane. Reverse osmosis removes most drinking water contaminants. Also used in wastewater treatment. Large-scale reverse osmosis plants are being developed.

Reversible effect: An effect which is not permanent; especially adverse effects which diminish when exposure to a toxic chemical stops.

RFI proposal: A document prepared by a company and approved by EPA which specifies the investigation to be performed during the RFI to identify and define the nature and extent of contamination.

Ribonucleic Acid (RNA): A molecule that carries the genetic message from DNA to a cellular protein –producing mechanism.

Rill erosion: The removing of thin layers of soil by surface runoff; can be easily smoothed put or obliterated by normal tillage.

Ringlemann chart: A series of shaded illustrations used to measure the opacity of air pollution emissions, ranging from light grey through black; used to set and enforce emissions standards.

Rio Conference: The United Nations Conference on Environment and riparian doctrine, under which those who own land next to water have rights to use the water.

Riparian habitat: Riparian habitat is characterized by vegetated areas along bodies of freshwater including stream, lakes and rivers with a differing density, diversity, and productivity of plant and animal species relative to nearby uplands.

Riparian rights: Riparian rights are the rights of landowners to use water that is on or adjacent to their property. Landowners have riparian only if their land touches somebody of water. There is no right to go over the property of others to take water, even if that intervening property is only a narrow strip such as a road or railroad. Landowners have rights in the groundwater underneath their property that are similar to surface water riparian rights. If there is not enough water to satisfy all users, allotments are generally fixed in proportion to frontage on the water source.

Risk: A measure of the probability that damage to life, health, property, and/or the environment will occur as a result of a given hazard.

Risk assessment: It is a common first step in a risk management process. Risk assessment is the determination of quantitative value of risk related to a concrete situation and a recognized threat.

Risk characterization: It is the final phase of the health risk assessment process. It integrates the three phases: Hazard Identification, Dose – Response Assessment, and Exposure Assessment. This phase determines the probability of an adverse effect to a human population by a toxic substance and outlines permissible exposure levels from which standards of exposure are set.

Risk communication: A process of providing the public with information about the risks associated with particular products, substances, activities, and technologies; the exchange of information about health or environmental risks among risk assessors and managers, the public, news media, interest groups, etc.

Risk estimate: A description of the probability that organisms exposed to a specific dose of a chemical or other pollutant will develop an adverse response, *e.g.*, cancer.

Risk evaluation: A determination of the acceptability of an identified health, safety, or environmental risk.

Risk factor: Characteristics (*e.g.*, race, sex, age, obesity) or variables (*e.g.*, smoking, occupational exposure level) associated with increased probability of a toxic effect.

Risk management: is a structured approach to managing uncertainty related to a threat, a sequence of human activities including: risk assessment, strategies development to manage it, and mitigation of risk using managerial resources. The process of evaluating and selecting alternative regulatory and non-regulatory responses to risk. The selection process necessarily requires the consideration of legal, economic, and behavioral factors.

Risk perception: Human perception of health, safety, or environmental risks.

Risk reduction: Actions that can decrease the likelihood that individuals, groups, or communities will experience disease or other health conditions.

Risk-based targeting: The direction of resources to those areas that have been identified as having the highest potential or actual adverse effect on human health and / or the environment.

Risk-benefit analysis: A comparison of the risks and benefits associated with a particular hazard to determine its acceptability.

Risk-specific dose: The dose associated with a specified risk level.

River basin: The land area drained by a river and its tributaries.

RIY BY: Recycling In Your Backyard.

Rodenticide: A chemical or agent used to destroy rats or other rodent pests, or to prevent them from damaging food, crops, etc. An effective rodenticide must be tasteless and odorless in lethal concentrations, and have a delayed effect.

Rotary kiln incinerator: An incinerator with a rotating combustion chamber that keeps waste moving, thereby allowing it to vaporize for easier burning.

Rough fish: Fish not prized for sport or eating, such as gar and suckers. Most are more tolerant of changing environmental conditions than are game or food species.

Route of exposure: The way people come into contact with a hazardous substance. Three routes of exposure are breathing (inhalation), eating or drinking (ingestion), or contact with the skin (dermal contact).

Routes of entry: Means by which material may gain access to the body.

Running losses: Evaporation of motor vehicle fuel from the fuel tank while the vehicle is in use.

Run Off: The portion of precipitation, snow melt, or irrigation water that runs off the land into streams or other surface – water. In broad terms, the water available for human use. It can carry pollutants from the land into receiving waters.

S

Sacrificial anode: is a metallic anode an easily corroded material deliberately installed in a container or intake pipe where it is intended to be dissolved (sacrifice it) to protect other metallic components or rest of the water supply facility from corrosion.

Safe water: Water (treated or untreated but uncontaminated water) that does not contain harmful bacteria, toxic materials, or chemicals, and is considered safe for drinking and bathing even if it may have taste, odor, color, and certain mineral problems.

Safe Yield: The annual amount of water that can be taken from a source of supply over a period of years without depleting that source beyond its ability to be replenished naturally in "wet years" and without producing unacceptable negative effects such as contamination of the aquifer water by induced infiltration, decreased river flows, and lowering of the water table.

Safe: Condition of exposure under which there is a practical certainty that no harm will result to exposed individuals; with acceptable risk.

Safener: These are additives for herbicides that accelerate the breakdown of the active substance in crop plants. In contrast, they do not hinder the required action against weeds, because these species apparently possess variants of enzymes that are different from those present in the crop. The safener principle was first shown by the German chemist Otto Hoffmann in 1947. The first (in 1987) safener was fenchlorazol-ethyl, used as an additive for fenoxaprop a post crop emergence herbicide applied in wheat.

Salinity: The dissolved salt content of water. The technical term for saltiness in the ocean is salinity, from the fact that halides – chloride specifically – are the most abundant anions in the mix of dissolved elements. Based on salinity water is classified as fresh (<0.05% salts), brackish (0.05-3%), saline (3-5%) and brine (>5%). Salinity in Australian and North American English may also refer to the salt in soil.

Salinity landfills: A method of ground disposal of solid waste in which waste material is spread in relatively thin layers, compacted and covered with clean earth by the end of each working day to eliminate odours, vermin and litter as required by current land disposal regulatory for Municipal solid waste.

Sahel: The Sahel is widely French-speaking, Islamic and takes its name ("shore") from Arabic. The Sahel or Sahel belt is a narrow band of arid region south of the Sahara Desert in West Africa in which food shortages often occur because of infrequent seasonal rains, determined by the movements of the Intertropical Convergence Zone. Strong correlation between rainfall in the Sahel region and intense hurricane activity in the Atlantic was reported.

Saint Andrew's Cross (X): Used in packaging for transport; It means harmful-stow away from foodstuffs.

Salinization: A process by which the salt content of the soil is increased. This is a common problem when crops are irrigated in arid regions; much of the water evaporates and salts accumulate in the soil. It typically is attributed to irrigation practices and often makes land useless for crop production.

Salt water intrusion: The invasion of fresh surface or ground water by salt water because of the differences in the hydrostatic pressure. This process occurs virtually in all coastal aquifers, where they are in hydraulic continuity with sea water.

Salvage: A property saved from destruction in a calamity (as a wreck or fire); something extracted from waste as valuable or useful.

Salvage value: is the estimated value of an asset at the end of its useful life.

Sampling frequency: The interval between the collection of successive samples or the number of samples per second or per other unit.

Sanborn maps: Detailed property and land-use records kept for fire insurance purposes. The Sanborn Company, USA began making fire insurance maps in 1867 when found by Daniel Alfred by Daniel Alfred Sanborn. Sanborn maps are included as part of the record search in a Phase I environmental audit.

Sanctions: Mechanisms of social control; punishment or permission depending on context; like a ban on trade (economic sanctions), punitive measures adopted by a country of group of countries against another nation for political reasons (international sanctions).

Sand filters: Devices that remove some suspended solids from water.

Sanitary landfill: is a disposal site that is carefully selected, constructed, and operated to protect the environment and public health where in refuse can be safely deposited.

Sanitary sewers: Underground pipes that carry off only domestic or industrial waste to treatment or disposal, not storm water.

Sanitary survey: An on-site review of the water sources, facilities, equipment, operation and maintenance of a public water system to evaluate the adequacy of those elements for producing and distributing safe drinking water.

Sanitation: It is the hygienic means of preventing human contact from the hazards of either wastes or agents of diseases (physical, microbiological, chemical) to promote development, health, or survival.

Saprolite: A soft, clay-rich, thoroughly decomposed rock autochthonously formed in place, by chemical weathering of igneous or metamorphic rock. Forms in humid, tropical, or subtropical climates.

Saprophyte: An organism (especially a plant) that lives on dead decaying organic matter that help natural decomposition of organic matter in water.

SARA: Superfund Amendments and Reauthorization Act. On October 17,1986, the SARA amended CERCLA by making numerous changes and expanding the scope of the Superfund program Title III of SARA is known as the Emergency Planning and Community Right-to Know Act of 1986. It provides citizens and local governments with information about potential chemical hazards in their communities.

Sarcoma: A tumor that is often malignant Sarcoma is a general class of uncommon cancers in which the cancer cells arise from or resemble normal cells in the body known as"connective tissues". Sarcomas are given a number of different names, based on the type of tissue from which they arise. For example, Osteosarcoma (arises from bone), Chondrosarcoma (cartilage), and Leiomyosarcoma (smooth muscle), Liposarcomas (fat tissue), Synovialsarcoma (cells in joints) Rhabdomyosarcoma (skeletal muscle).

Saturated zone: The area below the water table where all open spaces are filled with water under pressure equal to or greater than that of the atmosphere. Many hydrogeologists separate this zone into two subzones: the phreatic zone and the capillary fringe.

Savanna: A savanna is a rolling grassland scattered with shrubs and isolated trees, which can be found between a tropical rainforest and desert biome. This receives more rainfall but not enough to support forests. Savannas are also known as tropical grasslands. They are found in a wide band on either side of the equator on the edges of tropical rainforests.

SCBA: Self Contained Breathing Apparatus worn by first responders, firefighters, industrial worker and others to provide breathable air in environments that are Immediately Dangerous to Life and Health. The term "self – contained" differentiates SCBA from other apparatus connected to a remote supply by a long hose.

Science: (from the Latin scientia, meaning "knowing") An effort to discover or understand physical or social phenomena through replicable procedure, research, observation and analysis of information.

Sclerae: Tough, white, fibrous covering all of the eyeball except the cornea. Also called sclerotic coat.

Scrap metal processor: Intermediate operating facility where recovered metal is sorted, cleaned of contaminants, and prepared for recycling.

Scrap: Discarded materials from manufacturing operations and used products (for ex: parts of vehicles) that may be suitable for reprocessing or recycling. Often confused with waste, scrap, in fact has monetary value.

Screening level risk assessment: A risk assessment performed with few data and many assumptions to identify exposure estimates that should be evaluated more carefully for potential risk. They are not designed to provide definitive estimates of actual risk.

Screening: Use of screens to remove coarse floating and suspended solids from sewage.

Scrubber: An anti air pollution device that uses a spray of water or reactant or a dry process to trap pollutants like acid gasses and particulates in emissions or exhausted air.

S-curve: The "S" curve describes a paradox of growth in physical systems like the hysterises curve explaining why magnets and electrical coils behave, to biological systems of population growth, to the nature of organizations and interpersonal relationships. S-curve of population growth indicates that the population increases until it reaches the carrying capacity of the ecosystem and then levels off. The S-Curve is a well known project management tool and it consists in "a display of cumulative costs, labour hours or other quantities plotted against time". The "S" curve concept expands into a series of five curves in the successful growth of any organization-corporate, non-profit or government.

Second law of thermodynamics: is commonly known as the Law of Increased Entropy. A law holding that, when energy is converted from one form to another, some of the energy is degraded or lost.

Second world: May be defined as countries having developed planned economies, *i.e.*, those of Eastern Europe from 1948 to 1990. The phrase "Second World" was used to describe the communist states within the Soviet Union's sphere of influence. Additionally, the term is often used incorrectly, to describe a moderately developed country.

Secondary air pollutants: Substances created when primary pollutants combine with one another and/or other substances. Examples are sulfuric acid formed when the primary pollutant sulfur dioxide reacts with oxygen and moisture and photochemical smog.

Secondary air standards: National ambient air quality standards designed to protect public welfare including protection against damage to soils, water, crops and animals, decreased visibility, damage to property; transportation hazards (see primary air standards)

Secondary drinking water regulations: Non-enforceable regulations applying to public water systems regulating contaminants that may effect cosmetic (such as skin or tooth discoloration) or aesthetic (such as taste, odor, color, stained plumbing) qualities in drinking water.

Secondary effect: Action of a stressor on supporting components of the ecosystem, which in turn impact the ecological component of concern (see: primary effect). It is Indirect Effect for which the causal linkages to the action or inaction are not readily apparent.

Secondary forest: A forest or woodland area that has resulted from replacement of original trees (primary forest species) by new species.

Secondary materials: Materials that have been manufactured and used at least once and are to be used again.

Secondary pollutants: A general term applied to pollutants that are formed by chemical reactions in the environment. These reactions form compounds other than the substances released by the primary sources of pollution.

Secondary standards: The Clean Air Act established two types of national air quality standards. The Secondary standards – set limits to protect public welfare, including protection against decreased visibility, damage to animals, crops, vegetation, and buildings.

Secondary succession: Occurs on substrate that previously supported vegetation before a disturbance destroyed the plant life. It is a response to a disturbance, for example, forest fire, tsunami, flood, or an abandoned field.

Secondary treatment: A phase in the treatment of waste water. In this second phase, conditions are established to maximize the functions of bacteria in destroying or mineralizing dissolved or suspended organic materials that are not removed by the primary process. It is accomplished by bringing together waste, bacteria, and oxygen in trickling filters or in the activated sludge process. This treatment removes floating and settleable solids and about 90 per cent of the oxygen-demanding substances and suspended solids. Disinfection is the final stage of secondary treatment (see: primary, tertiary treatment).

Secure landfill: A solid waste disposal site lined and capped with an impermeable barriers to prevent leakage of leaching of dangerous substances into ground water. A piping network called a leachate collection system in placed beneath the facility to allow the removal of any liquid that accumulates in the land fill.

Secure maximum contaminant level: Maximum permissible level of a contaminant, at which it does not have any adverse effect on the health of the user, in water delivered to the free flowing outlet of the ultimate user, or of contamination resulting from corrosion or piping.

Sediment yield: The quantity of sediment arriving at a specific location. The amount of material eroded from the land surface by runoff and delivered to a stream system.

Sedimentation tanks: Wastewater tanks in which floating wastes are skimmed off and settled solids are removed for disposal.

Sedimentation: It is a physical water treatment process in which suspended solids settle out of wastewater by under the influence of gravity.

Sediments: Soil, sand, and minerals washed from land into water bodies, usually after rain or snow melt. Their accumulation may destroy fish and wildlife habitat, and cloud the water so that sunlight cannot reach aquatic plants. Sediments are commonly subdivided into three major groups – mechanical (derived from the erosion), chemical (from chemical reactions in seawater) and organic (peat and coal form).

Seed protectant: A chemical applied before planting to protect seeds and seedlings from disease or insects.

Seepage: Percolation of water through the soil from unlined canals, ditches, laterals, watercourses, or water storage facilities.

Selective pesticide: A chemical designed to affect only a narrow group of closely related pests. These are useful in IPM programmes as they have less impact on beneficial organisms and lower risk for humans and wild life.

Semi-confined aquifer: An aquifer partially underlain by an impervious layer and overlain by a semi-impervious layer through which recharge and discharge can still occur.

Semiperiphery: Or the semi-periphery countries refers to countries in the middle position in the three – strata world economic system, are the industrializing, mostly capitalist countries which are in the process of transforming from the periphery countries (low income) into core countries (high income) depend. Brazil, China, Taiwan, the former Eastern Bloc nations, and related countries fall in this category.

Semi-volatile organic compounds: Organic compounds that has a boiling point higher than water and volatilize slowly at standard temperature (20°C and 1 atm pressure). These include phenols and polynuclear aromatic hydrocarbons.

Senescence: The aging process. Sometimes used to describe lakes or other bodies of water in advanced stages of eutrophication. Also used to describe plants and animals.

Sensitization: In the context of allergies, state of immune-response reaction in which further exposure elicits an immune or allergic response. A person previously exposed to a certain material is more sensitive when he experiences further contact with it.

Sensitizer: Substance which, on first exposure, causes little or no reaction in man or test animals but which, on repeated exposure, may cause a marked response not necessarily limited to the contact site.

Sentinel organism: Sentinel organisms, or indicator species that accumulate pollutants in their tissues from the surrounding environment or from food, are important biomonitoring devices. Example is brown trout for PAH's in rivers.

Septic: The term "septic" refers to the anaerobic bacterial environment that develops in the tank and which decomposes or mineralizes the waste discharged into the tank.

Septic system: An on-site system designed to teat and dispose of domestic sewage. A septic system works by allowing waste water to separate into layers and begin the process of decomposition while being contained within the septic tank. Bacteria, which are naturally present in all septic systems, begin to digest the solids that have settled to the bottom of the tank, transforming up to 50 per cent of these solids into liquids and grasses. When liquids within the tank rise to level of the outflow pipe, they enter the drainage system. This outflow, or effiuent, is then distributed throughout the drain field through series of subsurface pipes. Final treatment of the effluent occurs here as the soil absorbs and filters the liquid and microbes break down the rest of the waste into harmless material. Systems cannot dispose of all the material that enters the system. Solids that are not broken down by bacteria begin to accumulate in the septic tank and eventually need to be removed periodically.

Septic tank: An underground storage tank for wastes from homes not connected to a sewer line. Waste goes directly from the homes to the tank. It holds the wastewater long enough to allow solids to settle out (forming sludge) and oil and grease to float to the surface (as scum). It also allows partial decomposition of the solid materials.

Service connector: The pipe that carries tap water from a public water main to a building.

Service line sample: A one-liter sample of water that has been standing for at least 6 hours in a service pipeline and is collected according to federal regulations.

Service pipe: The pipeline standing from the water main to the building served or to the consumer's system.

Set Back: Broadly Set Back means a problem, in the context of land use it means it means the distance a structure must be form the edge of a lot. In Setting a thermometer this refers to setting it to a lower temperature when the building is unoccupied to reduce consumption of heating energy. Also refers to setting the thermometer to a higher temperature during unoccupied periods in the cooling season.

Settleable solids: Material heavy enough to sink to the bottom of a wastewater treatment tank.

Settled sewage: The effluent from settling tanks.

Settling chamber: A series of screens placed in the way of flue gases to slow the stream of air, thus helping gravity to pull particles into a collection device.

Settling tank: A holding area for wastewater, where heavier particles sink to the bottom for removal and disposal.

Sewage: Domestic or municipal discharge containing cooking, cleaning, kitchen or toilet (human faeces and urine) waste discharged into sewers.

Sewage sludge: The solid component of sewage. Sludge produced at a Publicly Owned Treatment Works, the disposal of which is regulated under the Clean Water Act.

Sewer: A channel or conduit that carries wastewater and storm – water runoff from the source to a treatment plant or receiving stream. "Sanitary" sewers carry household, industrial, and commercial waste. "Storm" sewers carry runoff from rain or snow. "Combined" sewers handle both.

Sewerage: The entire system of sewage collection, treatment, and disposal.

Sex ratio: The number of males relative to females in a population.

Shading coefficient: It is a measure of the total amount of the sun's heat transmitted through a given window compared with that of a standard 1/8 – inch –thick single pane of glass under the same conditions.

Sievert: See also rem. One sievert is equal to 100 rem.

Sharps: Sharps represent about 1% of the total waste from health-care activities. These include hypodermic needles, syringes, Pasteur pipettes, scalpel blades, blood vials and culture dishes used in animal or human patient care or treatment, or in medical, research or industrial laboratories. Also included are other types of broken or unbroken glassware that were in contact with infectious agents.

Sheet erosion: The loss of thin layers of soil from unprotected land surfaces, commonly agricultural lands; accomplished primarily by wind and water.

Shock load: The arrival at a water treatment plant of raw water containing unusual amounts of algae, colloidal matter, suspended solids, turbidity, or other pollutants.

Short Circuiting: When some of the water in tanks or basins flows faster than the rest; may result in shorter contact, reaction, or settling times than calculated or presumed.

Sick building syndrome: Building whose occupants experience acute health and/or comfort effects that appear to be linked to time spent therein, but where no specific illness or cause can be identified. Complaints may be localized in a particular room or zone, or may spread throughout the building.

Siderosis: Pneumoconiosis caused by the inhalation of iron particles. Also, tissue pigmentation caused by contact with iron.

Sierra club: An environmental organization founded by John Muir in 1892 that promotes public education, litigation, and outing and conferences.

Sign: Abnormally in the body indicating poisoning or disease which is observable by another person.

Signal words: Distinctive words on a MSDS which serves to alert the reader to the existence and relative degree of a hazard. Signal words are limited to Danger: Materials that are highly toxic; corrosive to living tissue; extremely flammable; or are suspected human carcinogens. Warning: Materials that are moderately toxic; have severe skin irritation potential; cause allergic skin reactions; or are flammable. Caution: Materials that have a low order of toxicity; produce only slight to moderate skin irritation; or are combustion.

Significant deterioration: Pollution resulting from a new source in previously " clean" areas.

Significant municipal facilities: Those publicly owned sewage treatment plants that discharge a million gallons per day or more and are therefore considered by states to have the potentially affect the quality of receiving waters.

Significant potential sources of contamination: A facility or activity that stores, uses, or produces compound with potential for significant contaminating impact if released into the source water of a public water supply.

Significant violations: Violations by point source dischargers of sufficient magnitude or duration to be a regulatory priority.

Silicosis: Condition of massive fibrosis of the lungs causing shortness of breath because of prolonged inhalation of silica dusts.

Silviculture: It is the art and science of controlling the establishment, growth, composition, health, and quality of forest trees or crops to meet diverse needs and values of the many landowners, societies and cultures over the parts of the globe that are covered by dry land.

Single Breath canister: Small one-liter canister designed to capture a single breath. Used in air pollutant ingestion research.

Sink holes: Collapse of the ground surface as a result of the collapse of an underground channel or cavern; often triggered by groundwater withdrawal. These areas can be quite large, involving loss of roads, trees and homes.

Sink: A part of an environmental system that absorbs substances generated. The ocean acts as a sink for atmospheric carbon dioxide.

Sinking: Controlling oil spills by using an agent to trap the oil and sink it to the bottom of the body of water where the agent and the oil are biodegraded.

SIP call: EPA action requiring a state to resubmit all or part of its State Implementation Plan to demonstrate attainment of the require national ambient air quality standards within the statutory deadline. A SIP Revision is a revision of a SIP altered at the request of EPA or on a state's initiative (See: State Implementation Plan).

Site assessment program: A means of evaluating hazardous waste sites through preliminary assessments and site inspections to develop a Hazard Ranking System score.

Site control: The legally binding right to acquire the land on terms established in a legal agreement. Types of site control include purchase and sale agreements, options, earnest money contracts, or donation agreements.

Site inspection: A site inspection involves visually inspecting a property to determine its use and condition, and to identify or assess potential areas of environmental contamination.

Site safety plan: A crucial element in all removal actions, it includes information on equipment being used, precautions to be taken, and steps to take in the event of an on – site emergency.

SITE: Superfund Innovative Technology Evaluation Program.

Site: An area or place within the jurisdiction of the EPA and/or a state.

Siting: The process of choosing a location for a facility.

Skimming: Using a machine to remove oil or scum from the surface of the water.

SLAPP: Strategic Lawsuit Against Public Participation – is a civil complaint or claim filed against non governmental individuals or organizations or general population on a substantive issue of public concern, like real estate development, environment protection and NIMBY actions.

Slash – and – burn cultivation: *See* swidden agriculture.

Slope failure: A slope failure is a phenomenon that a slope collapses abruptly due to weakened self-retainability of the earth under the influence of a rainfall or an earthquake.

Slow sand filtration: Passage of raw water through a bed of sand at low velocity, resulting in substantial removal of chemical and biological contaminants. Slow sand filtration is inexpensive, does not require pretreatment, chemicals or extensive operator control.

Sludge: A general term used to designate residual semi-solid material waste products from industrial or sewage water treatment processor or muddy sediment that collects in a boiler used to produce steam. It contains a variety of substances, some useful and some toxic, dependent on their origins.

Sludge digester: Tank in which complex organic substances like sewage sludges are biologically digested. During these reactions, energy is released and much of the sewage is converted to methane, carbon dioxide, and water.

Sludge, activated: The sediment from biologically treated sewage, it contains the microbes for aerobic digestion of sewage.

Slum and Blight: Fire damaged, vacant and abandoned, and generally unsafe properties which constitute a menace to health, safety and overall deterioration of neighborhoods.

Slurry: A pourable watery mixture of insoluble matter resulting from some pollution control techniques.

Small Quantity Generator (SQG – sometimes referred to as "Squeegee"): Persons or enterprises that produce 100-1000 kg per month of hazardous waste; they are required to keep more records than conditionally exempt generators. SQGs may accumulate hazardous waste on site for 180 days without a permit (or 270 day if shipping a distance greater than 200 miles.). The quantity of hazardous on site waste must never exceed 6,000 kilograms.

Smelter: A facility that melts or fuses or, often with an accompanying chemical change, to separate its material content.

Smog: A term coined by Dr. H. A. Des Voeux in 1905 of England to describe the conditions of fog that had soot or smoke in it. Smog is a combination of various gases with water vapor and dust.

Smoke: is the collection of solid and liquid particulates and gases generated by combustion of a material combined with and suspended in the gases from combustion.

Social ecology: A radical ecological position championed by Murray Bookchin and maintaining that environmental problems can be traced to a hierarchical political/economic and social system that dominates humans and nature.

Social movement: A large number of people acting together to pursue some shared objective.

Social sciences: Sciences concerned with the study of human behavior, including anthropology, economics, psychology, political science, and sociology.

Social structure: Emphasizes the idea that society is grouped into structurally related groups or sets of roles, with different functions, meanings or purposes.

Socialist economy: An economic system in which the means of production and distribution are owned by the public authority or the Government.

Soft detergents: Cleaning agents that break down in nature.

Soft energy paths: Alternative energy sources. Soft energy paths supply appropriate energy in the most cost effective, environmentaliy efficient manner.

Soft water: Any water that does not contain a significant amount of dissolved mineral salts of calcium or magnesium, especially water containing less than 85.5 parts per million of calcium carbonate that interfere with the formation of lather with soap.

Soil adsorption field: A sub-surface area containing a trench or bed with clean stones and a system of piping through which treated sewage may seep into the surrounding soil further treatment and disposal.

Soil and water conservation practices: Control measures consisting of managerial, vegetative, and structural practices to reduce the loss of soil and water.

Soil borings: Holes drilled in the ground for the purpose of collecting subsurface soil samples for geologic and chemical analysis.

Soil erodibility: An indicator of a soil's susceptibility to raindrop impact, runoff, and other erosive processes.

Soil erosion: The process by which productive topsoil is eroded by wind or water action.

Soil gas: Gaseous elements and compounds in the small spaces between particles of the earth and soil. Such gases can be moved or driven out under pressure.

Soil moisture: The water contained in the pore space of the unsaturated zone.

Soil sterilant: A chemical that temporarily or permanently prevents the growth of some or all living matter in soils.

Solar energy: It is heat and light from the sun.

Solar pond: A solar pond is a large body of saline water in which a density gradient is deliberately introduced. When solar radiation (sunlight) is absorbed, the density gradient prevents heat in the lower layers from moving upwards by convection and leaving the pond. This means that the temperature at the bottom of the pond will rise to over 90°C while the temperature at the top of the pond is usually around 30 °C . A solar pond can be used for various applications, such as process heating water desalination, drying, generating electricity and thermal energy storage.

Solar power: The use of solar energy for heating purposes or for generating electricity.

Solder: Metal alloys used to seal joints between metallic surfaces. Alloys that melt between 180 and 190 °C are the most commonly used. Until recently, most solder contained 50 per cent lead. Use of solder containing more than 0.2 per cent lead in pipes carrying drinking water is now prohibited.

Sole – source aquifer: An aquifer that supplies 50 per cent or more of the drinking water of an area.

Solid waste disposal: The final placement of refuse that is not salvaged or recycled.

Solid Waste Management Units (SWMU's): Areas at a facility in which solid wastes have been placed at any time, whether for treatment, storage, disposal, or any other reason, and regardless of whether the area was originally intended for the management of solid waste.

Solid waste management: Supervised handling of waste materials from their source through recovery processes to disposal.

Solid waste: Non-liquid, non-soluble materials ranging from municipal garbage to industrial wastes that contain complex and sometimes hazardous substances. Solid wastes also include sewage sludge, agricultural refuse, demolition wastes, and mining residues.

Solidification and stabilization: Trapping or immobilizing contaminants within their "host" medium (*i.e.*, the soil, sand, and/or building materials that contain them) instead of removing them through chemical or physical treatment

Solidification and/or stabilization reduces the mobility of hazardous substances and contaminants in the environment through both physical and chemical means.

Solubility: The amount of mass of a compound that will dissolve in a unit volume of solution. Aqueous Solubility is the maximum concentration of a chemical that will dissolve in pure water at a reference temperature.

Solution: Uniformly dispersed mixture. Composed of a solvent and a dissolved substance, called the solute.

Solvent: Substance, usually liquid, in which other substances are dissolved. Water is the most common solvent.

Soot: Fine particles, usually black, formed by incomplete combustion consisting chiefly of carbon.

Sorption: The action of soaking up or attracting substances; process used in many pollution control systems.

Source characterization measurements: Measurements made to estimate the rate of release of pollutants into the environment from a source such as an incinerator, landfill, etc.

Source reduction: Source reduction is decreasing the amount of materials or energy used during the manufacturing or distribution of products and packages. Synonymous with waste reduction.

Source separation: Segregating various wastes at the point of generation (*e.g.*, separation of paper, metal and glass from other wastes to make recycling simpler and more efficient).

Sources and sinks: The places from which air pollutants emanate are called "Sources". The places where air pollutants disappear from the polluted air are called "sinks".

Source water assessment: Delineation of the area to be protected and inventory of the potential contaminants in that area.

South or southern: Less developed countries located in the Southern Hemisphere and often referred to as the Third World.

SOx: Oxides of sulfur where x equals the number of oxygen atoms.

Sparge or sparging: Injection of air below the water table to strip dissolved volatile organic compounds and/or oxygenate ground water to facilitate aerobic biodegradation of organic compounds.

Spasm: Involuntary, convulsive muscular contraction.

Special review: Formerly known as Rebuttable Presumption Against Registration (RPAR), this is the regulatory process through which existing pesticides suspected of posing unreasonable risks to human health, non-target organisms, or the environment are referred for review by EPA. Such review requires an intensive risk/benefit analysis with opportunity for public comment. If risk is found to outweigh social and economic benefits, regulatory actions can be initiated, ranging from label revisions and use – restriction to cancellation or suspended registration.

Special risk group: A group that is at high risk because of sensitivity or exposure to hazards. Children often are identified as a special risk group because their high sensitivity to pesticides and other hazardous substances.

Special waste: Items such as household hazardous waste, bulky wastes (refrigerators, pieces of furniture, etc.), tyres, and used oil. The new hazardous waste regime replaces the special waste regime.

Species: A reproductively isolated aggregate of interbreeding organisms having common attributes and usually designated by a common name.

Specific chemical identity: Precise chemical identity of a registry number assigned by chemical abstract service (CAS).

Specific conductance: Rapid method of estimating the dissolved solid content of a water supply by testing its capacity to conduct electrical current. It is an indirect measure of the presence of dissolved solids such as chloride, nitrate, salfate, phosphate, sodium, magnesium, calcium, and iron, and can be used as an indicator of water pollution.

Specific gravity: The specific gravity of a substance is a comparison of its density to that of water.

Specific yield: The amount of water a unit volume of saturated permeable rock or aquifer will yield when drained by gravity.

Spill prevention, Containment, and Countermeasures Plan (SPCP): Plan covering the release of hazardous substances as defined in the Clean Water Act.

Split sample: A sample collected and analyzed for comparison to another sample collected simultaneously at the same location. The two samples are analyzed at different labs so that the results from one can verify the results from the other.

Spoil: Dirt or rock removed from its original location – destroying the composition of the soil in the process-as in strip-mining dredging, or construction.

Sprawl: Unplanned development of open land.

Spray tower scrubber: A device that sprays alkaline water into a chamber where acid gases are present to aid in neutralizing the gas.

Spring Melt/Thaw: The process is hereby warm temperatures melt winter snow and ice. Because various forms of acid deposition may have been stored in the frozen water, the melt can result in abnormally large amounts of acidity entering streams and rivers, sometimes causing fish kills.

Spring: A spring is a location where groundwater naturally emerges from the Earth' s subsurface in a defined flow and in an amount large enough to form a pool or stream-like flow. Springs can discharge fresh groundwater either onto the ground surface, directly into beds of streams, or directly into the ocean below sea level.

Stability: Ability of a material to remain unchanged. A material is stable if it remains in the same form under expected and reasonable conditions of storage or use.

Stabilization ponds: Stabilization ponds consist of shallow man-made basins comprising a single or several series of anaerobic, facultative or maturation ponds. Are often referred to as oxidation ponds or lagoons, are holding basins used for secondary wastewater (sewage effluents) treatment where decomposition of organic matter is processed naturally, *i.e.* biologically. The activity in the WSP is a complex symbiosis of bacteria and algae, which stabilizes the waste, hencethe name.

Stabilization: Conversion of the active organic matter in sludge into inert, harmless material.

Stable air: A motionless mass of air that holds, instead of dispersing, pollutants.

Stack effect: Air, as in a chimney, that moves upward because it is warmer than the ambient atmosphere. One way to ventilate a building that is hotter or colder on the inside than outside is to use what is known as "stack". Because of the temperature difference, the air inside the building is either more or less dense than the air outside. If there is an opening high in the building and another low in the building, a natural flow will be caused. If the air in the building is warmer than the outside, this warmer air will float out the top opening, being replaced with cooler air from outside. If the air inside is cooler than that outside, the cooler air will drain out the low opening, being replaced with warmer air from outside.

Stack gas: Stack gas is anything that comes out of a burner stack in gaseous form.

Stack: A chimney, smokestack, or vertical pipe that discharges used air.

Stage II controls: Systems placed at gasoline dispensing facilities that pump over 400,000 gallons in a calendar year to control and capture gasoline vapors during refueling.

Stakeholder: Any organization, governmental entity, or individual that has a stake in or may be impacted by a given approach to environmental regulation, pollution prevention, energy conservation, etc.

Standard industrial classification code: Also known as SIC Codes, a method of grouping industries with similar products or services and assigning codes to these groups.

Standard of living: The quality of life or the extent to which basic human needs are met. Various indicators have been used to determine it, including infant mortality, life expectancy, literacy, and economic output measures such as gross national product.

Standard sample: The part of finished drinking water that is examined for the presence of coliform bacteria.

Standards: Norms that impose limits on the amount of pollutants or emissions produced. EPA establishes minimum standards, but states are allowed to be stricter.

State environmental goals and indication project: Program to assist state environmental agencies by providing technical and financial assistance in the development of environmental goals and indicators.

State Implementation Plans (SIP): EPA approved state plans for the establishment, regulation, and enforcement of air pollution standards.

State management plan: Under FIFRA, a state management plan required by EPA to allow states, tribes, and U.S. territories the flexibility to design and implement ways to protect ground water from the use of certain pesticides.

Static water depth: The vertical distance from the centerline of the pump discharge down to the surface level of the free pool while no water is being drawn from the pool or water table.

Static water level: Elevation or level of the water table in a well when the pump is not operating.

Stationary source: A fixed-site producer of pollution, mainly power plants and other facilities using industrial combustion processes.

Steady state economy: An economy with a constant population size and stock of capital goods.

STEL: Short-Term Exposure Limit.

Sterilization: The removal or destruction of all microorganisms, including pathogenic and other bacteria, vegetative forms, and spores.

Sterilizer: One of three groups of anti-microbials by EPA for public health uses. EPA considers an antimicrobial to be a sterilizer when it destroys or eliminates all forms of bacteria, viruses, and fungi and their spores. Because spores are considered the most difficult form of microorganism to destroy, EPA considers the term sporicide to be synonymous with sterilizer.

STEV: Short-Term Exposure Value.

Stockholm conference: The 1972 United Nations Conference on the Human Environment. Many scholars assert that this conference was the springboard for the modern environmental movement.

Storage: Temporary holding of waste pending treatment or disposal, as in containers, tanks, waste piles, and surface impoundments.

Storm drains: Separate drainage systems used for collecting and draining runoff from precipitation in urban areas.

Storm sewer: A system of pipes (separate from sanitary sewers) that carries water runoff from and land surfaces.

Strategic lawsuit against public participation: A lawsuit used increasingly by corporations to stifle environmentalists.

Stratification: Separating into layers. A composition, and sequence of sediments, whether consolidated or not.

Stratosphere: The portion of the atmosphere 10-to-25 miles above the earth's surface.

Stress: Physical, chemical or emotional factors that place a strain on an animal. Plants also experience physiological stress under adverse environmental conditions.

Stressors: Physical, chemical, or biological entities that can induce adverse effects on ecosystems or human health.

Strip-cropping: Growing crops in a systematic arrangement of strips or bands that serve as barriers to wind and water erosion.

Strip-mining: A process that uses machines to scrape soil or rock away from mineral deposits just under the earth's surface.

Structural deformation: Distortion in walls of a tank after liquid has been added or removed.

Stupor: It is unresponsiveness from which a person can be aroused only by vigorous, physical simulation. (Whereas Coma is unresponsiveness from which a person cannot be aroused).

Subcutaneous: Beneath the skin.

Sublimation: Change from the solid to the vapor phase without passing through the liquid phase.

Submerged aquatic vegetation: Vegetation lives at or below the water surface; an important habitat for young fish and other aquatic organisms.

Subsistence economy: An economy in which production meets a population's minimum needs but products no surplus.

Sub watershed: Topographic perimeter of the Catchment area of a stream tributary; a smaller basin of a larger drainage area that all drains to a central point of the larger watershed.

Succession: A sequential change in vegetation often in response to environmental change.

Sulfur Dioxide (SO_2): A pungent, colorless, gas formed primarily by the combustion of fossil fuels; sulphur becomes major air pollutant when present in large amounts.

Sulfuric acid: A strong, highly collusive acid, the acid is a colorless liquid that is a strong dehydrating and oxidizing agent. Sulfuric acid is a widely used industrial chemical and is a major formed in the atmosphere as a result of reactions produced by the emission of sulfur di oxide.

Sump: A pit or tank catches liquid runoff for drainage or disposal.

Super chlorination: Chlorination with doses that are deliberately selected to produce water, free of combined residuals so large as to require dechlorination. It is the addition of an extra dose of chlorine to bring the free available chlorine level to 5ppm. This will restore its ability to control algae and bacteria.

Supercritical fluid: Any fluid at a temperature and a pressure above its critical point; also, a fluid above its critical temperature regardless of pressure. Below the critical point the fluid can coexist in both gas and liquid phases, but above the critical point there can be only one phase. Supercritical fluids are of interest because their properties are intermediate between those of gases and liquids, and are readily adjustable.

Supercritical water: A type of thermal treatment using moderate temperatures and high pressures to enhance the ability of water to break down in large organic molecules into smaller, less toxic ones.

Superfund: It is the common name for the US Environmental Policy officially known as the Comprehensive Environmental Response, Compensation, and Liability Act (CERCLA). An economic fund earmarked for cleaning up major hazardous waste sites that have been abandoned to protect people, families, communities and others.

Superfund Innovative Technology Evaluation (SITE) Program: EPA program to promote development and use of innovative treatment and site characterization technologies in Superfund site cleanups.

Supplemental registration: An arrangement whereby a registrant licenses another company to market its pesticide product under the second company's registration.

Surface impoundment: Treatment, storage, or disposal of liquid hazardous wastes in ponds.

Surface runoff: Precipitation, snowmelt, or irrigation water in excess of what can infiltrate the soil surface and be stored in small surface depressions; a major transporter of non-point source pollutants in rivers, streams, and lakes.

Surface uranium mines: Strip mining operations for removal of uranium-bearing ore.

Surface water: Bodies of warm water on the earth's surface that are exposed to the air such as streams, rivers, lakes and oceans.

Surface water: All water naturally open to the atmosphere (rivers, lakes, reservoirs, ponds, streams, impoundments, seas, estuaries, etc.).

Surface-Water treatment rule: Rule that specifies maximum contaminant level goals for Giardia lamblia, viruses, and Legionella and promulgates filtration and disinfection requirements for public water systems using surface – water or ground-water sources under the direct influence of surface water. The regulations also specify water quality, treatment, and watershed protection criteria under which filtration may be avoided.

Surfacing ACM: Asbestos-containing material that is sprayed or troweled on or otherwise applied to surface, such as acoustical plaster on ceilings and fireproofing materials on structural members.

Surfacing material: Material sprayed or troweled onto structural members (beams, columns, or decking) for fire protection; or on ceilings or walls for fire proofing, acoustical or decorative purposes. Includes textured plaster, and other textured wall and ceiling surfaces.

Surfactant: A surface active agent. It is defined as a material that can greatly reduce the surface tension of water when used in very low concentrations.

Surplus: Production of goods and services beyond the minimum needed to sustain life or demand. Produces surplus and consumer surplus in economics, and operational surplus in natural accounts.

Surrogate data: Surrogate data is random data generated to have the same mean, variance, and autocorrelation function as the original data. In the analysis of real world data, the surrogate data test is often performed in order to investigate nonlinearity in the data.

Surveillance: is the monitoring of behaviour.

Systems surveillance: is the process of monitoring the behaviour of people, objects, or processes within systems for conformity to expected or desired norms for social or environment control.

Susceptibility analysis: An analysis to determine whether a Public Water Supply is subject to significant pollution from known potential sources.

Suspended load: Specific sediment particles maintained in the water stream/column by turbulence and are mechanically carried with the flow of water. In most streams, the suspended load is composed primarily of silt and clay size particles.

Suspended particulate matter: In air quality management, solid particles or liquid droplets suspended in air or carried by stream of air or other gases through a duct.

Suspended solids: Refers to small solid particles, with or without pollutants and pathogens on their surface, which remain in suspension in water as a coilloid or due to the motion of the water. It is used as one indicator of water quality.

Suspension culture: The cultivation of cells suspended in the medium rather than adhering to a surface. It is common practice for microorganisms but less so for multicellular organisms. This is useful for the maintenance of cell types, which do not adhere, including some types of blood cells, or in order to have cells express characteristics, which are not seen in the adherent form.

Suspension: Suspension the use of a pesticide when EPA deems it necessary to prevent an imminent hazard resulting from its continued use. An emergency suspension takes effect immediately; and under an ordinary suspension a registrant can request a hearing before the suspension goes into effect.

Sustainable agriculture: Agricultural practices that ensure long-term productivity with few harmful effects, integrating three main goals-environmental health, economic profitability and social and economic equity.

Sustainable development: The perspective emphasizing the need to reconcile present and future economic needs through environmental management. The term was coined by the Brundtland Commission and defined as "development that meets the needs of the present without compromising the ability of future generations to meet their own needs".

SVE: Soil Vapor Extraction.

Swamp: A type of wetland without appreciable peat deposits and with dry – land protrusions. The water of swamps may be fresh or salt water and tidal or non-tidal.

Swidden agriculture: Another term for slash-and-burn agriculture. Is a system of alternating clearance of forest and a short cultivation period with a long fallow period during which forest and soils recover.

Synergism: An interaction of two or more chemicals that results in an effect greater than the sum of their separate effects. An injury caused by exposure to two environmental factors together is greater than the sum of exposure to each factor individually.

Synonym: Another name or names by which a material is known.

Synthetic fuel or synfuel: is any liquid (less often gas) hydrocarbon fuel manufactured from something that has energy like from coal, natural gas, oil shale, tar sand, waste plastic, or biomass.

Synthetic Organic Chemicals (SOCs): Man-made (anthropogenic) organic chemicals. Some SOCs are volatile; others tend to stay dissolved in water instead of evaporating.

System with a single service connection: A system that supplies drinking water to consumers via a single service line.

Systemic effects: Acute or chronic adverse health effects that takes place at a location distant from the body's initial point of contact and presupposes absorption has taken place. Substances with systemic effects often have "target organs" in which they accumulate and exert their toxic effect such as arsenic effect on skin and benzene effects to the bone marrow.

Systemic pesticide: A chemical absorbed by an organism that interacts with the organism and makes the organism toxic to pests.

T

Tachypnea: Increased rate of respiration. This differs from hypernea in that tachypnea is rapid shallow breaths, while hypernea is rapid deep breaths.

Tail water: The runoff of irrigation water from the lower end of an irrigated field.

Tailings: Mining waste left after extraction of metal or mechanical/ chemical separation of minerals from ore.

Tailing ponds: Are areas of refused mining tailings where the refuse material is ponded to allow the settlement of solid particles from the water it is carried in.

Tailpipe standards: Emission limitations applicable to mobile source engine exhausts.

Take-Back: A concept commonly associated with product stewardship, placing responsibility on brand-owners, retailers, manufacturers or other supply chain partners to accept products returned by consumers once they have reached the end of their useful life. Products may then be recycled, treated or sent to landfill.

Tank tightness test: A test which measures, which detects a leak as small as 0.1 gallon per hour with certain probabilities of detection and of false alarm, in an underground. Different methods of testing include "precision," "volumetric", and "non-volumetric" testing.

Target organ toxin: Toxic substance that attacks a specific organ of the body.

Technical – Grade Active ingredient (TGA): A pesticide chemical in pure forms as it is manufactured prior to being formulated into an end-use product (*e.g.*, wettable powders, granules, emulsifiable concentrates). Registered manufactured products composed of such chemicals are known as Technical Grade Products.

Technology: A set of standardized operations that yields predetermined results.

Technology – Based limitations: Industry – specific effluent limitations based on best available preventive technology applied to a discharge when it will not cause a violation of water quality standards at low steam flows. Usually applied to discharges into large rivers.

Technology – Based standards: Industry – specific effluent limitations applicable to direct and indirect sources which are developed on a category – by – category basis using statutory factors, not including water-quality effects.

Technosphere: Synthetic and composite components and materials made by human activity. True technosphere materials, are generally non-renewable, not readily returned to cycles of nature, some like plastics are not biodegradable and will never re-enter the biosphere.

Tectonic plates also called lithospheric plates: Huge rocky plates of the earth's crust that slide around slowly, pulling apart to open new ocean basins or dashing ponderously into each other to create new, larger land masses.

Temperate: The changes in these regions between summer and winter are generally mild, rather than extreme hot or cold.

Temperate forests or forests in the temperate climate zones: They include ;Temperate forest-consists of that lose their leaves every year: Temperate broadleaf and mixed forests- a temperate and humid; Temperate coniferous forest is a terrestrial biome: temperate rainforests.

Temperate deciduous forest: The major biome found in regions with moderate temperatures, distinct seasons, and abundant rainfall (30-60 inches). It generally corresponds to some of the most densely populated regions of the globe and thus is often greatly modified. The forest has well-developed shrub and herb layers. In North America, this biome contains several important, distinct types of vegetation, including beech-maple and maple-basswoods of north-central states, oak-hickory forests in western and southern regions, oak (originally oak-chestnut) and mixed hardwood forests in western and southern regions, oak (originally oak-chestnut) and mixed hardwood forests of the Appalachians, forest.

Temperate rainforest: The conifer-dominated vegetation found along the coast from Alaska to central California, areas with high humidity (frequent fog), abundant rainfall (30 to 150 inches), and moderate temperatures restricted to a fairly small range. It has more under-story than the northern boreal forests and includes the redwood forest of California. Also called moist coniferous forest.

Temperature inversion: One of the weather conditions that are often associated with serious smog episodes in some portions of the country. In a temperature inversion, air doesn't rise because it is trapped near the ground by a layer of warmer air above it. Pollutants, especially smog and smog-forming chemicals, including volatile organic compounds, are trapped close to the ground. As people continue driving, and sources other than motor vehicles continue to release smog-forming pollutants into the air, the smog level keeps getting worse.

Teratogen: Substance or agent to which exposure of a pregnant female can result in malformation in the fetus.

Teratogenesis: The introduction of nonhereditary birth defects in a developing fetus by exogenous factors such as physical or chemical agents acting in the womb to interfere with normal embryonic development.

Terracing: Dikes built along the contour of sloping farm land that hold runoff and sediment to reduce erosion.

Terrestrial: Of or relating to the land rater than the water; the opposite of aquatic. Terrestrial organisms live or grow on the land.

Tertiary sector: It is one of the three economic sectors that involves the provision of services to businesses as well as final consumers, is also known as the service sector or the service industry.

Tertiary treatment: Advanced cleaning of wastewater that goes beyond the secondary or biological stage, removing nutrients such as phosphorus, nitrogen, and most BOD and suspended solids.

Tetraethyl lead: Tetraethyl lead is an organo metallic compound with the chemical formulae (C_2H_5) and Pb. Addition of a small amount of this compound to gasoline improves engine performance and reduce knocking.

Theoretical maximum residue contribution: The theoretical maximum amount of a pesticide in the daily diet of an average person. It assumes that the diet is composed of all food items for which there are tolerance-level residues of the pesticide. The TMRC is expressed as milligrams of pesticide/kilograms of body weight/day.

Therapeutic index: The ratio of the dose required to produce toxic or effects to the dose required to produce nonadverse or therapeutic response. It indicates the selectivity of the drug and consequently its usability.

Thermal pollution: An abnormal increase in heat in some part of the environmental, usually in aquatic systems which adversely affects the biota or water quality. For example discharges of heated water from industrial processes which may kill or injure aquatic organisms.

Thermal stratification: The formation of layers of different temperatures in a lake or reservoir.

Thermal System Insulation (TSI): Asbestos-containing material applied to pipes, fittings, boilers, breeching, tanks, or other interior structural components to prevent heat loss or gain or water condensation.

Thermal treatment: Use of elevated temperatures to treat hazardous wastes.

Ex: incineration, pyrolysis.

Thermocline: The middle layer of a thermally stratified lake or reservoir. In this layer, there is a rapid decrease in temperatures in a lake or reservoir.

Third Pipe System: A third pipe, in addition to the standard water supply pipe and sewer disposal pipe, which carries recycled water for irrigation purposes (non-potable water use).

Third world: Less developed countries that have low per capita incomes, large agricultural sectors, and a shortage of most kinds of capital:

Three mile island: A nuclear power accident that occurred at the Three Mile Island Unit 2 (TMI-2) nuclear power plant near Midde town, Pennsylvania, on March 28, 1979. Even though it led to no deaths or injuries to plant workers or members of the nearby community, it brought about sweeping changes in regulations enhancing safety, involving emergency response planning, reactor operator training, human factors engineering, radiation protection, and many other areas of nuclear power plant operations.

Three R's: Reduce, reuse, and recycle.

Threshold: The dose or exposure level below which a significant adverse effect is not expected.

Threshold level: The intensity level that is just barely perceptible. Time-weighed average pollutant concentration values, exposure beyond which is likely to adversely affect human health.

Threshold Limit Value (TLV): The concentration of an airborne substance to which average person can be repeatedly exposed without adverse effects. TLVs may be expressed in three ways: 1. TLV-TWA Time weighted average, based on an allowable exposure averged over a normal 8-hour workday or 40-hour work- week; 2. TLV_STEL-Short-term exposure limit or maximum concentration for a brief specified period of time, depending on a specific chemical (TWA must still be met); and 3. TLV-C- Ceiling Exposure Limit or maximum exposure concentration not to be exceeded under any circumstances. (TWA must still be met.)

Trophic levels: A functional classification of species that is based on feeding relationships (*e.g.* generally aquatic and terrestrial green plants comprise the first trophic level, and herbivores comprise the second).

Tidal marsh: Low, flat marshlands traversed by channels and tidal hollows, subject to tidal inundation; normally, the only vegetation present is salt-tolerant bushes and grasses. According to the salinity of the flooding water; freshwater, brackish and saline tidal marshes are distinguished.

Tillage: The use of mechanized means to loosen the soil, ploughing, seedbed preparation, and cultivation practices to improve growing conditions for crops.

Time-weighted Average (TWA): In air sampling, the average air concentration of contaminants during a given period. Time-weighted average exposure is the airborne concentration of a material to which a person is exposed, averaged over the total exposure time, generally the total workday (8 to 12 hours).

Tinnitus: Ringing sound in the ears.

Tipping fee: Or gate fee is the amount charged to dispose of garbage at a landfill or incinerator.

Tire processor: Intermediate operating facility where recovered tires are processed in preparation for recycling. Car and truck tires are recycled into rubber products such as trash cans, storage containers, rubberized asphalt or used whole for playground and reef construction.

Tolerance petition: A formal request to establish a new tolerance or modify an existing one.

Tolerances: Permissible residue levels for pesticides in raw agricultural produce and processed foods. Whenever a pesticide is registered for use on a food or a feed crop, a tolerance (or exemption from the tolerance requirement) must be established. EPA establishes the tolerance levels, which are enforced by the Food and Drug Administration and the Department of Agriculture. These limits on pesticides left on foods are called "tolerances" in the U.S (they are referred to as maximum residue limits, or MRLs, in many other countries).

Tonnage: The amount of waste that a landfill accepts, usually expressed in tons per month. The rate at which a landfill accepts waste is limited by the landfill's permit.

Topography: The physical features of a surface area including relative elevations and the position of natural and man-made (anthropogenic) features.

Topsoil: This layer contains large amounts of organic matter and is seat of biological soil activity. It often is viewed as a transitional region between the living and nonliving segments of the larger biosphere. A variety of soil mixtures are sold commercially as topsoil, usually for use us potting soil, for improving growing conditions in gardens and lawns.

Total dissolved phosphorus: The total phosphorous content of all material that will pass through a filter, which is determined as orthophosphate without prior digestion or hydrolysis. Also called soluble P. or ortho p.

Total Dissolved Solids (TDS): All material that passes the standard glass river filter; now called total filterable residue. Term is used to reflect salinity.

Total fertility rate: The number of children that, on overage, a woman would have in her lifetime at present age-specific fertility rates. Calculated as the average number of children born per woman of every give age in a particular year and totaled for all ages.

Total maximum daily loads (TMDL): The maximum quantity of a particular water pollutant that can be discharged into a water body from both point and non-point sources without violating a water quality standard.

Total petroleum Hydrocarbons (TPH): Measure of the concentration of mass of petroleum hydrocarbon constituents present in a given amount of soil or water.

The word "total" is a misnomer-few, if any, of the procedures for quantifying hydrocarbons can measure all of them in a given sample. Volatile ones are usually lost in the process and not quantified and non-petroleum hydrocarbons sometimes appear in the analysis.

Total Suspended Particles (TSP): A method of monitoring airborne particulate matter by total weight.

Total suspended Solids (TSS): A measure of the suspended solids in wastewater, effluent, or water bodies, determined by tests for "total suspended non- filterable solids."

Total water use in water accounting: Distributed water use+self-extracted water use+reuse water use cf. water consumption; here used to mean total direct and indirect water use.

Town water: Water supplied by government or private enterprise and known as the mains or reticulated water supply.

Toxaphene: Chemical that causes adverse health effects in domestic water supplies and is toxic to fresh water and marine aquatic life.

Toxic chemical use substitution: Replacing toxic chemicals with less harmful chemicals in industrial processes.

Toxic chemical: Chemicals that can cause harm to humans and the environment. Any chemical listed in EPA rules as "Toxic Chemicals Subject to Section 313 of the Emergency Planning and Community Right-to-know Act of 1986."

Toxic cloud: Airborne plume of gases, vapors, fumes, or aerosols containing toxic materials.

Toxic concentration: The concentration at which a substance produces a toxic effect.

Toxic dose: The dose level at which a substance produces a toxic effect.

Toxic pollutants: Materials that cause death, disease, or birth defects in organisms that ingest or absorb them. The quantities and exposures necessary to cause these effects can vary widely.

Toxic release inventory: An annual record of releases of toxic chemicals to the environment and the locations of waste handling and disposal sites. In U.S. information is supplied by companies to the USEPA and subsequently made available to the public covering the toxic emissions from certain manufacturing facilities.

Toxic substance: A chemical or mixture that may present an unreasonable risk of injury to health or the environment.

Toxic Substances Control Act of 1976 (TSCA): The Act authorities the environmental Protection Agency (EPA) to study the health and environmental effects of chemical already in the market and new chemical proposed for commercial manufacture.

Toxic Threshold: The level at which a substance causes poisonous effects or death.

Toxic wasters: Wastes that can cause harm to humans and the environment and can be found in the air, water, or soil.

Toxic: Poisonous; having the ability or property to produce harmful or lethal effects on humans or the environment. One of the four hazardous waste characteristics.

Toxicant: A harmful substance or agent that may injure an exposed organism.

Toxicity: The ability of a chemical substance to cause acute or chronic adverse health effects on exposure, ingestion, inhalation, or assimilation in animals, plants or humans.

Toxicity assessment: Characterization of the toxicological properties and effects of a chemical, with special emphasis on establishment of dose-response characteristics.

Toxicity testing: Biological testing (usually with an invertebrate, fish, or small mammal) to determine the adverse effects of a compound or effluent.

Toxicological profile: An examination, summary, and interpretation of a hazardous substance to determine levels of exposure and associated health effects.

Toxicology: The study of the impacts of toxic substances on living organisms including humans health and the pathways by which such substances reacts.

Toxins: Poisons (usually proteins) produced by living organisms, especially those capable of stimulating the production of antibodies. Botulism and tetanus are both caused by bacterially produced toxins. In vertebrates, toxins are broken down by enzyme action, mainly in the liver.

Trace elements: Elements essential to plant or animal life but required in only small amounts, such as the trace amounts of manganese, zinc, iron, molybdenum, and copper.

Trace gases: Gases that occur in only small amounts.

Trace metal: A term used to describe a metallic element that is present in environment/air, water, soil, or food chains in low concentrations.

Tradable permits: Pollution quotes or variances that can be bought or sold.

Trade secret: Any confidential formula pattern, process, device, information, or compilation of information used in employer's business and gives the employer an opportunity to obtain an advantage over competitors.

Traditional agriculture: Farming based on practices such as crop rotation, use of animal manures instead of chemical fertilizers, and use of animal power.

Tragedy of the commons: The tendency for people to overuse and degrade public resources to which they have free access, because it is in their individual short-term interest to use them in an unconstrained fashion.

Transboundary pollutants: Air pollution that travels from one jurisdiction to another, often crossing state or international boundaries. Also applies to water pollution.

Transfer station: Intermediate facility at which municipal solid waste is transferred from collection vehicles into larger trucks or rail cars for transport to its final disposal destination. Transfer stations have become an integral part of municipal waste management, playing an important role in materials recovery and improving transportation economics associated with municipal waste disposal.

Transfrontier pollution: Pollution that moves across national boundaries through natural forces such as rivers and air currents.

Transgenic plants: Possess a gene or genes that have been transferred from a different species by genetic engineering.

Transient water system: A non-community water system that does not serve 25 of the same nonresidents per day for more than six months per year.

Transmissible disease: A disease caused by a living organism transferred from one person to another through physical contact.

Transmission lines: Pipelines that transport raw water from its source to a water treatment plant, then to the distribution grid system.

Transnational corporation: A firm with substantial operations in many countries but controlled from its home country.

Transpiration: The process by which water vapor is lost to the atmosphere from living plants. The term can also be applied to the quantity of water thus dissipated.

Transportation Control Measures (TCMs): Steps taken by a locality to reduce vehicular emission and improve air quality by reducing or changing the flow of traffic; *e.g.*, bus and HOV lanes, carpooling and other forms of ride-sharing, public transit, bicycle lanes.

Trash: Material considered worthless or offensive that is thrown away. Generally defined as dry waste material, but in common usage it is a synonym for garbage, rubbish, or refuse.

Trash – to – Energy plan: Burning trash to produce energy.

Treatability studies: Tests of potential cleanup technologies conducted in a laboratory.

Treated regulated medical waste: Medical waste treated to substantially reduce or eliminate its pathogenicity, but that has not yet been destroyed.

Treated sludge: Solid organic material that has been removed from sewage and treated so that it is non-hazardous.

Treated wastewater: Wastewater that has been subjected to one or more physical, chemical, and biological processes to reduce its potential of being health hazard.

Treatment plant: A structure built to teat wastewater before discharging it into the environment.

Treatment: Any method, technique, or process designed to remove solids and/or pollutants from solid waste, waste-streams, effluents, and air emissions. Methods used to change the biological character or composition of any regulated medical waste so as to substantially reduce or eliminate its potential for causing disease.

Tremie: Device used to place concrete under water.

Trial burn: An incinerator teat in which emissions are monitored for the presence of specific organic compounds, particulates, and hydrogen chloride.

Trichloroethylene (TCE): A stable, low boiling – point colorless liquid, toxic if inhaled. Used as a solvent or metal degreasing agent, and in other industrial applications.

Trickle irrigation: Method in which water drips to the soil from perforated tubes or emitters.

Trickling filter: A coarse treatment system in which wastewater is trickled over a bed of stones or other material covered with bacteria that break down the organic waste and produce clean water.

Trihalomethane (THM): One of a family of organic compounds named as derivative of methane. THMs are generally by-products of chlorination of drinking water that contains organic material.

Triple Bottom Line (or "TBL", "3BL" ,or "People, Planet, Profit"): A form of sustainability accounting going beyond the financial 'bottom line' to consider performance of three criteria *i.e.*, the social and environmental as well as economic consequences or values of an organization's activity. Term coined by John Elkington in 1994.

Tropical: Occurring in the tropics (the region on either side of the equator); hot and humid with a mean annual temperature greater than 20 °C.

Tropical rainforest: A forest biome found in the tropics near the equator, in climates with continually warm to hot weather and very heavy rainfall. Tropical rainforests are green all year long, characterized by high species diversity, soil that are usually old and nutrient – poor, stratification into distinct layers of trees and shrubs, and many epiphytes and vines. The largest tropical rainforest is in the Amazon region, which covers a large portion of South America.

Troposphere: The lowest layer of the earth's atmosphere, extending from the earth's surface up to about 7 to 17 km. Within the troposphere, temperature usually decreases at a regular rate with increasing altitude. Weather is confined to this turbulent layer of the atmosphere.

True costs: Market and nonmarket costs associated with the production and use of goods and services. Market costs fail to account for environmental degradation and related externalities.

TSD: "Treatment, storage, and disposal" describes facilities where hazardous waste is treated, stored and/or disposed.

TSDF: Treatment, Storage, and Disposal Facility.

Tsunamis: A tsunami is an oceanic wave, also termed a seismic sea wave, generated most commonly by slippage (Vertical displacement) of the Ocean floor along a fault in association with an earthquake of a rock slide.

Tube settler: Device using bundles of tubes to let solids in water settle to the bottom for removal by conventional sludge collection means; sometimes used in sedimentation basins and clarifiers to improve particle removal.

Tuberculation: Development or formation of small mounds of corrosion products on the inside of iron pipe. These tubercules roughen the inside of the pipe, increasing its resistance to water flow.

Tundra: A type of treeless ecosystem dominated by lichens, mosses, grasses, and woody plants. Tundra is found at high latitudes (arctic tundra) and high altitudes (alpine tundra). Arctic tundra is underlain by permafrost and is usually water saturated.

Turbidimeter: A device that measures the cloudiness of suspended solids in a liquid; a measure of the quantity of suspended solids.

Turbidity: Any condition in air or water that reduces its transparency by the presence of particles, suspended silt, organic matter or other pollutants.

Turbine: A machine for converting the heat energy in steam or high temperature gas into mechanical energy. In a turbine, a high velocity flow of steam or gas passes through successive rows of radial blades fastened to a central shaft.

U

Ultra Clean Coal (UCC): Coal that is washed, ground into fine particles, then chemically treated to remove sulfur, ash, silicon, and other substances; usually briquetted and coated with a sealant made from coal.

Ultraviolet light: Ultraviolet (UV) light is electromagnetic radiation with a wavelength ranging between 400 nm and 10 nm with energy range of 3.10 – 124 eV. The electromagnetic spectrum of ultraviolet light can be subdivided into UV A (long wave of black light), Near, UV B (medium wave), Middle, UV C (short wave or germicidal), Far, Vacuum and Extreme,. The Sun emits ultraviolet radiation in the UVA, UVB and UVC bands, but because of absorption in the atmosphere's ozone layer, 98.7% of the ultraviolet radiation that reaches the Earth's surface is UVA.

Unconfined aquifer: An aquifer containing water that is not under pressure; the upper surface (water table) is open to the atmosphere through permeable overlying material.

Underground Injection Control (UIC): The program under the Safe Drinking Water Act that regulates the construction, operation, permitting, and closure of injection wells that place fluids underground for storage or disposal, such that contamination of underground drinking water resources is prevented.

Underground sources of drinking water: Aquifers currently being used as a source of drinking water or those capable of supplying a public water system. They have a total dissolved solids content of 10,000 milligrams per liter or less, and are not "exempted aquifers." (See: exempted aquifer.)

Underground Storage Tank (UST): A tank located partially that is at least 10 per cent of its combined volume underground and designed to hold gasoline or other petroleum products or chemicals.

Unequal exchange: A pattern describing trade relations between two or more countries when one country benefits more than another.

Uneven development: The tendency for some areas of a country or region to prosper, while other areas stagnate.

Upwelling: The ascent of deep nutrients rich ocean water to the ocean surface at some continental margins, due to ocean currents diverging from coastline. The effectiveness of upwelling and its ability to support abundant sea life is greatly dependent upon the depth of the thermocline, the transition layer between the mixed layer at the surface and the deep water layer.

United Nations Conference on Environment and Development (UNCED): Popularly called as "The Earth Summit" was held in Rio in 1992; twenty years after the UN first global environment conference in Stockholm. The UN sought to help Governments rethink economic development and find ways to halt the destruction of irreplaceable natural resources and pollution of the planet to ensure a healthy planet for generations to come. The Summit's message was "that nothing less than a transformation of out attitudes and behaviour would bring about the necessary changes."

United Nations Development Programme (UNDP): UNDP is the UN's global development network with the stated purpose of enhancing development worldwide.

UNDP does not represent any one approach to development; rather it assists partner governments in finding their own approaches, according to their own unique national circumstances.

United Nations Environment Programme (UNEP): A program headquartered in Kenya, conceived at Stockholm Conference in June 1972 with the purpose of raising environmental consciousness on a global level. It coordinates United Nations environmental activities, assist developing countries in implementing environmentally sound policies and encourages sustainable development through sound environmental practices.

Unsaturated zone: The area above the water table where soil pores are not fully saturated.

Upper detection limit: The largest concentration that an instrument can reliably detect.

Uranium (Mill) Tailings: The left over piles of radio-active materials (wastes) including thorium-230, radium-226, radon-222 (radon gas) and unrecovered uranium at uranium ore mining and processing sites.

Urban Runoff: Runoff of storm water/rainfall/snowmelt from city streets and adjacent domestic or commercial properties that carries substances/pollutants of various kinds into the sewer systems.

Urbanization: The process by which an increasing share of the population of a country lives in cities.

Use Cluster: A "Use Cluster" is a set of chemicals, processes, and/or technologies that can substitute for one another in performing a particular function. This information helps companies who are looking to see which chemical alternatives may compare favorably as substitutes when they design their products and processes.

Used oil: Spent motor oil from passenger cars and trucks collected at specified or certified locations for recycling (not included in the category of municipal solid waste). Certified center managers will not accept used motor oil that has been contaminated with other fluids such as antifreeze, solvents, gasoline, or water. In USA one is paid for the used oil taken to certified center for recycling.

User fee: Fee collected from only those persons who use a particular public service or facility, as compared to one collected from the public in general. People pay 'user fee' for the use of many public services and facilities. For example fees to drive into a National Park, toll charges for driving on highways. In international development, 'user fees' refer to system fees for basic health care, education, and/or other services which a developing country implements in order to make up for the costs of these services.

V

Vadose Zone: The zone between land surface and the water table within which the moisture content is less than saturation (except in the capillary fringe) and pressure is less than atmospheric. Soil pore space also typically contains air or other gases. This zone is unsaturated zone + the capillary fringe zone.

Value addition: Increasing the value of commodities by some means like for *e.g.,* in agriculture breeding wheat with high protein content; or grinding wheat into flour or processing of perishables into much storable and enriched form.

Valued Environmental Attributes/Components: The components of ecosystems, human health, and environmental welfare which are potentially at risk from human activity or natural hazards and are considered in environmental impact assessment.

Vapor recovery nozzles: Special gas pump nozzles that will reduce release of gasoline vapor into the air when people put gas in their cars.

Vapor recovery system: A system by which the volatile gases are captured instead of being released into the atmosphere.

Variance: Government permission for a delay or exception in the application of a given law, ordinance or regulation.

Vector: An organism that carries disease (often an insect or rodent) or transport genes (plasmids, viruses, or bacteria) into a host cell.

Vegetation: All kinds of the plants growing in and characterizing a specific area or region.

Vegetative controls: Non-point source pollution control practices that involve vegetative cover to reduce erosion and minimize loss of pollutants.

Ventilation/Suction: The act of admitting fresh air into a space in order to replace stale or contaminated air; achieved by blowing air into the space. Similarly, suction represents the admission of fresh air into an interior space by lowering the pressure outside of the space, thereby drawing the contaminated air outward.

Venture scrubbers: Air pollution control devices that use water to remove particulate matter from emissions.

Vermicomposting: Use of worms to compost organic waste.

Vinyl: A type of plastic (usually PVC) used to make product such as fruit juice bottles, credit cards, pipes and hoses.

Vinyl chloride: A colorless flammable and toxic liquid when under pressure, but under ordinary conditions it is a gaseous organic compound of carbon, hydrogen and chlorine. Vinyl chloride monomers are used to make the polymer called polyvinyl chloride, is an ubiquitous plastic produced in enormous quantities. Occupational exposure to vinyl chloride is linked to a rare live cancer, angiosarcoma.

Virgin materials: Materials extracted from natural resources in their raw form, such as timber or metal ore.

Virtual water: It is now more widely known as embedded (embodied) water. It is the volume of water required to produce a commodity or service.

Viscosity: The molecular friction within a fluid that produces flow resistance.

Volatile: Any substance that evaporates readily.

Volatile liquids: Liquids which easily vaporize or evaporate at room temperature.

Volatile organic compounds (VOCs): A category of major air pollutants present in the air in the vapor form. Organic chemicals that evaporate readily at room temperature and normal atmospheric pressure, and exist as gases in the air. The category includes fragments of hydrocarbon fuels from incomplete combustion and evaporated organic compounds. Volatile organic chemicals include gasoline, industrial chemicals such as benzene, solvent such as toluene and xylene, and tetrachloroethylene (perchloroethylene, the principal dry cleaning solvent). Many volatile organic chemicals are also hazardous air pollutants; for example, benzene causes cancer. These are major factors in the formation of photochemical smog.

Volt (symbol V): The S.I. unit of electric potential or electromotive force. It is represented as one joule of energy per coulomb of charge. It is named after Alessandro Volta (1745-1827), who invented the voltaic pile, the first chemical battery. The volt is defined as the potential difference across a conductor when a current of one ampere dissipates one watt of power.

Volume reduction: Processing waste materials to decrease the amount of space they occupy, usually by compacting, shredding, incineration, or composting.

Vulnerability analysis: Assessment of elements in the community that are susceptible to damage if hazardous materials are released.

Vulnerable species: Natural race, organisms or species that has a high risk of extinction in the wild in the medium term future. The probability of extinction in the wild within 200 yrs may be 10%.

W

Wadin: Arabian term used to designate the channel of water course which is dry except during the time of rainfall.

Waste: A product or by-product or residue, which can not be used profitably. Any unwanted materials (liquid, solid or gaseous) left over from a manufacturing process. It also refers to refuse from places of human or animal habitation.

Waste assessment: Observing, measuring, and recording data and collecting and analyzing waste samples.

Waste exchange: Arrangement in which companies exchange their wastes for the benefit of both parties.

Waste generation: It represents the amount of unwanted materials including recyclables generated by a given source or category of sources.

Waste load allocation: The maximum load of pollutants each discharger of waste is allowed to release into a particular waterway. Discharge limits are usually required for each specific water quality criterion being, or expected to be, violated.

Waste minimization: It refers to source reduction and recycling activities; excludes treatment and energy recovery. Techniques to keep waste generation at a minimum level and thereby reduce the cost of waste collection, handling and disposal to landfill; recycling and other efforts made to reduce the amount of waste going into the waste stream.

Waste stream: The steady flow of varied wastes of gas, liquid or solids from domestic garbage or yard wastes or the by-products of a treatment operation/industrial process. Depending on the context they are called "residential waste stream" or the "recyclable waste stream".

Waste to energy: A process of combustion of solid wastes that also generates electrical energy.

Waste treatment: Is processing of particular waste to increase its degradability or compositability or reduce its toxicity.

Waste water: The spent or used water discharged from homes, business and industries that contain dissolved or suspended matter. This can not be used directly for drinking. It contains <1% by weight of total organic carbon and <1% by weight of total suspended solids.

Water cycle: See hydrological cycle.

Water footprint: It is the total volume or freshwater that is required in a given period to perform a particular task or to produce the goods and services consumed at any level of the action hierarchy. Country water footprint is a consumption-based indicator of water use in a country – the volume of water needed to produce the goods and services consumed by the inhabitants of a country.

Water harvesting: See rainwater harvesting.

Water intensity: The quantity of water required to produce a given outcome of production or service delivery; this is generally further reduced to monetary unit return per given volume of water used. Essentially equivalent to water productivity.

Water pollution: Degradation of the natural quality of water because of the presence of enough harmful or objectionable material to damage the water's quality.

Water productivity (WP): WP – field relates to crop output *e.g.*, kg of wheat produced per m^3 of water. WP – basin relates to water productivity in the widest possible sense as including crop, fishery yield, environmental services etc.

Water quality standards: State-adopted and EPA-approved ambient standards for water bodies.

Water Quality-Based permit: A permit with more stringent limits to protect the designated use of receiving waters *i.e.*, recreation, irrigation, industry or water supply compared to those based on technology performance.

Water resources: Water in various forms, such as groundwater, surface water, snow and ice, at present in the land phase of the hydrological cycle.

Water table: The upper level of groundwater that is the level below which the ground is saturated with water.

Water treatment: Purification of well water or surface water or contaminated water or sewage water in treatment plants to the extent that it could be re-used for irrigation or even for drinking purposes to meet the water shortage.

Water treatment lagoon: An impound for liquid wastes designated to accomplish some degree of biochemical treatment.

Watershed: The land area that drains into a common point.

Watershed approach: A coordinated framework for environmental management that focuses public and private efforts on the highest priority problems within hydrologically defined geographic areas taking into consideration both ground and surface water flow.

Weather: The hourly/daily change in atmospheric conditions of a particular place, which over a longer period constitute the climate or a region cf. climate.

Well field: Area containing one or more wells that produce usable amounts of water or oil.

Well injection: The subsurface emplacement of fluids into a well.

Well plug: A watertight, gastight seal installed in a bore hole or well to prevent movements of fluids.

Well: A bored, drilled or a dug hole whose depth is greater than the largest surface dimension and whose purpose is to reach underground water supplies or oil, or to store or bury fluids below ground.

Wellhead protection area: A protected surface and subsurface zone surrounding a well or well fields supplying a public water system to keep contaminants from reaching the well water.

Wetlands: Areas of permanent or intermittent inundation, whether natural or artificial, with water that is static or flowing, fresh, brackish or salt, including areas of marine water not exceeding 6 m at low tide. In urban areas wetlands are protected because they purify water, prevent floods and are called 'the kidney of a city'.

Wet meadows: Grazing land adjacent to wetlands which is flooded at peak water levels.

Wet scrubber: A device used to remove pollutants like particles, poisonous vapors from an exhaust gas by passing the exhaust through a shower of water alone or water that contains an agent to react with the material to be removed.

Whole-Effluent-Toxicity tests: Tests to determine the toxicity levels of the total effluent from a single source as opposed to a series of tests for individual contaminants.

Why In My Backyard (WIMBY): The question of why hazardous activities and substances are placed in a particular location.

Wilderness: An area uninhabited by humans.

Wildlife: All undomesticated organisms in an area, especially animals.

Wildlife refuge: An area set aside to shelter, feed and protect wild life, within which hunting and fishing are either prohibited or strictly controlled.

Wildness: An area of undeveloped land affected primarily by the forces of nature where humans are visitors only.

Wind Energy/Wind power: It is the kinetic energy present in the motion of the wind. Wind energy can be converted to mechanical or electrical energy. For example conversion of the force of the wind to electrical energy by modern electrical wind turbine for on-site consumption and/or supply to the electricity grid, is a renewable source.

Wood – Burning – Stove pollution: Air pollution caused by emission of particulate matter, carbon monoxide, total suspended particulates, and polycyclic organic matter from wood-burning stoves.

Working level (WL): A unit of measure for documenting exposure to radon decay products. One working level is equal to approximately 200 picocuries per liter.

World Bank: The popular name for the International Bank for Reconstruction and Development, which was established on December 27, 1945. It encourages private (rather than public) investment. The World Bank is an internationally supported bank that provides financial and technical assistance to developing countries for development programs (*e.g.*, bridges, roads, schools, etc.) with the stated goal of reducing poverty.

World Commission on Environment and Development (WCED): Also known as 'Brundtland Commission' by the name of its chair, Norwegian Prime Minister Gro Harlem Brundtland, was convened by United Nations in 1983, popularized the notion of 'sustainable development'. The commission was created to address growing concern about the accelerating deterioration of the human environment and natural resources and the consequences of that deterioration for economic and social development.

World environment day: June 5 of each year; designated by the 1972 UN Conference on the Human Environment as a day to focus on environmental problems.

World Health Organization (WHO): A United Nations agency established on 7th April 1948, to promote cooperation among nations in controlling diseases or on international public health.

World Resources Institute (WRI): It is a Policy research center set up in 1982 to address environmental issues on a global level. It is an independent, non-partisan and nonprofit organization supporting research, conferences, education, training, and policy initiative for environmental stewardship and improving people's lives. WRI is best known for its biennial publication, the world resources report, a well-regarded collection of data and in-depth analysis on current environmental issues.

World Trade Organization (WTO): Is the only global international organization dealing with the rules or trade between nations, established in Geneva, Switzerland on 1 January 1995. It is an association of 153 countries as on 23 July 2008. At its heart are the WTO agreements, negotiated and signed by the bulk of the world's trading nations and ratified in their parliaments. The goal is to help producers of goods and services, exporters, and importers conduct their business in order to improve the welfare of the peoples of the member countries.

World Watch Institute (WWI): It is an independent research organization founded in 1974 to track world's environmental and social problems. It is known around the world for its accessible, fact-based analysis of critical global issues and for sustainability analysis.

World Wildlife Fund (WWF): Is an international conservation organization founded in 1961 and working globally to protect endangered wildlife and wetlands. Its projects and activities play a part to stop the degradation of Earth's natural environment, and to build a future in which humans live in harmony with nature.

World view: A set of beliefs and perceptions regarding the manner in which the world operates.

X

Xenobiotic: Xenobiotic derived from the Greek roots "zeno", meaning "strangers" or "foreign" and "bio" meaning "life". It designates a foreign chemical or organism in a biological system.

Xeric: Refers to dry environment. It refers to an organism that requires little moisture or a habitat containing little moisture.

Xerophytes: Plants that grow in xeric or desert conditions.

X-rays: The deep penetrating highly ionizing electromagnetic radiations or rays and as such can be dangerous. X-rays have a wavelength in the range of 10 to 0.01 nanometers, corresponding to frequencies in the range of 30×10^{15} Hz to 30×10^{18} Hz and energies in the range 120 eV to 120 keV. X-radiation is called Rontgen radiation after one of its first investigators, Wilhelm Conard Rontgen. X-rays are primarily used for diagnostic radiography etc. As a result, the term "X-ray" is used to refer to a radiographic image produced using this method.

Y

Yard wastes: The part of solid waste composed of grass clippings, tree trimmings, other plant materials and other organic wastes that are typically generated in garden refuse and outdoor residential setting.

Yeast: Single-celled micro-organism that converts its food (sugar or starch) into alcohol and carbon dioxide through fermentation. Yeast are used for making beer, wine, cheese and some breads. When making bread, the carbon dioxide produced by yeast makes the dough rise. Yeast needs sugar or starch and a warm environment in order to grow.

Yellow – Boy: Clumps of Iron oxide in waste or water; usually observed as orange-yellow deposits in surface streams with excess iron content. When the pH of acid rock drainage is raised past 3, either through contact with fresh water or neutralizing minerals, previously soluble Iron (III) ions precipitate as Iron (III) hydroxide, a yellow-orange solid colloquially known as Yellow boy. Yellow boy discolors water and smothers plant and animal life on the streambed, disrupting stream ecosystems. The process also products additional hydrogen ions, which can further decrease pH. Research is currently being conducted as to the feasibility of using Yellow boy as a commercial pigment.

Yellow fever: It is an infection caused by the yellow fever virus. It has caused large epidemics in Africa and the America. Although an effective vaccine has been available for 60 years, the number of people infected over the last two decades has increased and yellow fever is now a serious public health issue again.

Yoga: It means "union" in Sanskrit, the classical language of India. The word yoga is derived from the Sanskrit root Yuj (To join, to concentrate one's attention on) meaning to bind, join, attach and Yoke to direct and concentrate one.

Z

Zero air: The atmospheric air which is purified to contain less than 0.1 ppm total hydrocarbons.

Zero discharge: Describing a facility that does not release any waste water to the environment but recycles and reuses it internally. Also called zero waster water discharge systems or closed circuit systems.

Zero population growth (ZPG): A lack of population growth caused by a balance between births and immigration, and deaths and emigration.

Zero waste: Turning waste into resource; the design of resource-use such that waste can ultimately be reduced to zero; ensuring that by-products are used elsewhere and goods are recycled, in emulation of the cycling of wastes in nature.

Zinc fume fever: Caused by inhalation of zinc oxide fumes and is characterized by flu-like symptoms, a metallic taste in the mouth, coughing, weakness, fatigue, muscular pain, and nausea, followed by fever and chills.

Zooplankton: Animal species of plankton in freshwater or marine environments; that do not produce their own energy but feed on phytoplankton or smaller animal plankton. Their movements are largely dependent upon currents.